THE GEOGRAPHY OF THE THIRD WORLD: PROGRESS AND PROSPECTS

PROGRESS IN GEOGRAPHY SERIES
Edited by Michael Pacione, University of Strathclyde, Glasgow

PROGRESS IN URBAN GEOGRAPHY
Edited by Michael Pacione

PROGRESS IN RURAL GEOGRAPHY
Edited by Michael Pacione

PROGRESS IN POLITICAL GEOGRAPHY
Edited by Michael Pacione

PROGRESS IN INDUSTRIAL GEOGRAPHY
Edited by Michael Pacione

PROGRESS IN AGRICULTURAL GEOGRAPHY
Edited by Michael Pacione

POPULATION GEOGRAPHY: PROGRESS AND PROSPECT
Edited by Michael Pacione

MEDICAL GEOGRAPHY: PROGRESS AND PROSPECT
Edited by Michael Pacione

HISTORICAL GEOGRAPHY: PROGRESS AND PROSPECT
Edited by Michael Pacione

SOCIAL GEOGRAPHY: PROGRESS AND PROSPECT
Edited by Michael Pacione

The Geography of the Third World:

PROGRESS AND PROSPECT

EDITED BY
MICHAEL PACIONE

ROUTLEDGE
London and New York

First published in 1988 by
Routledge
11 New Fetter Lane, London EC4P 4EE

Published in the USA by
Routledge
in association with Routledge, Chapman & Hall, Inc.
29 West 35th Street, New York, NY 10001

Printed and bound in Great Britain by Mackays of Chatham Ltd, Kent

British Library Cataloguing in Publication Data

The geography of the third world: progress
and prospect. — (Croom Helm progress in
geography series).
1. Developing countries — Economic conditions
I. Pacione, Michael
330.9172'4 HC59.7

ISBN 0-415-00467-5

Library of Congress Cataloging-in-Publication Data

ISBN 0-415-00467-5

CONTENTS

PART II: REGIONAL STUDIES

TABLES

FIGURES

To Christine, Michael John and Emma Victoria

PREFACE

The term Third World is a convenient shorthand description for that part of the world dominated by a particular combination of social and economic disadvantages, most evident in widespread poverty, overpopulation, and weak economic structures. The term was first employed as part of a three-fold division of the world which recognised the First World of industrialised market-economy states, the Second World of centrally planned economies, and the economically infirm and supposedly politically neutral Third World. The latter zone covers 60 per cent of the world's land surface and encompasses 70 per cent of the global population. A more recent four-fold classification of countries within the Third World (Figure 1.1) serves to underline the high degree of generality attached to the earlier term. Beneath the superficial unity, however, the Third World constitutes a multitude of varied social, economic, cultural, political and environmental settings.

This collection of original essays is designed to encapsulate the major themes and recent developments in a number of areas of central importance in the geography of the Third World. The twin themes of unity and diversity are acknowledged in the dual-level structure of the book, which first provides readers with an introduction to a number of key systematic issues (e.g. the global setting of the Third World, agrarian structure, urban economy, housing, welfare, power and politics, and regional development). Attention then focuses on the regional scale, to consider the contemporary situation in a number of major zones within the Third World (Latin America, Tropical Africa, the Middle East, South Asia, East Asia, and the island micro-states).

Figure 1.1: Countries of the world grouped according to type of economy. Low-income economies less than $400 per inhabitant in 1983. Countries with fewer than 1 million inhabitants are not shaded. Projection: Peter's equal area.

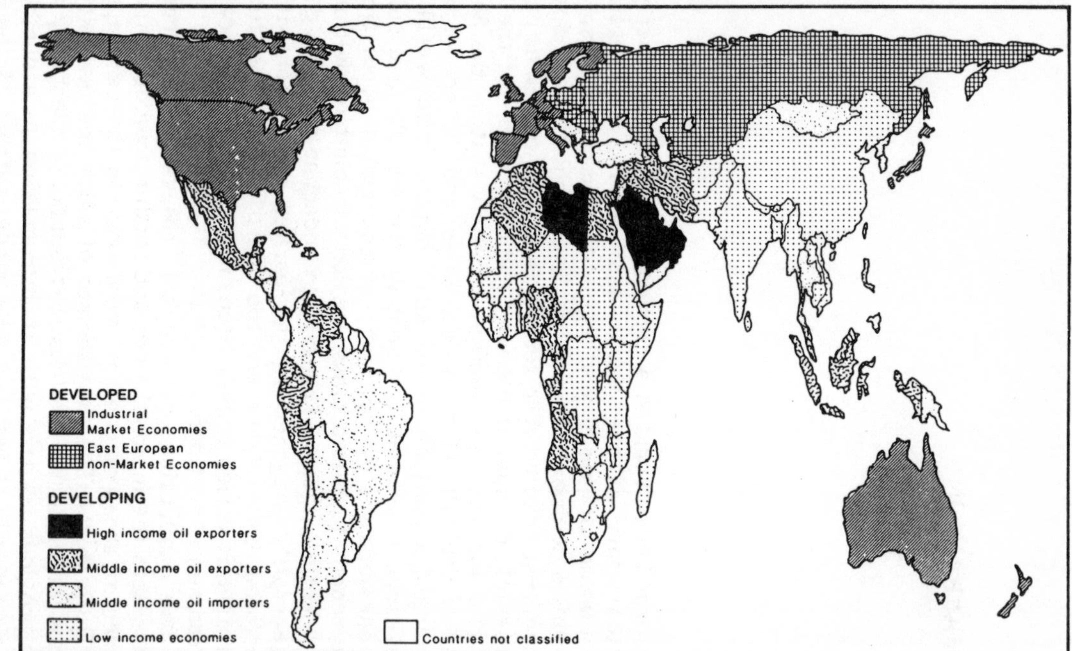

Source: World Development Report (1985), pp.170-1

INTRODUCTION

Inequality and uneven development are characteristics of contemporary society at all spatial scales ranging from the intra-urban to the international. Over the last three decades growing attention has been afforded to the global pattern of inequality and in particular to the marked differences in quality of life experienced by people in the First and Third Worlds (Figure 1.1). The imbalanced distribution of the world's resources is demonstrated by a range of social and economic indicators such as per capita income, GNP per head, literacy levels, per capita calorie intake and infant mortality rates. How ever it is measured, the gap between developed and underdeveloped, North and South, rich and poor, is a dominant feature of the geography of the modern world.

While the diversity within the unity implied by the term Third World is self-evident, the countries of the Third World do share some common characteristics. Most widespread is poverty, both relative compared with the advanced industrialised nations, and absolute, with millions of people lacking the resources to provide even a nutritionally adequate diet. In the absence of state welfare systems, hunger and undernourishment are the daily experience of the Third World poor. Poverty, seen as relative deprivation, also relates to higher-order human needs including self-esteem, and the right to uncensored information, freedom of speech and to organise trade unions and political parties. Not all of these are available in all Third World countries. In addition to their poverty, the majority of Third World states have weak economic structures characterised by a high proportion of population employed in the primary sector, low agricultural productivity, fragmentary industrialisation, limited application of technology, limited purchasing power, overdependence on a small number of export cash crops which place the economy at risk from fluctuations in world prices and terms of trade, dependence on foreign investment capital and on imports of capital goods, such as machinery, and extraction of profits by foreign multinational companies.

The third main distinguishing feature of the less developed countries is demographic, with generally low life-expectancy at birth, large family-size and a high proportion of dependent young children. In Kenya, for example, 21 per cent of the population in 1983 was under five years of age. Poverty, malnutrition and overpopulation contribute to the disease problems which afflict the region, with nutrition-related diseases such as rickets, kwashiorkor and pellagra reducing working capacity and heralding an early death. Reduced resistance to infection contributes to endemic plagues such as typhoid, cholera, malaria and yaws, while poor standards of hygiene and lack of medical facilities compound the problem. Although death-rates are being reduced by medical advances, the birth-rate shows no sign of slackening and there is every chance that the five billionth citizen of the world will be born in one of the Third World states. While rural-urban migration has improved the relative poverty of some, the rapid increase in population ensures the continued poverty of the masses in both town and country.

A final common feature of most Third World states concerns their previous colonial status. Some see this as the fundamental determinant of their disadvantaged position, since although now politically independent, in economic terms many remain closely tied to former colonial powers. Most countries of the Third World are entangled in a dependency relationship of unequal exchange with the countries of the developed realm. A major stumbling block for attempts to reduce the quality-of-life gap between rich and poor states is the belief by many of the former that global economic interaction is a zero-sum game in which improvement in the well-being of the Third World can only be at the expense of some decline in that of the First World. A similar rationale operating at a different scale underlies the efforts of powerful interest groups within Third World states, for example the urban elite and large landowners, to prevent radical change. On the world stage, for the political

Right this view results in opposition to any stronger global redistribution mechanisms, while for the Left it underlines the need for fundamental structural change in the nature of world trade.

In terms of praxis, structuralist interpretations of the 'development gap' postulate radical action as the only real means of resolving the problems of poverty, and castigate traditional planning and spatial policies as counter-revolutionary and supportive of the status quo. More recently, although radical change is still advocated by many Marxist geographers there has been an increased acceptance of the need to marry consideration of the global processes that are restructuring social, economic, and political reality with an understanding of 'the specifics of what is happening to individuals, groups, classes and communities at particular places at certain times'. In consequence, geographers have increasingly been drawn into the field to complement a global-system perspective with local and regional investigations. The geographer is well equipped to describe, explain and evaluate problem situations at the national and sub-national scales and to identify potential remedial strategies.

The communality of disadvantage which integrates the countries of the Third World must not, however, be allowed to disguise the great diversity that exists both between and within Third World states. Marked differences exist in terms of physical environments, resource endowments, historical experience, cultural traditions, societal structures, economic organisation and political practices. This internal diversity means that, to be most effective, policies must be tailored to suit particular situations. The geographer's traditional concern with the relationship between humanity and the environment, along with expertise in locational issues, and skills in regional analysis, make him or her ideally qualified to participate in not only the study but the making of the geography of the Third World.

CONTENTS OVERVIEW

The inter-connectivity of national economic systems within the global economy is of primary importance in the geography of the Third World. In Chapter 1, Stuart Corbridge provides an explanatory critique which traces the progress of development theory through the liberal, neo-

Marxist and Marxist traditions to the more recent regulationist perspective. Within the liberal formulations, comparative advantage theory has featured prominently both in a Ricardian framework and, since World War 1, within the structure proposed by Heckser, Ohlin and Samuelson. According to this model free trade is generally regarded as a precondition for successful economic development. A second liberal strand, modernisation theory, assumes that all societies lie along the same socio-economic continuum and advocates a process of Westernisation in pursuit of Third World economic development. The conceptual and empirical limitations of liberal theory were emphasised by proponents of the neo-Marxist underdevelopment model which assumed:

(1) an underconsumptionist view of the dynamics of capital accumulation;
(2) a structure of metropolis-satellite relationships which facilitates the centralisation of surpluses;
(3) a belief that the nature of the social formation in the Third World is dependent upon their position in the capitalist world economy; and
(4) that the prospects for industrial development are inversely related to the Third World states' ties to the capitalist world system.

The third major theoretical tradition is represented by Marxist theories of development. These seek to explain particular important events in the world economy, such as the rapid industrialisation of East Asia, which lay beyond the explanatory scope of dependency formulations. The several strands in this paradigm range from the pessimistic scenario portrayed by Luxemburg to the equally polarised optimistic interpretation posited by Warren.

None of these individual theoretical stances provides a complete understanding of the Third World's position in the global economy. In the present discussion, the extent to which the various liberal, neo-Marxist and Marxist theories successfully explain observed world events is assessed by examining the global geographies of trade, industry and finance. This analysis illustrates both the strengths and weaknesses of the major theoretical traditions: for example, while liberal theorists identify the important role of human agency in the development process, neo-Marxist theory illuminates the structural preconditions inherent in the capitalist mode of production. Taken to extremes, however, each perspective can result in voluntarism and functionalism, respectively. An acceptable model of development and underdevelopment must avoid the reification of concepts such as 'capitalism', recognise the changing spatial and temporal setting of capital accumulation in the world economy, and develop meso-level concepts to adjudicate between macro-scale structural forces and micro-scale agency factors. It is suggested that this goal may be approached by the post-Marxist regulationist school which:

(1) recognises the dynamic impermanence of global economic patterns;
(2) posits the need for a theory which acknowledges the variety of forms and processes which are consistent with the 'logic of capital accumulation';
(3) in terms of scale of analysis, emphasises the key importance of the national level, with the world economy theorised as a system of interacting national social formations; and
(4) proposes a set of meso-concepts, such as the regime of accumulation and mode of regulation, which may be employed to construct a model of capitalist development and crisis in the twentieth century.

Agriculture in Third World countries often forms the largest sector of the economy, yet paradoxically, it contributes relatively little to national economic growth. Although generalisation is difficult, this situation often arises as a result of a combination of environmental conditions, the low priority attached to agriculture in development planning, and world market conditions, such as recession and tariff barriers which adversely affect the export of primary products. In many countries the weak position of agriculture has meant that declining food production has had to be offset by imports in order to avoid famine. Government attempts to increase agricultural productivity have major implications for agrarian structure.

In Chapter 2, Bill Morgan describes the nature of traditional and modern market-oriented agricultural systems and examines the effects of recent trends in land tenure, holding structure, land reform and agricultural development policy. Six main classes of land tenure are identified. First are usufruct systems, such as those in tropical Africa, where land is held communally with individuals having user-rights only; secondly, estates of tenant farms are the most widespread tenure form in North Africa, Middle East and Asia, with the development of a land-owning class facilitated by colonial rule; thirdly, the latifundia systems of Latin America, characterised by absentee ownership; fourthly, socialist forms of tenure, including collective, co-operative and state farms; fifthly, plantations and ranches which are centrally managed, employ wage-labour and produce specialised crops for urban or export markets; and finally, owner-farmers, normally associated with commercial agricultural production linked to a well-developed marketing system.

A widespread problem in the Third World is the fragmentation of land holdings which imposes significant costs on farmers. This can arise from the traditional method of land allocation by a communal holding group, through purchase of extra land by owners, or through subdivision of existing holdings due to inheritance practices. Plot consolidation programmes can only produce a long-term improvement in productivity if supported by measures to prevent re-fragmentation (e.g. a change in inheritance laws or provision of alternative employment opportunities). Land reform is another structural measure which involves both the redistribution of land and reform of tenancy arrangements. In practice the motives behind this strategy are often mixed, with the social objective of redressing an inequitable distribution of land and wealth accompanied by economic (e.g. to increase agricultural productivity) and political (to enhance legitimation of the state) reasons. The major arguments in favour of a land-reform approach include the view that:

(a) it is an essential component of the development process supplying labour, capital and produce for the industrial sector;
(b) under- and unemployment are reduced by the labour-intensive nature of small-scale farming;
(c) greater use of the labour resource would reduce the need for mechanisation; and
(d) higher yields per acre are obtainable on intensive family farms.

Assessment of the efficiency of a land reform is difficult

however, not least because of the problem of separating out the effects of the policy from that of other contemporary social trends.

Apart from debate over the economic and social merits of the policy, a major obstacle for land reform is the entrenched position of landed interests and the backwardness of the rural peasantry in many Third World states. Land is only one component of the agricultural system and the process of agricultural development must also consider the role of capital and labour. It is often suggested that overpopulation is a general problem in Third World economies, yet in several countries rural-urban migration and the consequent reduction in the rural labour force has led to a decline in agricultural productivity. The availability and cost of labour is a vital factor which has yet to be adequately addressed in agricultural development policies. Strategies must also aim to expand the availability and use of capital for innovation or expansion. In most Third World countries this will require some redress of the concentration on industrial capital. Related social improvements include education and training schemes. An integrated rural development strategy with co-ordination of sectoral policies at the regional level is clearly imperative for efficient agricultural development.

The Third World city is a fragmented labour market in which conditions of, and opportunities for, employment are highly variable. Even the mass of the population categorised as the urban poor display considerable heterogeneity in their employment characteristics, with the principal income earners from poor urban households represented in virtually all types of employment including waged-, family-, and self-employment, and in both the public and private sectors. Contrary to popular belief, the activities carried out by the poor are often closely linked to those of the modern sector. As well as those in low-paid jobs within the modern sector, others are related through sub-contracting, a need for capital, or government taxation of street vendors. The Third World urban economy must be viewed as a complex interrelated system of activities and an appreciation of this diversity is essential.

In Chapter 3, Ray Bromley and Chris Birkbeck examine the nature of economy and employment in the Third World city by addressing the fundamental question of the appropriate conceptual framework and terminology for research. The dual-economy concept, which emphasises the divisions between the formal and informal sectors, is rejected as an oversimplification of limited value for either academic analyses or policy-making. The preferred Marxian-humanist approach seeks to understand the relationships between the modern and traditional, or large- and small-scale activities, since study of these links more clearly illustrates the factors which underlie revealed patterns of socio-economic inequality. The validity and analytical ability of a number of commonly used terms are also questioned. Underemployment and the accompanying assumptions that those engaged in certain activities, such as petty trading, work less hours and are less productive is regarded as an inaccurate description since those so categorised tend to work hardest for least rewards. Furthermore, undervaluing the efforts of the urban poor effectively disguises the exploitative nature of the prevailing economic system.

The terms production, work and employment also require careful definition when applied to the economic geography of the Third World. The concept of production, for example, may realistically have to include illegal and parasitic activities such as prostitution or begging. Similarly, it is recommended that the common distinction between wage-work and self-employment should be replaced by the idea of a continuum of employment relationships ranging between the extremes of 'career wage work' and 'career self-employment' (both having high levels of stability and security), with intermediate relationships described collectively as 'casual work'. This includes short-term wage work, disguised wage work, dependent work, and precarious self-employment. These terms are described in detail, and a conceptual model is presented as an organising framework for research into the nature of the social relations of production in the Third World city. Analysis of the employment interrelationships of individual workers and the links between workers and employers is complemented by an examination of enterprises and enterprise structure. This requires definition of terms such as enterprise, firm and company in a Third World context.

The practice of partially decentralised production is of particular importance in the Third World city, and it is in the analysis of this activity that the significance of sub-contracting and casual work in the production process is most evident. A minority of workers constitutes the 'protected' labour force, characterised by minimum wages, regular working hours, paid holidays and sickness benefits. Maintenance of this labour aristocracy is costly, however, and employers often use the 'reserve army' of casual workers when additional production is required. There is, in essence, a conflictual relationship between 'career wage-workers' and the mass of casual workers, and each may be used against the other to further the capitalist mode of production. While the former can be co-opted by governments and large companies to support continuation of the system, the latter may be used as a strike-breaking weapon.

Clearly, an appreciation of the diverse inter-relationships among the many different components of the Third World economic system is a prerequisite for a proper understanding of economy and employment in the Third World city.

The ongoing process of rural-urban migration has ensured that the key element of the housing question in the Third World is the provision of shelter for the poor attracted to the burgeoning urban areas. Within this general field, two main themes have received attention. Until recently, investigations were dominated by a concern with housing conditions, but over the last decade the multi-functional role of housing as shelter, provider of construction jobs, contributor to GDP, and aid in the reproduction of labour has been acknowledged by neo-Marxist interpretations of the urban development process.

In Chapter 4, David Drakakis-Smith first reviews the major influences on housing policy and programmes including, for example:

(1) the political philosophy of the state and in particular the priority attached to social, political and economic objectives in national planning;
(2) central-local state relations; and
(3) the role of international agencies and institutions.

The difficulty of accurately assessing housing need is considered. Doubt is cast on the validity of employing the number of squatters as a surrogate measure since this ignores the squatters' perceptions of their position and neglects the positive aspects of squatter life. A typology of the principal forms of low-cost housing is presented as an aid to further discussion. It is suggested that greater research attention should be given to the role of the private sector in low-cost housing provision. The basic components of low-cost housing production (land, labour, finance, infrastructure, materials and design) are then analysed.

The importance of land tenure for the urban poor remains open to debate, with some writers emphasising its role in encouraging investment in housing improvement and others stressing the relevance of factors such as perceived security, family finances and political climate. The operation of the commercial land market and the ways in which land becomes available for low-cost shelter are discussed. Consideration of the issue of building materials and house-design has identified significant problems relating to inappropriate building regulations, standards, high-cost imported materials, and technology (e.g. high-rise blocks). Clearly, greater attention must be given to the use of intermediate and small-scale technology and to the views of consumers.

In terms of housing finance, the poor exist outwith the formal credit system yet few studies have been undertaken of housing finance schemes within 'peripheral capitalism'. In practical terms what is required is an initiative which acknowledges the position of the urban poor and provides, for example, a flexible repayment regime. A major recent innovation in the provision of low-cost housing has been the suggestion of aided self-help schemes. These seek to mobilise the resources of the poor in an attempt to improve their living conditions. A major criticism, however, has been that the cost of housing has placed it outwith the reach of the poorest households. Reasons cited for the gap between housing cost and household income include a combination of global factors (e.g. reduced wages and local job opportunities due to world recession) and local factors (e.g. the complexity and inefficiency of administrative frameworks, and minimal public participation in planning). More radical objections to such schemes are that they serve to fragment the solidarity of the poor and seek to placate them without changing their lowly social position.

The political-economy perspective underlines the fact that the housing problem cannot be resolved in isolation but is inter-connected with the need for increased employment opportunities for the poor. The nature of the impact of global trends and the current world recession on housing provision has attracted particular attention. While negative effects can include a reduction in employment in construction, increased crowding in existing stock and reduced government budgets for housing, there is also evidence to suggest that capital may actually flow into the construction sector in times of recession in manufacturing. There is, in effect, no universal reaction to global economic forces. What is required are detailed studies of how these general processes are manifested in particular locations at specific times.

Social commentary of all kinds is replete with references to quality of life and human welfare. The meaning of these terms differs as they are variously used, but generally they refer to either conditions of the environment in which people live and work (e.g. housing standards or pollution levels) or to attributes of the individual (such as health or educational achievement). While the great majority of people in the developed world satisfy their basic needs for shelter and food as of right these fundamental components of life quality are denied to a large proportion of the population of the Third World.

In Chapter 5, Morag Bell focuses on the relationship between the concepts of welfare and development, and critically examines national and international efforts to enhance human well-being. Two general conclusions are, firstly, that the normal division of responsibility for issues (such as medicine and public health engineering) in development policy and administration has obscured the interrelationships among the different elements of human welfare. Secondly, this Western-based sectoral approach generally overlooks indigenous welfare systems which remain of importance in Third World societies. Welfare strategies in indigenous socio-cultural systems are characterised by an adaptive flexibility embodied in a variety of community-based systems, such as risk-minimising agricultural practices, mutual support, and food redistribution to the disadvantaged in times of need. Such evidence refutes the suggestion that only 'modern' societies are capable of administering a wide-ranging system of social welfare. External influences on indigenous welfare provision include the centralisation and professionalisation of welfare services in line with the Western model. This reduces local responsibility and subordinates indigenous knowledge to technical expertise.

This perspective on welfare is incorporated into the Western definition of development which, until the early 1970s, focused on economic criteria of progress. In welfare terms, this meant that social services were largely urban-oriented and based on a 'top-down' imported model. The broadening of the index of development to include social indicators has promoted consideration of other welfare goals such as the reduction of inequality and intra-national variations in levels of provision. Greater attention has been afforded to rural areas, to urban slums, and to small-scale projects targeted on the poor. The earlier view of welfare as charity is being replaced by a developmental perspective which views welfare as contributing to general social advancement and involving the active participation of beneficiaries. This 'bottom-up' approach has stimulated a pluralist attitude to welfare which acknowledges the potential benefits of intermediate technologies and traditional knowledge. A major structural constraint on the implementation of 'bottom-up' strategies is represented by the power of vested interests, such as multinational corporations, to maintain the status quo (e.g. capital-intensive curative medicine as opposed to locally oriented preventive measures).

The inappropriateness of many imported welfare strategies has been illustrated with references to issues of health care, water-supply and environmental sanitation. It is essential that solutions be culturally sympathetic. Social conventions over food preparation and distribution within a household can determine the efficacy of a feeding programme, while beliefs about the causes of illness will affect the acceptability of treatments. Indigenous welfare systems co-exist with imported systems and, in many areas, sustain a greater population than the latter. Coming to terms with these indigenous sytems and conventions is a major challenge for national and international agencies charged with improving the quality of life of those who live in the Third World.

Inequality is an inherent characteristic of economic development both under the capitalist mode of production and in theoretically more egalitarian economic regimes. The spatial concentration of different levels of income and well-being highlights inequalities in the distribution of resources within any society and if pronounced at an intra-national scale, may threaten the legitimacy of the state. For a variety of social, economic and political reasons, therefore, most Third World governments have included regional planning objectives in their development strategies.

In Chapter 6, Graham Hollier first identifies the changing bases of regional development theory, and then examines the practical efficacy of the major strategies. In the early post-war period greater attention was given to the North-South divide and to national economic growth, and regional policies were afforded comparatively little attention. Subsequent efforts were stimulated by industrial location theory, drainage basin studies and in particular the growth pole model. However, the debate over whether a non-interventionist economic strategy would produce convergence or divergence in regional development was not resolved until the mid-1970s, when the need for government intervention was generally acknowledged. The major philosophical re-orientation to affect regional development strategy has been the decline in influence of positivist neo-classical economic models which dominated the concept of development until the late 1960s, and the rise of the notion of culturally specific development incorporating social objectives in a broader less economically deterministic definition of the concept. This shift in emphasis can be summarised as a move away from the goals of functional integration (development of regional resources for their contribution to the national and international economy) to those of territorial integration (use of a region's resources to satisfy the needs of its inhabitants). This functional-territorial dichotomy is reflected in the distinction between 'top-down' and 'bottom-up' planning strategies.

Three forms of regional development policy have characterised spatial planning in Third World countries: (1) river basin programmes, (2) growth pole strategies, and (3) 'territorial' initiatives. The first drew on the example of the Tennessee Valley Authority established in the USA in the inter-war period. It has been widely applied, though many schemes have failed to achieve the objective of comprehensive development of a drainage area's resources. Growth pole strategies which have dominated Third World regional planning since the early 1960s have three main concerns: stimulation of economic growth in lagging regions, relief of pressure in congested metropolitan areas, and development of the urban hierarchy. Supporters of the growth pole concept attribute the lack of success inappropriate application of the theory and, in particular, to the insufficient time perspective employed. The focus of attention however has now turned to the range of territorial approaches. These include:

(1) decentralised concentration based on the promotion of rural development through the strategic location of urban functions;
(2) selective regional closure to reduce the outflow of local surpluses; and
(3) agropolitan development which seeks to increase local participation in the development of a self-reliant region.

As yet such 'neo-populist' schemes have not been widely employed, and a major issue at the research frontier is to assess the potential and the impact of such strategies in reducing regional inequalities.

A fundamental structural problem confronting practitioners and academics is the gap between theory and practice with most developing countries still committed to macro-scale national goals. Regional policy in general, and equity issues in particular, are afforded much less attention, with Third World governments seemingly unable or unwilling to enact strategies which would radically alter the distribution of power in society.

It should not be surprising, given the enormous differences in the distribution of wealth and power, both globally and within Third World states, that the geographical study of development should be one of the early adopters of a structuralist perspective. However, while the analytical rigour of the power-oriented neo-Marxist view has illuminated our understanding of the processes underlying the contemporary geography of the Third World, the concentration on the macro scale and the relative neglect of the spatial dimension is a serious deficiency. It is axiomatic that in order to achieve a full understanding of any society it is necessary to incorporate investigation of both the social and the spatial, infrastructure and superstructure.

Such a view is stressed by Rob Potter and Tony Binns. In Chapter 7, they first define the concepts of power, politics and social organisation, and then adopt a hierarchical framework to examine the operation of these variables at the national, regional, inter-household and intra-household levels. The polarised distribution of income and life quality within Third World states is characterised by the existence of elite political, economic and social groups. Under some regimes the state merely acts to maintain the position of the ruling group, and outcomes include maladministration, corruption and the unmitigated continuation of slum and squatter areas. In other instances discrimination may exist as a legacy of colonial rule. These social inequalities exhibit a spatial character with major life quality differences evident between regions, urban and rural areas, and different levels of the settlement hierarchy. Several examples of such socio-spatial variations are discussed. It is argued that in order to understand the extent to which the people of the Third World can influence decisions which impact upon their quality of life, attention must be directed to the local scale. Practical support for a micro-scale approach is provided by the failure of many 'top-down' development programmes which neglected to consider local circumstances.

Recent work on farming systems and on urban informal sector employment suggest that the household, which is the basic unit of production and consumption in many Third World societies, is increasingly becoming a focus of research attention. The major problems associated with conducting household research in the Third World are identified before the discussion turns to consider the question of power at the inter-household and intra-household levels. Social stratification at the inter-household level is long established in some communities. In the Indian caste system, for example, more powerful groups have used their position for self-advancement, thereby translating a traditional social division into a modern economic or class-based society. Generally, a major factor in household status and wealth is access to land, with land owners effectively controlling the political economy. Another key factor is the importance of political and economic patronage, the operation of which is illustrated by the power relationship built up between squatters and political leaders, by the biased allocation of public service jobs, and by the practice of entrepreneurs employing people from their extended family or home area.

At the intra-household level, the question of gender relations is of central importance, since this underlies a clear division of labour and decision-making responsibility. Clearly, greater attention should be afforded to the role of women in Third World society, since in some areas they are the primary food producers, and where male migration occurs, women assume a predominant position within the family. Similarly, since the problems and pressures on a family change with stage in the life cycle, further empirical investigation of household social characteristics would assist the preparation of effective aid policies.

In conclusion, it is emphasised that a multi-level perspective which accommodates macro- and micro-level analyses is essential both for extending academic knowledge of power relationships in Third World society and for the development of programmes designed to alleviate the disadvantaged position of the mass of the population.

Part 2 of the book explicitly acknowledges the diversity encompassed by the term Third World. The approach changes from consideration of systematic issues to focus on regionally defined issues, thus providing vignettes of the major sub-areas within the Third World. This shift of perspective enables us to complement the general situation uncovered in Part 1 with detailed discussions of regional priority issues.

In Chapter 8, Bob Gwynne identifies contemporary issues and research questions in Latin America, focusing particularly on the field of economic and social geography. A world-economy perspective underlies the analysis, which firstly examines the international forces affecting development patterns in the region, and then considers the economic growth processes at a national level, before discussing a number of rural and urban issues. The discussion of Latin America's position in the world economy is predicated upon an examination of trade and capital flows. In terms of the former, with the exception of Brazil, the long-standing distinctive regional pattern is one of export of primary products and import of manufactured goods. Disadvantages of this dependency on primary exports are exposed by the fall in the Third World's share of world agricultural markets, partly due to trade barriers; and the decline and instability of prices. In terms of capital flows the prime characteristic is that of indebtedness as a result of heavy borrowing in the 1970s and increased interest rates in the 1980s. These factors have led some to suggest that industrialisation is the only means of reducing the effect of external constraints. However, the fact that Latin American industry is generally characterised by restricted domestic markets and dependence on imported technology questions the possibility of increasing manufacturing exports. The capital-intensive nature of industrialisation has also done little to absorb the rapidly increasing labour pool, and policy emphasis is turning to the opportunities for generating employment in small-scale enterprises. Evaluating the success of such initiatives represents a key research area. The pronounced concentration of manufacturing activity in the major cities also serves to draw migrants from rural areas where poverty is endemic.

The major policy initiative of agrarian reform has had mixed results, with the social goal of redistributing land conflicting with the economic objective of increased productivity. While social equity has been advanced in some localities, there has not been an expansion of food production due to the small scale of farms, low market-prices and limited government support. This also represents a research issue with a significant applied component. The influx of the rural poor into cities leads to a range of social (e.g. housing) and economic (e.g. unemployment) problems.

With reference to housing, Turner's thesis of improvement through aided self-help suggests several research questions, including:

(1) the relationship between successful consolidation of squatter settlements and the national economic situation;
(2) the role of the rented sector in the provision of low-cost accommodation;
(3) the mobility patterns of inner city residents; and
(4) the operation of the legal and extra-legal land market.

More general issues which demand attention include the nature and effectiveness of government social welfare policies and programmes.

The difficulties and deficiencies of life for the mass of the population in the Third World are exposed most starkly in tropical Africa where general Third World problems such as poverty and national indebtedness are compounded by regional issues including civil war, an exploding population, malnutrition and famine. In Chapter 9, Tony O'Connor examines contemporary issues in what may validly be characterised as the problem continent of the Third World. Research on tropical Africa is hampered, however, by a lack of statistical information, with the reliability of even national censuses open to question. Major demographic problems include:

(1) the concentration of population is densely settled areas and a consequent need for planned redistribution;
(2) the accelerating rate of growth with little sign of demographic transition (fertility decline) occurring;
(3) the high dependency ratio due to the preponderance of children in the population; and
(4) increased urbanisation fuelled by rural-urban migration.

All are areas in which geographers can make a contribution.

In the field of cultural and political geography, the implications of the strength of tribalism and the colonial subdivision of Africa across ethnic lines provide clear research opportunities for geographers to advance the study of national integration and disintegration, and the causes and consequences of refugee movements. Famine is perhaps the most pernicious manifestation of the economic infirmity and political fragility of tropical Africa. Detailed analysis of agrarian structure and local farming practices would be of considerable benefit, as would an atlas of drought designed to indicate the nature, extent and intra-regional impact of this key environmental factor. The fundamental importance of the land for the livelihood of most inhabitants of tropical Africa and the need to husband natural resources indicate a major applied role for geographers in devising appropriate strategies to combat soil erosion, desertification and deforestation. In relation to health-related environmental hazards, insufficient information is available on the ecology of many debilitating tropical diseases and medical geographers are well placed to contribute towards further understanding. Finally macro-scale issues such as the logistics of famine relief, the development impact of multinational corporations, and the distribution of the benefits and costs of international aid projects all represent critical research areas which would benefit from the application of a geographical perspective.

The stereotype image projected by the Middle East centres upon civil and international conflict, a predominantly Arab culture, and the importance of oil for national economies. As Dick Lawless points out in Chapter 10, oil resources have had the greatest single influence on the region's position within the global economy, with the economic well-being of many Middle Eastern states tied directly to the international price of this commodity. The unprecedented accumulation of financial resources by the oil-exporting countries was the pre-eminent regional trend of the 1970s. Integration of the region into world trade, based on oil and the replacement of socialist economic policies by more liberal strategies, resulted in a two-way exchange, with foreign capital, commodities and culture entering the region. In general, however, the bulk of oil revenues have been invested in non-productive goods and services (e.g. infrastructure), with little being directed into the development of alternative economic activities in the agricultural and industrial sectors. A major socio-economic issue on which little information is available is the extent of income and welfare variations within the Arab states.

One of the major demographic features of the region over the last decade has been the pattern of labour migration between oil-poor and oil-rich states. Despite advantages for migrants, home areas (via remittances) and host states, the loss of skilled labour from exporting countries, together with the volume of the influx into others, has created significant problems in some parts of the region. Further investigation of the impact of labour migration would be of benefit.

During the 1970s, increases in domestic food production barely kept pace with population growth and the value of agricultural imports rose markedly. The emphasis placed on urban-industrial development as the path to modernisation has been at the expense of agriculture, from which resources have been effectively transferred (e.g. by price-fixing). In some states, rural-urban migration has produced acute shortages of agricultural labour. Agrarian reform, stimulated by a variety of political, economic and social aims, has been introduced in most Middle Eastern countries. However, the results of land redistribution, in terms of individual advancement and increased productivity, have proved disappointing, with insufficient land to satisfy demand and inadequate funding being directed into complementary training and marketing institutions. Though as yet spared the ravages of sub-Saharan famines, the problems of how to preserve agricultural land under threat from urban expansion or ecological deterioration and how to feed a growing population are of increasing relevance for the region. The most enduring issue within the political geography of the Middle East is the Arab-Israeli conflict and the indeterminate position of the Palestinians, and a host of ensuing social, economic, demographic and environmental issues pose significant research questions.

For economic geographers, a key issue is whether, in the light of past experience, the oil-producing states will be capable of directing revenues into productive investments which will diversify national economies and reduce dependence on imports. Particularly problematic is the future for the oil-poor states of the region which are heavily reliant upon intra- and extra-regional aid.

One-fifth of the world population lives in the South Asian sub-continent, a region of physical and social diversity bound by a common climatic, political and cultural heritage. In Chapter 11, John Soussan identifies a range of specific research issues relating to environmental maintenance, international trade relations, overpopulation, health questions, migration, political instability, energy supplies, and the suite of problems consequent upon rapid urbanisation. Particular attention is focused on the key issues of poverty and inequality and on the prospects for agricultural and industrial development of a scale sufficient to alleviate these problems.

In marked contrast to countries of the developed world, agriculture is the most important economic activity in South Asia, in terms of GNP, export earnings and employment. Remarkably, although many rural areas are characterised by all the symptoms of a backward economic and social structure (polarised land ownership, fragmented holdings, usurous tenure systems, indebtedness, limited availabilty of credit, and primitive technology) plus an undependable rainfall regime, in the period since the Green Revolution food output has outpaced population growth in the region. Despite such achievements in terms of productivity, assessments of the Green Revolution are controversial. A chief criticism surrounds the unequal spatial distribution of benefits. While the programme has achieved national self-sufficiency in food production in many states, the essential selectivity of the exercise has increased disparities in welfare between regions, and between social groups within regions. The fundamental question is whether the dynamic growth generated in the successful regions can be reproduced elsewhere.

Industrial development in South Asia is generally concentrated in the urban areas of the larger states. Growth rates have not mirrored those of East Asia, but the region has also avoided the excessive debt and dependance on foreign capital and markets which accompanied rapid growth in regions such as Latin America. In the smaller states of South Asia, industry accounts for less than 15 per cent of GNP, and progress is hampered by major structural inadequacies including capital shortages, poor infrastructure and inefficiency. Sri Lanka, Pakistan and especially India have fared better. However, the particular structure of the industrial sector in each, together with the volatility of world markets, means that prediction and prognosis must be based on detailed study of individual countries. This would include examination of government development policies (ranging from protectionist to economic liberalism), the public-private mix, including the role of state firms, and the contribution of small-scale enterprise, marketing and financing. The question of regional development policy within states is also of central importance in efforts to reduce poverty and inequality. The variety of research opportunities for the geographer in South Asia serves to underline the importance of recognising the links between individual themes, such as agricultural change, population structure, migration and urban problems, and clearly demonstrates that the question of Third World development is a complex multi-faceted problem.

East Asia, despite limited natural resources, high population density and restricted people:land ratios, is arguably the most developed part of the Third World. Over the last three decades the region has achieved an exceptional rate of economic growth, narrowed the gap in living standards with the West, and introduced a dynamic new centre of advanced technology into the world economic system. The economic success of the Far East is easier to describe than to explain, with free enterprise systems like Hong Kong juxtaposed with China's brand of communism.

As Jack Gray points out in Chapter 12, the pattern of

economic advancement in the region challenges

(1) socialist contentions that capitalist industrialisation of surplus-labour peasant societies inevitably increases inequality;
(2) conservative support for the necessity of individual enterprise free from government intervention; and
(3) the liberal view that successful development requires a pluralist political system with a strong role for organised labour.

There is no single model to explain the economic distance of the region from the remainder of the Third World. East Asian economies exhibit as much structural diversity as those in any other world region with, for example, varying degrees of government intervention, union power, and multinational investment. Possible regionally specific factors include:

(1) a general sense of national insecurity which may have provided a stimulus to attain rapid growth and to reduce disruptive social inequalities;
(2) political patronage and economic aid from allies in the post-war period, and in particular access to Western markets; and
(3) the innovative role of Japan in the region, especially the marketing skills of the general trading companies (Sogo Shosha) whose primary aim is to increase trade flows.

The experience of Taiwan is symptomatic. Post-war economic policy was initially based on import-substitution accompanied by protection for the fledgling domestic industry. The limited size of the home market led to a re-orientation towards manufacturing for export, based at first on labour-intensive products. When faced with increasing real wages, competition from other labour-rich developing countries and the threat of protectionism in world markets, the emphasis was switched to higher-value exports and to an intensification of import-substitution in response to the increasing purchasing power in the domestic market. Agriculture was seen as a source of surplus capital for industrialisation (extracted, for example, by state monopolies), but in contrast to other Third World states, farmers were not deprived of the opportunity to increase productivity and incomes. In addition, once industrial 'take-off' had been achieved, policy has changed to one of agricultural support. Sound forward planning and flexibility have been key features of economic success.

Another common thread running through the regional diversity is that of culture. It has been suggested that a major explanatory factor may be the inherent qualities of a 'Confucian economy' characterised by:

(1) an unparalleled achievement/work ethic;
(2) a capacity for entrepreneurship;
(3) a willingness to accept economic activity within government rules;
(4) a tradition of mutual co-operation and obligation in industry; and
(5) the extensive support afforded to enterprise by national bureaucracy in pursuit of economic growth.

Only by tying these cultural elements to the individual characteristics of each state can the economic achievements of East Asia be fully appreciated. This is not to argue for 'cultural determinism', but rather, explicates a model, the transferability of which is a question of considerable relevance for other parts of the Third World.

In the final chapter, John Connell examines the position of island micro-states in the Third World. Small in size and often isolated, their increasing incorporation into the world economy poses serious questions for their political status, which in principle can range from full independence, through free association with a former colonial power, to dependence. A fundamental difficulty for those seeking to achieve an independent status is the limited resource base of most island micro-states. Political independence implies a degree of economic self-reliance and among a host of obstacles to be overcome are:

(1) reliance on a few primary export products;
(2) a small domestic market and therefore limited industrialisation and dependence on imports;
(3) limited technical expertise;
(4) dependence on foreign capital;
(5) disproportionately high expenditure on administration and welfare services; and

(6) high transport costs.

Unfortunately, most structural factors are outwith the state's control and the prospects for self-reliant economic development are slight. In many states, the impact of 'technological modernisation' has diluted local cultures, raised expectations, and stimulated extra-regional emigration. Although draining the most able population, this operates as a safety valve and as a source of remittances, and some states actively encourage emigration. Links with a colonial power insures the flow of foreign earnings, loss of which would diminish the attraction of independence. Urbanisation focused on the primate city has been accompanied by growing unemployment and the spread of squatter settlements and has exacerbated the problems of rural peripheries. In some island groups, this has led to the rise of separatist movements.

The agricultural sector, generally characterised by large estates and plantations as a legacy of the colonial period, presents several problems for the governments of island micro-states. These include the management of nationalised systems, introduction of land reform measures, diversification of cash crops, and stimulation of domestic food production to overcome a situation of food dependency. Apart from basic import-substitution and food-processing industries, industrialisation is limited. Major sources of income and foreign investment centre on the issue of postage stamps, tourism and provision of financial services (tax havens). The impact of tourism on local economies represents a major research question in which the benefits in terms of income and employment must be weighed against factors such as the loss of agricultural land and cost of subsidies and tax concessions. Despite the existence of regional organisations in the Caribbean and Pacific, they have been relatively unsuccessful in attempts to promote trade or sustained economic growth, and intra-regional linkages are generally weaker than those between individual states and the outside world.

Remote island micro-states are highly dependent elements of the global political economic system while in some instances their small size in physical and resource terms is offset by a strategic location, these states are generally economically vulnerable, dependent upon imports, aid, remittances and tourist revenue, and consequently, they tend to be politically relatively powerless. These factors will ensure a continued pattern of dependent development for island micro-states. Quite simply, for most, independence is a luxury they cannot afford.

Part I

SYSTEMATIC ISSUES

Chapter One

THE THIRD WORLD IN GLOBAL CONTEXT

S. Corbridge

In the years since Yalta and Bretton Woods there has been a remarkable transformation in the world economy, in the position of the Third World within it, and in a development studies literature which exists to explain these changes. In this chapter I want to convey a sense of the connectivity of these theories and events.

The chapter comprises three main sections. The first examines three traditions of development studies, each of which proposes a distinctive account of the relationship between capitalism, development and underdevelopment. A first tradition can be described as liberal, although I include within it a Keynesian variant whose optimism regarding the pace of modernisation sits uneasily with the logic of comparative advantage theories. In the schemas of underdevelopment theory (and some versions of dependency theory - see Palma, 1978), the development of capitalism in the core is said to depend upon the capitalist underdevelopment of the periphery. The underdevelopment of the Third World is here located in a set of 'external' factors and attention is directed to the unequal and asymmetrical insertion of satellite economies in a world economy dominated from abroad. Finally, more orthodox Marxian perspectives present capitalism as a dynamic regime of accumulation, which is marked by contradictions in the process of production and realisation and by a necessary tendency to uneven development. In some quarters a grudging tribute is paid to the role of capital in transforming natural economies and in bringing to them the fruits of development. This is most evident in the work of Bill Warren and in the associated debates on the new international division of labour. Another variant of Marxism directs attention to the role played by metropolitan capitalism in reproducing pre-capitalist modes of production in the periphery. This tradition speaks also of a 'blocked transition' in the Third World (Kay, 1975).

Section 2 appears to retreat from theory in favour of the real world. Our purpose here is to present a series of statistics which help place the Third World in global context. Given the concerns of other chapters our coverage is limited to the (changing) global geographies of trade, industrialisation and finance. The divorce from theory is never complete, however - nor could it be. Section 2 strives to examine the claims of competing traditions of development studies as they are illuminated by evidence from the world economy. (That this evidence is in turn constituted by theoretical practice goes without saying).

Section 3 brings development theory and development practice together in more explicit union. Whereas sections 1 and 2 develop an 'internal critique' of existing paradigms of study, section 3 looks forward to another mode of analysis. The inspiration for this new account - if such it is - comes from a number of sources; from Hindess and Hirst (1977) and Cardoso (1977), and, more latterly, from the work of the Regulationist school of French 'post-Marxists'. The strengths of the fourth approach are as follows. Firstly, the new analysis moves us away from totalising conceptions of development. No place is found here for the tyranny of 'progress' or of 'iron laws of development' as organising ideas in the world economy. Secondly, we are reminded of the geographical variation of global development and of the part played in the world economy by nation-states and sub-national agents and institutions. Room is also found for the part-structuring role of space in the development process, and for theories of the production of nature. Finally, the Regulationists alert us to the part played by contingency and chance in the production of social 'laws and tendencies'. As Alain Lipietz describes it, the best that social science can hope for is in an 'a posteriori or almost metaphoric functionalism' (Lipietz, 1987, p. 16).

1. THEORIES OF THE WORLD ECONOMY

Liberal Theories of Development

The model of economic development proposed by liberal social theory draws upon two traditions of analysis which at times support conflicting conclusions (Hoogvelt, 1982). The more long-standing of these traditions is that of economic liberalism. As formulated by Smith, Ricardo and Mill, and as later refined by neo-classical economics, this model suggests that (capitalist) production and exchange is carried out by discrete economic agents whose purposeful interaction in the market secures fair profit for themselves and a full employment equilibrium in the economy. The patterns of production and exchange thus instituted also effect a rational use of society's scarce resources.

Applied to the international arena, this model has some insistent implications. It suggests, firstly, the *dirigisme* for development is unhelpful. This is clearly the view of Deepak Lal, who believes that:

> Despite the current problems of the world economy, the best service the North can give to the Third World is to ensure that the post-war *liberal* international economic order is maintained by refusing to surrender to the blandishments of either the Southern *dirigistes* of the New International Economic Order or the Northern advocates of the 'new protectionism' (Lal, 1983, p. 131, original emphasis; see also Bhagwati, 1979; Little, 1982).

It is a perspective which also finds favour in the offices of the International Monetary Fund and in parts of the World Bank (World Bank, 1981). Secondly, the basic Smithian framework extends easily to the arena of international trade, where it is reproduced as comparative advantage theory. In the nineteenth century, the work of David Ricardo was used by some to defend a stark division of labour between the primary-producing colonies and the industrial heartlands of North America and Western Europe. In a grotesque display of circular reasoning, it was argued that the colonial powers, logically and eternally, had a comparative advantage in the production of manufactures *because* they were the industrial powers. *Pari passu* the colonies held a natural advantage in the production of foodstuffs and raw materials (to which they were to be confined). Such a division of labour was assumed to work to the advantage of both sets of countries because each shared in the greater surplus for trade created by specialisation. (As Lord Cromer, Governor of Egypt, 1883-1907, put it:

> The policy of the government may be summed up thus: (1) export of cotton to Europe ... (2) imports of textile products manufactured abroad ... nothing else enters the government's intentions, nor will it protect the Egyptian cotton industry because of the dangers and evils that arise from such measures ... Since Egypt is by her nature an agricultural country, it follows logically that industrial training can only lead to neglect of agriculture while diverting the Egyptians from the land - quoted in Hayter, 1981, p.49).

Since World War I, this Ricardian framework has been modified by the work of Heckscher (1919) and Ohlin (1967) (and later Samuelson). The Hecksher-Ohlin-Samuelson (HOS) model of international trade assumes that technologies of production are equally and freely available in all parts of the world. This being so, comparative advantages are forged only on the basis of different factor endowments, with relatively labour-abundant countries specialising in relatively labour-intensive goods. Thus, if rice is a labour-intensive good relative to apparel, and if labour is relatively abundant in India, then India will specialise in rice production. As in Ricardo's model, specialisation is assumed to increase the surplus for trade, but it does something else besides. The HOS model suggests there is a tendency for wage rates (and other factor prices) to be equalised following the development of trade. The reasoning behind this 'factor price equalisation' is as follows (after Edwards, 1985, p. 23): as India specialises in the production of rice, its pattern of production becomes more labour-intensive. As a result, India's relative abundance of labour is reduced, the marginal productivity of labour in India rises and wages also rise. Conversely, in the 'industrial' country, a movement into the production of apparel causes labour to become less scarce, with the result that marginal productivities and wage rates fall. In sum, the HOS model holds out to the developing countries the prospect of development on two fronts. In the short run they will gain relative wage rises by producing primary commodities for the world economy. In the long run it seems clear that less developed country (LDC) 'factor endowments' will shift away from a surplus in labour and towards a comparative advantage in semi-manufactured goods (see Haberler, 1961; the shift is observed first in export-processing or free-trade zones).

The notion that free trade is a precondition for successful economic development is accepted by most liberal theorists. Since 1945, however, a second strand of liberal social theory has emerged, which adds to this insight a more interventionist account of the process of Third World development. In part, this new tradition draws upon a body of Keynesian economics. In the work of Harrod (1948) and Domar (1957) especially, and in that of Rostow (1960), the development process is effectively reduced to an exercise in maximising the savings of a given national economy and ensuring that this surplus is invested in manufacturing industry. Where the local economy cannot support the required level of savings, it is up to the developed nations to plug the 'gap' with disbursements of foreign aid and direct transfers of foreign capital. In this schema, the West is offered a role in the development process akin to that played by the US in the reconstruction of war-torn Europe and Japan, and results are expected quite as swiftly. The promulgation of a Development Decade, in the 1960s, suggests that a timetable for development of ten years was not considered amiss.

International Keynesianism, however, forms only one element in the reorientation of liberal social theory. In some ways more pressing is the rediscovery of the idea of 'necessary progress' which marks Parsonian sociology. In broad terms, the Parsonians argue that societies are akin to organisms and exhibit similar tendencies to integration and adaptation over time. More pointedly, both Parsons (1948) and Smelser (1959) argue that the mechanics of integration becomes more complex as the process of modernisation takes hold. In traditional ('backward/underdeveloped') societies social integration is mechanical in form and is articulated through local, place-specific, kinship systems, mythologies and simple divisions of labour. In modern societies social integration is organic. Social cohesion is maintained over time and across space through the impersonal, inter-personal, relations forged in the market-place. Likewise, societies become more stable as their level of structural differentiation increases.

Given this insight, the project of modernisation theory

becomes straightforward. Once it is accepted that all societies lie along the same continuum between tradition and modernity, the political task reduces to spurring on a necessary progression (or, if you will, the developing countries are enjoined to copy the West, to abjure communism - which is deviant - and to remove local obstacles to progress). The fact that this message can be phrased in different ways is not of primary concern. (In demographic circles birth control becomes the new panacea for development (Davis, 1949); in geography attention fixes on the diffusion of technologies down a central-place hierarchy (Friedmann, 1966; Gould, 1970); in psychology much is made of X-achievement scores as a development indice (McClelland, 1961)). What matters is that modernisation theory combines (uneasily) with liberal economics to support a distinctive account of the process of capitalist development in the Third World. In effect, this account equates the diffusion of capital with Third World development, with Westernisation and (later) with industrialisation. In so far as it makes predictions and proposals, the liberal model leads us to expect a continuing and quickening pace of development in the ex-colonies. It further suggests that development will proceed with most vigour where countries pursue free-trade policies and where the door is left open to foreign assistance and capital (Shenoy, 1971; Cheung, 1986).

The Development of Underdevelopment

The claims of liberal social theory have not always been accepted in academe (Dutt, 1901; Naoroji, 1901), and the prescriptions of liberal economics have not been widely welcomed in the Third World. Throughout the 1950s the United Nation's Economic Commission for Latin America (ECLA), headed by Raoul Prebisch, challenged the bases of comparative advantage theory and bemoaned the slow rate of progress being made towards a new international economic order. Prebisch noted, firstly, that the Panglossian logic of free-trade models works only under the most restrictive assumptions. The Hecksher-Ohlin-Samuelson model, for example, assumes that initial factor endowment ratios, on a country-by-country basis, are not too dissimilar (Edwards, 1985, p. 29). It further assumes that the same technologies of production are available in all countries (which seems unlikely: Vernon, 1966), that there is global full employment (which is mischievous), and that factors of production can be aggregated unambiguously into the categories of capital and labour (a proposition disputed by Sraffa, 1960). Turning to empirical matters, Prebisch noted that in the 1950s, the Third World was still locked into a service role in the world economy, whereby it suffered a secular decline in its (primary commodity-based) income terms of trade. This decline was enforced, allegedly, by the oligopolistic nature of commodity and factor markets in the core. The First World exacted a rent upon trade (Nurske, 1953; Prebisch, 1959).

To snap this frozen division of labour, ECLA proposed a strategy of import-substitution industrialisation (ISI). Third World countries would be justified in protecting their infant industries behind tariff walls until they could compete freely in world markets. More generally, ECLA looked to policies of 'healthy protectionism, exchange controls, the attraction of foreign investment into Latin America and the adoption of wage policies aimed at boosting effective demand' (Palma, 1978, p. 907) to secure the industrialisation which would bring with it the developmental externalities absent from primary commodity-based growth.

For some years ISI became a new orthodoxy in the Third World, where its tinge of economic nationalism echoed the ambitions of populist regimes (Kitching, 1982). Not everyone thought that it went far enough, however, or that it escaped the liberal bias of international Keynesianism. At the time that Prebisch was attacking comparative advantage theory, Paul Baran, a leading US radical, was writing his great work on The political economy of growth (first published in 1957). According to Baran, it was not sufficient to urge reforms upon the core power of monopoly capitalism or to seek to reform the system from within. Activists in the Third World had to recognise, instead, that their countries were trapped in a capitalist world economy which was, in its fundamentals, asymmetrical, unequal and repressive. In Baran's judgement, economic development in the Third World will be resisted because it is 'profoundly inimical to the dominant interests in the advanced countries' (Baran, 1973, p. 120). The bases of this zero-sum account are argued with vigour in Baran's work and they have since been built upon to provide a second account of the relationship between capitalism, development and underdevelopment. Simplifying greatly, we can refer to an underdevelopment tradition which is distinguished by four linked propositions.

The underdevelopment model depends, firstly, on an underconsumptionist account of the dynamics of capital accumulation. Paul Baran's work makes this point most clearly. According to Baran, capitalism presents itself in two main forms. In its youth, it is competitive, and at this time, market forces are more or less untrammelled and local forces of competition are set free to produce an actual economic surplus which approximates a hypothetical 'potential economic surplus'. The problem is that competitive capitalism carries within it the seeds of its own destruction. Precisely because capitalism favours the strong, a process of monopolisation occurs wherein economic production and exchange is centralised in a small number of oligopolistic concerns. These enterprises are able to generate an enormous economic surplus, but they are unable to distribute it beyond a small group of controlling capitalists. As a result, says Baran, monopoly capitalism is prone to crises of underconsumptionism which force its comptrollers to shore up demand by spending on arms, by deficit financing and by extending the reach of capital into the periphery of the world system. The Third World, for Baran, becomes an investment outlet for metropolitan capitalism and a major (and very cheap) source of raw materials. For its part, the Third World underdevelops and begins to exhibit a peculiar 'morphology of backwardness'.

In the 1960s this insight was 'extended' by Andre Gunder Frank, who popularised an analysis of 'the development of underdevelopment'. According to Frank, the development of capitalism in the core has from the very beginning depended upon the transfer of a (physical) surplus from the periphery. As Frank explains:

> (From) the time of Cortez and Pizarro in Mexico and Peru, Clive in India, Rhodes in Africa, the 'Open Door' in China - the metropolis destroyed and/or totally transformed the earlier viable social and economic systems of these societies, incorporated them into the metropolitan dominated worldwide capitalist system, and converted them into sources for its own metropolitan capitalist accumulation and development. The resulting fate for these conquered, transformed or newly established societies was and remains their decapitalization, structurally generated unproductiveness, ever-increasing poverty for the masses - in a word

> their underdevelopment (Frank, 1969, p. 225).

In this fashion Frank detaches an emerging neo-Marxism still further from its classical roots. For Frank (and later Wallerstein, 1974, 1979, 1984), the economies of the periphery have been capitalist since they first produced for exchange in a world market. Again, it matters not that this production is carried on according to several different systems of 'labour control' - for example, free wage labour, serfdom and slavery. For Wallerstein

> the relations of production that define a system are the 'relations of production' of the whole system and the system at this point in time is the European world economy. Free labour is indeed a defining feature of capitalism, but not free labour throughout the productive enterprises (Wallerstein, 1974, p. 127).

A second element of the underdevelopment model describes the structure of metropolis/satellite relationships which facilitates the transfer of a surplus from the 'bottom to the top' of the world system. Frank's early work describes this structure in terms both graphic and geographical. He tells of a chain of metropolis/satellite relationships wherein:

> at each stage along the way the relatively few capitalists above exercise monopoly power over the many below, expropriating some or all of their economic surplus and, to the extent that they are not expropriated in turn by the still fewer above, appropriating it for their own use ... at each point the international, national and local capitalist system generates economic development for the few and underdevelopment for the many (Frank, 1969, p. 7-8).

Later scholars added a more precise account of unequal exchange to this imagery. In Wallerstein's model of surplus transfer, actors in the core (the metropolitan capitalists) call on their state machines to manipulate an economic system which is geared otherwise to the geographical equalisation of profits. In effect they use state power deliberately and persistently to weaken (underdevelop) the periphery - by conquest, by monopoly pricing, by protectionism, and so on; but not so the semi-periphery. Wallerstein implies that it suits the core states to preserve a (variable) semi-periphery as a sort of buffer between themselves and the periphery. More pointedly, Arghiri Emmanuel (1972) provides a complex theory of unequal exchange which hinges upon the power of trade unions in the core to raise real wages in a manner not open to Third World workers. According to Emmanuel, this underpins an unequal exchange of goods in which a greater quantity of 'embodied labour time' flows from the periphery to the core than vice versa. (For a critique, see Bacha, 1978; Brewer, 1980.)

A third element of underdevelopment theory concerns itself with the effects in and upon the Third World of their dependent insertion into a capitalist world economy. Again, there are several variations upon a theme. In the work of Frank and Wallerstein the morphology of dependent societies seems to be determined entirely by the 'logic and needs' of metropolitan capitalism. It is not just that production for exchange takes place in a capitalist world system: for Wallerstein, the class systems of the Third World - or modes of labour control - take shape according to their ability to service this grand global machine. Thus: 'free labour is the form of labour used for skilled work in core countries whereas coerced labour is used for less skilled work in peripheral areas. The combination thereof is the essence of capitalism' (Wallerstein, 1974, p. 127). In other accounts of dependencia (Dos Santos, 1970; Sunkel, 1973; Evans, 1979), more attention is paid to the role played by comprador elites in enforcing local geographies of production. Even in the work of Sunkel and Dos Santos, however, the determination of internal 'factors' by external forces is never far from the surface. For Dos Santos, 'Dependence is a conditioning situation in which the economies of one group of countries is conditioned by the development and expansion of others' (Dos Santos, 1970, pp. 289).

A fourth proposition of underdevelopment theory suggests that (industrial) development within the capitalist world system is at best unlikely and at worst unthinkable. Among those taking a stagnationist line are Frank, Wallerstein and Dos Santos. Frank, indeed, holds out a hostage to fortune when he declares that 'the satellites experience their greatest economic development ... if and when their ties to the metropolis are weakest' (Frank, 1969, p. 9-10). In Frank's judgement, the political choice in the Third World lies between socialism or barbarism: either a country breaks from the capitalist world system or it does not. In the work of Sunkel and others in the (reborn) ECLA tradition of dependency analysis, this stagnationism is neatly sidestepped. Sunkel prefers to speak of certain obstacles to development which are induced by dependent patterns of local market constriction. At this point, however, we are slipping towards more orthodox Marxian perspectives (the dividing line is always very thin). Although Frank's work is now regarded as an extreme case within the dependency paradigm, and thus of limited interest in itself, it remains true that underdevelopment theory 'exemplifies perfectly a form of analysis in common use' (Booth, 1985, p. 762, emphasis added). It remains to see what competency it displays as a guide to events.

Marxist Theories of Development

A third account of the dynamics of global capitalism is rooted in the concepts of classical Marxism first developed by Marx, Lenin and Luxemburg. The bases of this new account will be well known. By the mid-1970s it was clear to many on the Left (and not just the Left: see Lall, 1975) that dependency theory was unable to account for certain important developments in the world economy (and not least the rapid industrialisation of parts of the Third World). Some scholars judged this failure as emanating from the attachment of neo-Marxism to a 'circulationist' or 'neo-Smithian' conception of capitalist development. According to Laclau (1979) and Brenner (1977) (and many others besides: Bath and James, 1976; Slater, 1977; Wolf, 1982), dependency theory from Baran onwards has accepted (unwittingly) Adam Smith's thesis that capitalist economic growth depends only on the extension of an ever more efficient global division of labour. To this, the neo-Marxists have added a radical twist, in the form of a theory of unequal exchange. For Baran, Frank and Wallerstein the extension of an uneven and exploitative world market makes possible the development of core capitalism through the underdevelopment of a capitalist periphery (which loses its economic surplus).

In the eyes of more orthodox Marxists this position is signally flawed. To begin with, the equation of capitalism with a system of production for world market exchange does not stand up. As Laclau explains, if this is all that capitalism amounts to, it must have shaped the lives of 'the slave on a Roman latifundium or the glebe serf of the European Middle Ages, at least in those cases - the overwhelming majority - where the lord assigned part of the economic surplus extracted from the serf for sale' (Laclau, 1979, p. 23). Indeed, by this logic 'we could conclude that from the neolithic onwards there has never been anything but capitalism' (ibid.). More importantly, the world systems perspective fails to grasp the true uniqueness of capitalism as a system of qualitatively expanding commodity production based upon the prior separation of the workers for their means of production and of enterprise from enterprise. Lacking this insight, the neo-Marxists are condemned to draw two false inferences. Firstly, they conceive of 'changing class relations as emerging more or less directly from the (changing) requirements for the generation of surplus and development of production, under the pressures and opportunities engendered by a growing world market' (Brenner, 1977, p. 27). Instead of seeing in local class structures a context for the formation of a 'world market', the neo-Marxists present local 'modes of labour control' as the functional outcomes of this grand world system. Secondly, the neo-Marxists are charged with mistaking the effects of an inflow of wealth from the periphery to the core. This will stimulate a systematic development of the core's productive forces,

> only when it expresses certain specific social relations of production, namely a system of free wage labour where labour-power is a commodity. Only where labour has been separated from possession of the means of production, and where labourers have been emancipated from any direct relation of domination (such as slavery or serfdom) are both capital and labour-power free to make possible their combination at the highest level of technology. Only where they are free, will such combination appear feasible and desirable. Only where they are free will such combination be necessitated (ibid, p. 32, original emphasis).

This last statement directs us to what is distinctive about more orthodox Marxian theories of capitalist development. Unfortunately, there is not space here to do justice to a rich and absorbing literature. Put simply, however, we can say that the 'new Marxism' recovers from Marx an account of the (autocentric) dynamics of capitalist accumulation. In place of the zero-sum logic of unequal exchange models, we learn now that capital is everywhere driven to exploit labour-power, in part through the creation of those fixed capitals in which some see the trappings of 'development'. It follows that there can be no question of capitalism promoting the development of underdevelopment (at least not in the sense that Frank understands it, and certainly not through local systems of production which are in fact pre-capitalist). Marxists must instead explain the continuing underdevelopment of parts of the Third World in terms of a failure of capitalism to take root there (or to take root in forms which demand the production of relative surplus-value). To date, this task has been undertaken in several ways, of which three stand out.

(a) Taking a lead from Rosa Luxemburg is a group of scholars concerned to theorise the articulation of modes of production. This tradition follows Luxemburg in suggesting that capitalism is driven to invade the non-capitalist world. This is not because the extraction of a 'surplus' is a precondition for capitalist development in the core - pace Frank - nor must it lead, at once, to the promotion of capitalism in the periphery. Metropolitan capitalism expands because its domestic markets are incapable of realising an expanded surplus. Once established in the periphery, capitalism comes into conflict with pre-capitalist relations of production. According to Luxemburg (1972), capitalism must 'win' its struggle with 'natural economy' if it is to secure the liberation of labour-power and its coercion into the service of capital. To this end metropolitan capitalism calls up the full force of colonial violence, using military might, oppressive taxation and cheap imported goods to drive the peasants from the land and into the mines and plantations. Victory proves to be pyrrhic, however, for once capitalism engulfs the periphery, its external escape route is lost and the system is condemned to perish in the mires of overproduction and proletarian revolution.

Later theorists have sought to soften Luxemburg's conclusion while making use of her analysis. In the work of P.P. Rey (1971, 1973) attention is directed to a three-stage process of articulation wherein

(1) an initial link is forced in the sphere of exchange, where interaction with capitalism reinforces the pre-

capitalist mode;
(2) capitalism takes root, subordinating the pre-capitalist mode but still making use of it;
(3) there follows the total disappearance of the pre-capitalist mode, even in agriculture.

Others have moved further from Luxemburg. In the 1970s a less deterministic model of 'articulation' was advanced which suggests that the preservation of pre-capitalist modes of production in the periphery can long continue to be 'in the interest' of metropolitan capitalism. This is not just because the capitalists are content to exploit their 'partners' through exchange, nor even is it because they face resistance in the periphery (although they surely do). It is because the reproduction of pre-capitalist systems removes from the agents of metropolitan capitalism the expense of providing 'real wages' to a fully proletarianised labour force. Instead, some of these costs are offset to the pre-capitalist sector where a small cash wage helps to supplement the traditional social wage of the villager. In this way the pre-capitalist society bears the (major) cost of reproducing the labour force of capitalism. The articulation of modes of production ensures that there is 'a process of transfer of labour value to the capitalist sector through the maintenance of self-sustaining domestic agriculture' (Hoogvelt, 1982, p. 179). It also ensures that Third World countries are marked by an extraordinary dualism of form, with pre-capitalist 'underdevelopment' and capitalist 'development' existing side by side in supposed symbiosis (Wolpe, 1980; Armstrong and McGee, 1985).

(b) A second strand of Marxist theory claims to take its lead from Marx (and to stand in opposition to the later heresies of Lenin and the neo-Marxists). According to Bill Warren (1973, 1980), the emergence of dependency theories in the 1960s must be traced back to a betrayal of classical Marxism first breached by Lenin (1970) in his pamphlet on imperialism. In 1928, says Warren, the 'traditional Marxist view of imperialism as progressive ... was sacrificed to the requirements of bourgeois anti-imperialist propaganda and, indirectly, to what were thought to be the security requirements of the encircled Soviet state' (Warren, 1980, p. 8). The Comintern now endorsed two theses that were at best only implicit in Lenin's original pamphlet: firstly, that imperialism actually retarded the industrialisation of the colonies, and secondly, that as a consequence, the Soviet Union and the industrial bourgeoisies of the colonial countries were natural allies in the fight against imperialism. Warren rejects both claims. He reaffirms that, for Marx, capitalism, and indeed imperialism, is always progressive and is everywhere associated with an increase in democracy, individual freedom, scientific rationality and undreamed of technological advance. As Marx himself puts it: 'the bourgeoisie cannot exist without constantly revolutionising the instruments of production, and thereby the relations of production, and with them the whole of society' (Marx and Engels, 1967, p. 83). Nor does this stop at the borders of Europe and North America. As Warren argues: 'Since Marx and Engels considered the role of capitalism in pre-capitalist societies progressive, it was entirely logical that they should have welcomed the extension of capitalism to non-European societies' (Warren, 1980, p. 39) - exploitation and all.

Having thus reclaimed Marx, Warren concludes by 'demonstrating' the vitality of capitalist development in the Third World. Warren acknowleges that the pace of development before 1945 was not especially fast. At this time the full flowering of peripheral capitalism was hindered by the politics of colonialism and imperial preference. Since 1945, however, Warren sees only progress in the Third World. He dismisses the view that GNPs have not grown rapidly or that income inequalities have widened significantly. He also rejects the claim that 'marginalisation' is endemic in the Third World and that people have not gained in health, nutrition and education. Most vigorously, Warren disputes the claim that metropolitan capitalism has acted to prevent the industrialisation of the Third World. Looking in turn at statistics on national average rates of growth in manufacturing industry and on the percentages of national GDP earned by industry, Warren declares that 'the underdeveloped world as a whole has made considerable progress in industrialisation during the post-war period' (ibid, p. 241). For Warren, this is as it must be. His reading of Marx leads him to expect, and to welcome, the capitalist development of the Third World, if only as a precursor to global revolution and the transition to socialism.

(c) A third strand of Marxism stands firm against the

conclusions of Bill Warren. Like Warren (and _contra_ Frank), a group of Marxists concerned to theorise the internationalisation of capital, or a new international division of labour (NIDL), accepts that capitalism is driven to expand into the periphery and that it is there promoting a selective industrialisation. McMichael _et al._ are not alone, however, in dismissing Warren's claims that such industrialisation is 'developmental', or even widespread. They argue that Third World industrialisation is still small in volumetric terms and that it consists, very often, 'in the simple elaboration of raw materials (or) the assembly of parts' (McMichael _et al._, 1978, pp. 110-11). Later authors have developed this complaint. According to Frobel _et al._ (1980), the recent industrialisation of South-East Asia, and parts of Latin America, is 'an institutional innovation of capitalism itself', the NIDL is not being established in response to the changing needs or strategies of Third World countries. Further, the emergence of a NIDL does not alter the fundamental structures of inequality which exist between core and periphery, rather the reverse: the NIDL is said to be based on the exploitation of cheap (mainly female) labour (Ross, 1983); to be guaranteed by repressive Third World regimes (Lamb, 1981); to be directed by foreign transnational corporations (Landsberg, 1979); and to produce local enclave economies unconnected by positive multipliers to a still dependent periphery (Raj, 1984). In short, the internationalisation of capital is promoting growth without development; it is turning 'banana republics (into) pyjama republics' (Adam, 1975, p. 102).

Between them, these three strands of Marxism (which themselves overlap) describe a model of capitalist development and underdevelopment which departs significantly (although again, not absolutely) from the accounts of neo-Marxism and liberal social theory. Once again, it remains to see how it serves as a guide to events.

2. THE THIRD WORLD IN THE WORLD ECONOMY

The position of the Third World in the world economy has changed markedly since 1945. At the end of World War II, it was common to refer to three major blocs in the world economy. In North America and Western Europe (and including, perhaps Australia, New Zealand and Japan) were the countries of the First World. This group of nation-states was characterised by its high per capita incomes, by its strong manufacturing base, by its low birth and death-rates, by its high levels of literacy and educational provision and by its positive 'attitudes to growth and entrepreneurship'. A Second World was found in Eastern Europe (and later, less certainly, in China and Cuba). The economies of this bloc were organised according to non-market principles and were thought to be less developed than their First World counterparts. Finally, there was a residual category: a Third World. Describing itself as a Tiers Monde distinguished by its political non-alignment, this bloc of Latin American, African and Asian countries more often appeared to others as one scarred by poverty, where predominantly agricultural economies were worked by ill-fed, ill-clothed, ill-educated, but highly fertile, 'backward households'. Quite explicitly, the 'Third World' was measured against the First World and as a bloc found wanting (Jalée, 1969).

By the 1960s, and more so the 1970s and 1980s, this stark trinity was less easily recognised. The different dynamics of Third World countries, together with their different insertion in the world economy, helped split the presumed unity of the Third World. The 1986 Report of the World Bank gives some indication of this fragmentation when it presents statistics not on the First, Second and Third Worlds, but on Low-income Countries, Lower Middle-income Countries, Upper Middle-income countries, High Income Oil-exporting countries, Industrial Countries and Non-market Countries.

But what does all this change amount to and to what should we trace its dynamics? Statistics alone cannot answer this question, but data on the emerging (global) geographies of trade, industrialisation and finance can help us interrogate the theories of the world economy outlined in section 1. Moreover, any tension seen to emerge between theory and events can point us towards a 'set of conditions' against which we might judge prospective theories of development (cf. section 3).

Trade

The growth of world trade since 1945 appears to lend support to liberal theories of development. Under a managed

Table 1.1: Growth in world merchandise exports, 1955-84 (billions of U.S. dollars, current prices)

Country group	1955	1965	1970	1980	1984
All developing countries	26	38	58	426	403
Low-income Countries	5	6	8	38	48
Middle-income Countries	21	32	50	388	355
High-income Oil Exporters	2	5	7	203	88
Industrial Non-market Economies	8	20	32	158	180
Industrial Market Economies	58	123	216	1208	2000
Total world exports	94	186	313	1995	2672

Sources: Lal, 1983, p. 21; World Bank: World Development Report, 1986

Table 1.2: The structure of LDC exports: selected years, 1955-78 (percentages)

	1955	1960	1970	1978
Total exports	100.0	100.0	100.0	100.0
Food	36.5	33.6	26.5	16.4
Agricultural raw materials	20.5	18.3	10.0	4.8
Minerals, ores	9.9	10.6	12.3	4.6
Fuels	25.2	27.9	32.9	52.8
Manufactures	7.7	9.2	17.7	20.9
Total non-fuel exports	100.0	100.0	100.0	100.0
Food	48.9	46.7	39.5	34.8
Agricultural raw materials	27.4	25.3	14.9	10.1
Minerals, ores	13.3	14.6	18.3	9.7
Manufactures	10.4	12.8	26.4	44.4

Source: Lal, 1983, p. 35

Table 1.3: The regional destination of the exports of the USA, Japan and the EEC, 1948-82 (percentages)

From North:	To: Developed market economies	Developing market economies: Non-OPEC	Developing market economies	Developing market economies: OPEC	Centrally planned economies
<u>USA</u>					
1948	56.4		40.4		0.5
1955	61.9		37.3		0.05
1960	63.9		35.8		0.8
1965	64.5	25.0		4.9	0.5
1970	69.5	24.8		4.8	0.8
1978	60.7	24.1		11.6	2.3
1982	63.5	23.0		12.1	2.2
<u>Japan</u>					
1948	38.5		57.7		1.5
1955	39.6		58.2		0.5
1960	47.7		50.9		1.5
1965	51.4	36.4		6.4	5.6
1970	54.6	34.9		5.1	5.4
1978	47.2	31.5		14.5	5.8
1982	49.4	31.9		13.1	5.6
<u>EEC</u>					
1948	63.9		32.2		2.9
1955	62.3		31.8		5.3
1960	72.5		22.7		3.3
1965	78.9	13.1		4.1	3.4
1970	80.1	10.7		3.4	3.7
1978	76.8	10.2		8.6	4.2
1982	77.4	9.3		8.5	4.4

Source: United Nations International Trade Yearbooks, 1960, 1978 and 1982.

regime of free trade and fixed exchange rates up to the early 1970s, and floating rates thereafter, the Third World has seen its volume of exports rise in line with a general increase in global exchange (see Table 1.1). The structure of Third World exports has also diversified. Table 1.2 reveals that 'whereas manufactures accounted for only 10% of (LDC) non-fuel exports in 1955, that share had risen to over

Table 1.4: Change in export prices and in terms of trade, 1965-85 (annual percentage change)

Country group	1965-73 average	1973-80 average	1981	1982	1983	1984	1985
Change in export prices							
Developing countries							
Food	5.0	9.6	-8.2	-8.8	5.6	2.0	-8.1
Non-food agriculture	4.2	10.5	-14.4	-8.6	5.7	-2.0	-10.0
Metals and minerals	2.4	4.8	-7.6	-8.5	-0.1	-1.7	-4.9
Fuels	7.9	27.2	12.5	-3.2	-12.4	-2.1	-2.5
Manufactures	7.2	8.1	0.2	-3.2	-2.5	-1.9	1.3
Industrial countries							
Manufactures	5.4	11.0	0.5	-1.4	-2.6	-1.8	1.3
Change in terms of trade							
Low-income Countries							
Africa	0.1	-1.8	-11.8	-0.9	4.8	5.0	-5.6
Asia	3.2	-2.4	1.1	1.2	-1.2	1.5	-1.9

Table 1.4: continued

Country group	1965-73 average	1973-80 average	1981	1982	1983	1984	1985
Middle-income Countries							
Oil exporters	-0.4	8.5	5.4	0.2	-7.7	0.3	-2.9
Oil importers	0.0	-3.0	-4.4	-0.6	2.3	0.1	-0.1
All developing countries	0.8	1.5	-1.0	-0.1	-1.3	0.4	-1.1

Note: Data are based on a sample of 90 developing countries.

Source: World Bank, World Development Report, 1986.

40% by 1978' (Lal, 1983, p. 34). Furthermore, the World Development Report for 1986 records that in Upper-middle Income Countries the share of 'industrial' goods in total exports is now 55 per cent. The Report also documents the increasing penetration of developed countries' import markets achieved by some developing countries.

These figures should not be lightly dismissed, but they do need to be seen in a context not always highlighted by liberal theorists. Thus, for example, the growth of Third World exports in volumetric terms needs to be set against the growing relative closure of developed nations' trading patterns. Table 1.3 confirms that US, EEC and Japanese exports are increasingly being shipped to developed market economies. (The growth of exports to OPEC countries is one exception to this trend and it is significant that Japan still depends more on Third World markets than either the US or the EEC.) Nor is it the case that Third World countries have shared equally in the post-war growth of global exchange. The trading success of OPEC and the NICs, and of some Middle-income Countries, has not been matched by the Low-income Countries (and especially those in Sub-Saharan Africa), with the result that a united Third World perspective on trade is now more difficult to attain (UNCTAD, 1983).

With respect to the terms of trade figures, the evidence is (at best) equivocal. Even if the link between a declining terms of trade and slow development is not yet proven (Morawetz, 1977), the evidence clearly points to a significant and continuing loss of relative earnings suffered by food and non-fuel mineral exporting countries (Table 1.4). In the 1980s this loss of earnings extended to other commodity exports as the world trading system entered a period of prolonged crisis and rapidly falling commodity prices. Third World countries are also faced with creeping protectionism in the markets of industrial economies and it is unlikely that the slowdown in this circuit of international trade is being offset by a rapid growth in intra-firm trade (which now accounts for 30-40 per cent of the total world trade: United Nations, 1983).

Industry

There can be no doubt that parts of the Third World have industrialised apace over the past 30 years. According to

Bill Warren,

> for the LDCs as a whole manufacturing accounted for 14.5% of GDP in 1950-4; the figure rose to 17.9% in 1960 and 20.4% in 1973. In the developed capitalist countries manufacturing contributed 28.4% to GDP in 1972. The difference is therefore becoming rather small (Warren, 1980, p. 244, original emphasis).

This rate of progress was maintained throughout the 1970s until, in 1980, the corresponding manufacturing:GDP ratio for the LDCs stood at 23 per cent (Cody et al., 1980). Moreover, in selected Newly Industrialising Countries (NICs) the average rate of growth of manufacturing has been outstanding and seems to have risen in tandem with a growth in manufacturing employment. Table 1.5 reveals that South Korea, Taiwan and Singapore have each maintained annual average rates of growth of manufacturing in excess of 12 per cent (compared with annual average rates of 2-3 per cent during the height of Britain's 'Industrial Revolution').

The 'success' of these NICs has been warmly welcomed by liberal economists, and it has encouraged one of their number to develop a new 'transition theory' of global economic change. According to Michael Beenstock, the industrialisation of the Third World must be traced back 'to the more open economic policies that were implemented in the so-called "newly industrialising countries"' (Beenstock, 1984, p. 14). In broader terms, Beenstock contends that the industrialisation of the South has altered the balance of power between the First and Third Worlds, so that the principal dynamic in the world economy comes today from the South and not from the North. As far as the North is concerned, it has suffered from the lower relative prices of manufactures on world markets consequent upon their greater supply. As Beenstock puts it, 'The rise in the relative price of raw materials towards the end of the 1960s was brought about by supply shocks in the market for raw materials' (ibid, p. 15). The fall in the relative prices of manufactures has in turn provoked a set of market signals which causes resources to shift from the manufacturing sector to other sectors of the economy in the bloc of developed countries. This is called the de-industrialisation effect because industry suffers while the rest of the economy expands. Meanwhile, the South has been gaining.

Table 1.5: The 'Newly Industrialising Countries': selected statistics

(a) Annual average rates of growth of manufacturing, 1951-69, 1965-74 and 1973-84, for selected countries (percentages)

	1951-69	1965-74	1973-84
Brazil	7.8	11.2	4.9
Jordan	15.2	-	12.9
Korea (Rep. of)	16.9	24.4	11.5
Malaysia	6.4	9.6	8.7
Mexico	7.4	7.6	5.0
Panama	14.2	8.5	2.1
Philippines	8.5	5.8	4.3
Singapore	14.8	15.3	7.6
Taiwan	16.1	-	-
Thailand	8.7	11.4	10.0
Venezuela	10.5	5.0	3.4

(b) Percentage of labour force in industry, (a) 1965 and 1980, for selected countries

	1965	1980
Brazil	20	27
Jordan	26	26
Korea (Rep. of)	14	27
Malaysia	13	19
Mexico	22	29
Panama	16	18
Philippines	16	16
Singapore	27	38
Taiwan	-	-
Thailand	5	10
Venezuela	24	28

Note: (a) Industry: mining, manufacturing, construction, electricity, water and gas.

Sources: Warren (1980); World Bank, World Development Report, 1986.

On the one hand, it gains from the rise in worker militancy that is induced by the first round of de-industrialisation in the North; and on the other, the South reaps the rewards of greater foreign investment. As the rate of return on capital improves in the South relative to the North, so investment begins to flow to the Third World (and the market is vindicated).

Not everyone takes this sanguine view of events. Gunder Frank (1982), not unexpectedly, tries hard to discount the significance of the NICs. In his judgement, the 'miracles' of growth in South-East Asia are but 'mirages'. On the one hand, because their rates of growth have been so extraordinary, they stand condemned as exceptions to the rule of stagnationism and not as models for the rest of the Third World to copy. On the other hand, such growth as has occurred is dismissed as non-developmental. In two telling phrases, Frank complains that 'the inability of much of the world to follow (the NICs) is that this development or ascent has been misperceived as taking place in particular countries when it has really been one of the processes of the world system itself'; and 'The new dependent export-led growth of manufacturing and agribusiness production for the world market are in no way significantly different from the old raw-materials export-led growth that underdeveloped the Third World in the first place' (Frank, 1982, p. 22). In short, it is all down to the transnational corporations (TNCs), or the Americans, or global capitalism, or the world system, no credit (or development) reflects upon the peoples, institutions and governments of the countries concerned (see Corbridge, 1986b).

Others accept some parts of Frank's account while avoiding the glum circularity of his logic. Thus, it surely is the case that much of the recent industrialisation of the NICs has been led by Western TNCs (but by no means all - Dicken, 1986). It is also true that many TNCs are attracted to the Third World by the low wages paid there (especially to young women, many of whom are discarded at 30), that many have transfer priced significantly against host countries (Lall, 1980), that early TNC operations in the Third World frequently involved a local assembly of parts (Taylor and Thrift, 1982), and that management and R&D decisions often remain the preserve of aliens (Teulings, 1984). (Again, however, this is not the whole story. The direction of transfer pricing on a global scale is much debated (Auty, 1983; Lall, 1984); most TNCs have not

Table 1.6: Manufacturing industry in the Low-income Countries: selected statistics

(a) Structure of production: distribution of gross domestic product (per cent), 1965 and 1984

	Percentage GDP from							
	Agriculture		Industry		Manufacturing		Services	
	1985	1984	1965	1984	1965	1984	1965	1984
All Low-income Countries	42	36	28	35	14	15	30	29
China and India	42	36	31	38	15	15	27	26
Other Low-income	43	36	16	20	11	15	41	44
Sub-Saharan Africa	43	39	16	18	9	10	41	43

(b) Annual average rates of growth of manufacturing, 1965-73 and 1973-84: selected Low-income Countries (percentages)

	1965-73	1973-84
Burma	3.2	6.1
Central African Republic	5.4	-4.3
Ethiopia	8.8	3.5
Ghana	6.5	-6.9
Guinea	-	-2.0
Haiti	3.0	5.4
India	4.0	5.9
Kenya	12.4	6.0
Senegal	6.2	1.9
Zaire	-	-5.0

Source: World Bank, <u>World Development Report, 1986.</u>

relocated to the Third World, but have opened up additional operations there (Thrift, 1986); product diversification is a feature of recent Third World industrialisation (Chudnovsky, 1979); and TNCs are attracted to the NICs as much by skilled labour forces and buoyant local markets as by cheap wages (Becker, 1984)).

Finally, some scholars have called into question the temporal and spatial significance of 'Third World industrialisation'. Not unreasonably, attention has been drawn to the continuing 'absence' of manufacturing industry in many Low-income Countries (Table 1.6), where a dependence upon primary commodity production (and the simple elaboration of parts) still prevails. To talk, as Warren does, of 'the underdeveloped world as a whole (making) considerable progress in industrialisation during the post-war period' (Warren, 1980, p. 241) is quite misleading. The selective industrialisation of the periphery is enforcing a further fragmentation of one Third World. Temporally, too, the fate of manufacturing industry in the semi-periphery is not yet certain. Since 1980 the annual average rate of growth of manufacturing industry in Upper Middle-income Countries has dropped to below 2 per cent (and in some cases is negative). There is evidence also that new production technologies are being developed which may allow the re-importation of some manufacturing to the core countries (Jenkins, 1984). To some scholars this signals the fragility and continuing dependence of Third World industrialisation. According to Frobel (1983) and Harris (1983), the internationalisation of capital in the 1970s was an attempt to solve a crisis of accumulation in the core, and was made possible by the extension of cheap credits and syndicated bank loans to the NICs. The collapse of these financial circuits in the 1980s, together with the 'failure' of market-widening strategies in the South, has sabotaged this attempt to regulate the world economy through 'Global Fordism' (see section 3). In future, it is possible that new forms of regulation will be sought, or will emerge, within the countries of core capitalism.

Finance

The emergence of a 'new international division of labour' has been matched by recent developments in the internationalisation of finance. In the managed economic

Table 1.7: Composition of net capital flows to developing countries, 1960-2 and 1978-80 (percentages)

Net capital flows	1960-2	1978-80
Official development assistance	59	34
Other non-concessional flows, mainly official	7	13
Private non-concessional flows	34	53
Direct investment	20	14
Export credits	7	13
Financial flows	7	26
Total	100	100
Memorandum item		
Total amount (billions of dollars)		
Current prices	9	84
1978 prices	25	76

Source: Lal, 1983.

order of the 1950s and 1960s, net capital flows to the Third World mainly took the form of official development assistance and private direct investment (Brett, 1985). Together, these two categories of flow comprised 79 per cent of the total net transfer in 1960-2, with just 7 per cent being transferred as private financial flows (Table 1.7). Since then developments in the international monetary system - and in the Bretton Woods system itself - have secured a reversal of this pattern, with private financial flows accounting for 26 per cent of the total net transfer of capital to the Third World in 1980 (and over 50 per cent of the net flow to Middle and Upper Middle-income Countries: Cline, 1983).

The reasons for this reversal are three-fold. In the first place, the demise of fixed exchange rates and the dollar-gold standard allowed the Americans, in the 1970s, to fund a growing balance of payments deficit by exporting dollars to the rest of the world (Parboni, 1981; Williamson, 1985). This had the effect both of fuelling global inflation and of placing Eurodollars in the coffers of the relatively unregulated Eurobanks. (A Eurodollar - or Xenodollar, as Lipietz more accurately describes it (Lipietz, 1987, p. 142) -

is a dollar, or any currency unit, deposited in a country other than that in which it is issued). To this growing pool of credit money was soon added the fund of petrodollars recycled through the Eurobanks by the OPEC countries. Following the 'oil price shock' of 1973-4 an estimated $80 billion was recycled in this fashion. Finally, the banks themselves adopted new structures of portfolio management (and new financial instruments: Llewellyn, 1985) which disposed them towards recycling some of these dollar deposits as syndicated loans to the Third World (and to the NICs and non-OPEC oil-exporting countries especially).

For a time, the process of recycling - and of deregulated private credit creation - appeared to meet the needs both of the developing world (for finance) and of the developed world (for export markets). In some quarters a return to market-led financial transfers to the Third World is still lauded, the recent emergence of a debt crisis notwithstanding. According to Lal (1983) and Beenstock (1984), the so-called debt crisis is much exaggerated and its causes and consequences widely misunderstood. Both authors point out that it was Brazil and Mexico, South Korea, Argentina and Venezuela - the big five debtor nations - which in the 1970s enjoyed some of the most rapid rates of growth in GDP, in exports and in manufacturing industry (Table 1.8). At this time real rates of interest were often negative. Lal and Beenstock also claim that the debt:export ratios of the major Latin American countries are not high by historical standards and can be swiftly brought down by a combination of world recovery based on sound money policies and by 'sensible' strategies of economic management inside the debtor countries. For Beenstock, 'what is going on at present is a liquidity crisis rather than a solvency crisis which is therefore likely to be a temporary problem' (Beenstock, 1983, p. 23).

The outlook of many others is less optimistic. Even within the liberal camp there are major disagreements, with Keynesian economists voicing alarm at the size of the debt overhang in the world economy (see Table 1.9). Doubt is also expressed at the wisdom of private financial flows substituting for aid (which hurts the Low-income Countries cut off from bank finance), and at the rectitude of real transfers of income flowing from the South to the North (Brandt, 1983; Griffith-Jones and Lipton, 1984). Critics further insist that the debt crisis is not of the Third World's making, but was triggered off by the financial 'experiments'

Table 1.8: Third World debt and development, 1970-80

Country	Average annual growth rates, %		Debt service (a) as a percentage of:			
			GNP		Exports of goods and services	
	Output 1970-80	Exports 1970-80	1970	1980	1970	1980
Mexico	5.2	13.4	2.1	4.9	24.1	31.9
Brazil	8.4	7.5	0.9	3.4	18.9	22.9
Argentina	2.2	9.3	1.9	1.4	21.5	16.6
Venezuela	5.0	-6.7	0.7	4.9	2.9	13.2
South Korea	9.5	23.0	3.1	4.9	19.4	12.2
Chad	-0.2	-4.0	1.0	3.1	3.9	n/a
Niger	2.7	12.8	0.6	2.2	3.8	2.3
El Salvador	4.1	1.5	0.9	1.2	3.6	3.5
Ghana	-0.1	-8.4	1.1	0.6	5.2	6.0
Ethiopia	2.0	-1.7	1.2	1.1	11.4	7.6

Note: (a) Debt service is the sum of interest payments and repayments of principal on external public and publicity guaranteed medium and long-term debt.

Source: World Bank, World Development Report, 1982, Tables 2, 8 and 13.

Table 1.9: Composition of debt outstanding: major borrowers, 1970-84

Country	Total gross external liability (billions of dollars)		Debt from official sources (per cent)		Debt from private sources (per cent)		Debt at floating rates (per cent)	
	1970	1984	1970	1984	1970	1984	1970	1984
Argentina	-	45.8	12.6	9.2	87.4	90.8	13.9	37.5
Brazil	-	104.4	29.7	13.8	70.3	86.2	43.5	79.1
Chile	-	19.9	47.2	8.8	52.8	91.2	9.6	81.2
Egypt	-	23.2	66.0	80.8	34.0	19.2	3.1	1.7
India	-	30.7	95.2	79.6	4.8	20.4	0.0	7.9
Indonesia	-	32.5	71.5	48.1	28.5	51.9	10.2	23.6
Korea (Rep. of)	-	43.1	37.8	32.3	62.2	67.7	15.6	46.8
Mexico	-	97.3	19.5	8.8	80.5	91.2	46.9	83.0
Turkey	-	22.3	92.1	68.0	7.9	32.0	0.8	28.5
Venezuela	-	34.2	28.5	0.7	71.5	99.3	20.6	93.8
Yugoslavia	-	19.8	37.3	25.7	26.7	74.3	7.6	56.0

Source: World Bank, World Development Report, 1986.

of the US Treasury, which sought to combat inflation (and balance its budget) through the medium of high interest rates (for an interesting account of 'Reaganomics', see Stockman, 1986). More telling still is the observation that the mechanisms of financial regulation in the world economy have broken down. Since the collapse of the Bretton Woods system the private creation of credit has been much expanded and this has added to the growth of moral hazard in international banking. It remains possible that a measure of certainty will be restored to the international monetary system by a revamped International Monetary Fund (the de facto lender of last resort) and by new agreements which bring into force target exchange rates and a multiple currency system (this last displacing the disturbing and asymmetrical role now assumed in international monetary affairs by the dollar). However, this is only one scenario among many (Aglietta, 1982; Parboni, 1986). Equally plausible scenarios must include:

(1) general stagflation amidst defensive protectionism and the formation of regional trading alliances;
(2) a debt moratorium, financial collapse and the fragmentation of the world economy;
(3) debt cancellation, the recycling of Japanese trade surpluses and renewed efforts to extend a regime of global Fordism - or some combination thereof.

It is hardly necessary to add that scenarios (1) and (2) are not conducive to economic development in the Third World.

3. DEVELOPMENT STUDIES: A PROSPECTUS

Our inability to foretell the future should not alarm us, or make us unduly sceptical of the claims of development theory. As Lenin used to say, history has infinitely more imagination than we have (cf. Lipietz, 1987, p. 11). Nevertheless, development theories should be able to 'predict the past', and against this criterion it seems that we must treat with caution each of the competing sets of claims advanced in section 1. (Do note, however, that these sets of claims are presented here as 'paradigms' or 'archetypes' and as such their differences are exaggerated and important nuances are lost: Binder, 1986.)

Consider, firstly, the liberal account of development.

A strength of this tradition is its willingness to invoke the 'idea of progress', all previous evidence to the contrary notwithstanding. By this I mean the liberals do not surrender to the teleology of underdevelopment so beloved by neo-Marxists. Liberal theorists are unimpressed by the claim that 'because capitalism promoted the underdevelopment of the periphery in the eighteenth and nineteenth centuries', so 'capitalism must continue to underdevelop the periphery in the twentieth and twenty-first centuries'. This claim would be rejected by Keynesian economists as much as by free-market theorists, the major difference being that free-market theorists would challenge both clauses in the argument, while Keynesians would reject the argument as a non-sequitur. Both versions of liberal theory would also wish to highlight the role played by actors and policies in the development process. Nation-states, on this account, are not simply at the mercy of a (malevolent) world system, but can make the system work to their advantage by pursuing 'rational' development policies. This is the core of the liberal account of differential development in the Third World (Balassa, 1981).

However, the strengths of liberal theory are also its Achilles heel. To the extent that liberal theory escapes one version of teleology, it imposes its own in the idea of 'necessary progress' (Taylor, 1987). Moreover, while liberals are right to insist that the past is no guarantor of the future - in the sense that Third World countries are not forever the prisoners of their imperial past - they go too far in promoting a voluntarist account of 'development policy' which finds no place for antecedent conditions or for the structural constraints upon action. This failing is evident in liberal accounts of the success of the NICs. It may be that today, Taiwan and South Korea operate open-market economies which like to respond to undistorted price signals. It must be emphasised, however, that the logic of such a policy, from a developmental perspective, was only vouchsafed by a series of prior structural reforms in each economy (including land reform, fiscal interventionism and protectionism), and that these reforms were themselves made possible by the particular and evolving class structures (and international position) of each economy. To extend the same logic to Bangladesh could be disastrous. In an economy almost totally lacking in physical and educational infrastructure, and in which patterns of commodity demand and labour supply are 'distorted' by a grossly unequal distribution of resources and massive rural to urban migration, a reliance on open market prices will only reinforce a comparative advantage in agricultural exports and a commodity import profile geared to elite goods (Toye, 1983).

A similar failure to think through the political context of economic policy is evident in Beenstock's model of the World Economy in Transition. The weakness of Beenstock's model is that it fails to develop even a rational-choice account of national action in an interdependent world (cf. Carling, 1986). Beenstock's work is marked by a naive assumption that each country will act benignly and for the greater good. The possibility that Northern governments might pursue policies of protectionism against a second generation of NICs - for reasons of self-interest - is simply not considered. Once again, the realm of politics does not impinge upon the 'pure' world of economics.

If liberal accounts of development are marked by an other-worldliness, the theory of underdevelopment is scarcely more credible. To be fair, scholars in this tradition have served us well by highlighting the flaws of comparative advantage theory, by challenging the equation of development with Westernisation and by drawing attention to the powerful asymmetries which structure the interdependent world-system. What the neo-Marxist account has been unable to anticipate is the rapidly evolving geography of the capitalist world-system. The selective industrialisation of the Third World since the 1950s has called into question the very bases of an account which insists that economic development in the periphery is inimical to the interests of the developed world. Moreover, the tension between underdevelopment theory and the events it so fails to describe highlights a series of conceptual flaws at the heart of neo-Marxism. Some of these flaws are well known and bear upon the quantitative model of the capitalist world system proposed by underdevelopment theorists (and by followers of Wallerstein today). More damning are the flaws of circularity and teleology. By defining capitalism as a zero-sum system in which the development of some countries depends upon the underdevelopment of others, Frank sets dependency theory *en route* to a 'pessimistic functionalism' (Lipietz, 1987, p. 18) wherein all developments in the Third World are decried as 'dependent developments' *because* they do not approach the model of autocentric development, which is by definition arrogated to the core countries of metropolitan capitalism (Bernstein, 1982). Furthermore, by equating capitalism with a system of production for unequal exchange, the neo-Marxists build out of their models any conception of the changing dynamics of capital accumulation (in both core and periphery).

In the 1950s and early 1960s this definition appeared to work to the advantage of neo-Marxism. For some years a regime of accumulation based on 'central Fordism' ensured that First World countries developed together through a system of internationalised production, exchange and consumption. At this time the Third World was 'marginalised' and the differences between core and periphery were re-emphasised. From the mid-1960s, however, a series of crises in central Fordism prompted a renegotiation of the 'logic of capital accumulation' which involved, in part, the internationalisation of capital and the selective industrialisation of the Third World. By failing to acknowledge the changing geography of capital accumulation, the theory of underdevelopment finds itself at odds with the historical geography of capitalism itself. In the curious world of neo-Marxism, capitalism is at once transcendental and insensitive to the rich national geographies of class and accumulation.

Marxist theories of development may also be charged with an unseemly formalism, although their greater sensitivity to time and space (under capital) warrants a less guarded welcome. Thus, on the one hand, the recovery of classical Marxism has helped direct the new radicalism 'to where the main dimension of development-underdevelopment lies: the economy, the forces and relations of production' (Mouzelis, 1978, p. 44). This, in turn, has fostered a more rigorous account of the dynamics of capital accumulation and a greater sensitivity to its changing rationale and differential spatial impress (see Taylor, 1979). On the other hand, where the 'new Marxism' has dressed itself in the garb of 'structuralism' it is still burdened by an unhealthy historicism (not to mention functionalism).

In the work of the articulation theorists an impression is given that changes and developments in the Third World are directed - in the last instance - by and for the First World (or metropolitan capital). On this account the existence and preservation of peripheral pre-capitalist modes of production (PCMPs) (or not, as the case may be) can simply be read off from the needs of metropolitan capital. If the PCMP survives (as in the Bantustans), then that is evidence of its functionality for capitalism; and if it does not (as in the plantations of Latin America), then that too is evidence of capitalism's functional requirements. In each case the possibility that the particular peripheral social formation might be the result of an unhappy compromise between two 'modes of production' is swept away beneath the structural causality of capitalism's laws of motion. Such latent functionalism is evident, too, in the work of some theorists of the 'new international division of labour'. Too often, we are presented with accounts of the selective industrialisation of the Third World which find no room for local class structures and policy decisions (and so fail to explain why particular countries are 'chosen' by capital to become NICs and not others), and which imply that because the extension of global Fordism did in some respects 'solve' the crises of central Fordism, so it was designed to function in this way and to this end. As Lipietz points out, this is to confuse a set of results 'with causes of existence ... A body of partial regularities which "forms a system" is not the same thing as a system which "unfolds"' (Lipietz, 1987, p. 18).

The need to avoid finalism and functionalism, tautology and teleology, in our accounts of the development process cannot be overemphasised. All of the development theories reviewed here have their merits, but each tends to mistake one conjuncture, one configuration, within the capitalist world system for the 'essence' of that system. Similarly, there is a tendency to privilege either structure over agency, or agency over structure, to the extent that the reproduction of each is guaranteed only by the dogmatism of concepts. To escape this Scylla and Charybdis, development geography must forge a less stylised account of development and underdevelopment. This account will be sensitive to:

(1) the constant and shifting production of space under the rule of capital;
(2) the changing sites and temporalities of capital accumulation and crisis formation in the world economy; and
(3) the fragile (economic and non-economic) conditions of existence of national and international regimes of accumulation.

Our fourth account would also need, firstly, to develop a clutch of 'meso-concepts' which split open the determinism of those theories seeking to read off particular empirical developments from the 'logic of capitalism'; and secondly, to eschew forms of reasoning which conceive of capitalism as a totality with functional requirements and/or necessary laws of motion.

Happily, there are signs that such an account is emerging. In geography, David Harvey (1982) has provided an extraordinary model of 'The Limits to Capital' which, if apocalyptic in its conclusions, is richly suggestive in its theory of crisis formation and displacement under the rule of capital (see also Neil Smith's account of Uneven Development: Smith, 1984). Dean Forbes (1984) has also argued cogently for development geography to take on board Giddens's theory of structuration, and in the work of Bell (1986) and Benton (1984) the role of culture and gender, respectively, is taken seriously and not simply read off from the 'economy'. Finally, Corbridge (1986a) suggests that the tyranny of rationalism may be undone by the adoption of less deterministic conceptions of capitalism and its effects. He endorses the argument of Cutler et al., 1977, p. 336) that, 'while specific social conditions and practices always presuppose definite social conditions of existence, they neither secure those conditions through their own action nor do they determine the form in which they will be secured'. Corbridge further insists that 'there is nothing in the concept of capitalism itself which should lead us to expect that it must have X, Y or Z development (or underdevelopment) effects. Such contingencies are not forged at this macro-theoretical scale' (Corbridge, 1986a, p. 67).

However, where are these contingencies forged? The most insistent answer to this question is now coming from a group of scholars who make up the so-called Regulationist school of French post-Marxists. Again, this is not the place for a full exposition of this school's views (which draw heavily on Marxism and which are known already to many industrial geographers: cf. the essays in Scott and Storper, 1986). Nevertheless, we can close this 'prospect' by pointing up three features of the new approach which together come close to meeting the criteria of adequacy for development theory outlined above.

The work of the Regulationists is distinguished, firstly, by the challenge it presents to various 'ideologies of globalism' (Aglietta, 1985). For Aglietta the national dimension is primary and the world economy is theorised as a system of interacting national social formations. This is an important point. Clearly, only a fool or a knave would deny that we live today in an interdependent world, and that the power of nation-states is being slowly eroded and transferred to the markets and to circuits of international capital (Coakley and Harris, 1983; Radice, 1984; Hamilton, 1986). Nevertheless, what Petras and Brill (1985) call the 'tyranny of globalism' can be pressed too far, with the result that the changing constitution and dynamics of the 'world system' are lost amidst a welter of platitudes about core and periphery. As Lipietz puts it:

> Something which 'forms a system' and which we intellectually identify as a system precisely because it is provisionally stable must not ... be seen as an intentional structure or inevitable destiny because of its coherence. Of course it is relatively coherent: if it were not, we would have international conflict and there would be no more talk of systems. But its coherence is simply the effect of the interaction between several relatively autonomous processes, of the provisionally stabilized complementarity and antagonism that exists between various national regimes of accumulation (Lipietz, 1987, pp. 24-5; see also Corbridge, 1987).

The Regulationist account is marked, secondly, by its understanding of the processes of accumulation and crisis under capitalism. In place of more orthodox Marxian formulations which stress the continuity of these processes, the Regulationists offer a set of 'meso-concepts' which help us to see the history of capitalism in terms of a theory of discontinuous equilibria (and within which the regime of accumulation and the site of crisis change periodically). A little detail is unavoidable here. In the work of Lipietz and Aglietta, we are introduced to the concepts of a <u>regime of accumulation</u> and a <u>mode of regulation</u>. A regime of accumulation 'describes the fairly long-term stabilization of the allocation of social production between consumption and accumulation ... (both) within a national economic and social formation and between the social and economic formation under consideration and its "outside world"' (Lipietz, 1987, p. 14). A mode of regulation 'describes a set of internalized rules and social procedures' which ensure the unity of a given regime of accumulation and which 'guarantee that its agents conform more or less to the scheme of reproduction in their day-to-day behaviour and struggles' (ibid.).

Thus defined, these concepts can be put to work to build up a four-stage model of capitalist development and crisis in the twentieth century. (Needless to say, much is lost in this simple schema: see Aglietta, 1982; Boyer and Mistral, 1983; Lipietz, 1985). Until the early twentieth century, the dominant regime of accumulation in the advanced capitalist countries was 'extensive'. This regime centred upon the expanded reproduction of means of production and involved both a sharp international division of labour and a relative orientation to external markets. The corresponding mode of regulation was 'competitive', which means, in part, that national regimes of accumulation were forced to adjust to one another through their balance of payments (and ultimately in response to rates of interest for short-term capital set in London, the financial centre of the hegemonic power). By the 1920s this combination of extensive accumulation/competitive regulation had entered a period of 'major crisis'. According to Lipietz, the dominant regime of accumulation now shifted to a system of Fordism - centred upon the United States - which sponsored a growth in output beyond that which could be realised under a competitive mode of regulation. Put simply, a system of competitive regulation demands 'the a posteriori adjustment of the output of the various branches to price movements ... and of wages to price movements' (Lipietz, 1987, p. 34). Accordingly, wages are able to rise only slowly - if at all - and capitalism falls into a crisis of overaccumulation.

After World War II, a regime of Fordist - or 'intensive' - accumulation came to be matched by a 'monopolistic' mode of regulation. Within nation-states this mode of regulation 'incorporated both productivity rises and a corresponding rise in popular consumption into the determination of wages and nominal profits a priori' (ibid., p. 35). Internationally, a system of regulation emerged which acknowledged the United States as the new hegemon and which installed the dollar as the accepted international unit of account. This system proved stable so long as the Americans had a trade surplus with Europe and Japan, and so long as Europe and Japan had funds to buy American producer goods (Triffin, 1960). Since the mid-1960s, this equation has become less assured and we are now living through a second 'major crisis' in twentieth-century capitalism. The difference this time is that demand is holding up well - thanks to the international credit economy (Strange, 1986) - but profits have fallen amidst generalised inflation and/or stagflation. According to Aglietta (1985), the origins of the present crisis are to be found in a crisis of intensive accumulation itself - the profits squeeze - and in the collapse of the Bretton Woods system, which was an embodiment of monopolistic regulation at the international level. Increasingly, the financial power of the US is at odds with its declining real economy, and this state of non-correspondence has encouraged the Americans to pursue monetary policies which now threaten the process of capital accumulation throughout the world system (Corbridge, 1984). The Third World has borne the brunt of this onslaught, with the tentative expansion of a regime of global Fordism for the present being curtailed by a US-inspired 'debt crisis'.

The model of capital accumulation and crisis formation proposed by the Regulationists is enhanced, finally, by the philosophical stance of this school. Lipietz, especially, is scathing in his critique of theoretical 'finalism and functionalism'. With regard to regimes of accumulation and modes of regulation, he insists that:

> Whilst no immanent destiny condemns a particular nation to a particular place within the international division of labour, a provisional solution for the immanent contradictions of capitalism can at times be found (and I insist that it is a matter of chance discovery) in deviations and differences between regimes of accumulation in different national social formations. In such periods, a field of possible positions ... does exist, but positions within it are not allocated in advance (Lipietz, 1987, p. 24, original emphasis).

Put bluntly, the emergence of Fordism, and its extension as global Fordism, must not be seen as preordained 'solutions' to capitalist crisis, neatly identified and invented by a controlling class of capitalists. They are, rather, one of many 'experiments' thrown up by capitalism, and survive only as 'successful mutants' on probation.

4. CONCLUSION

The work of the Regulationists is more complex than I can convey here and refinements continue to be made to its leading concepts (De Vroey, 1984; Tokman, 1985). Nevertheless, it is not unreasonable to propose this body of work as a theory both of and for these 'post-modern' times (Gregory, 1986). The development of the world economy over the past 30 to 40 years has shattered the unity of the Third World and presents a challenge to those theories which see in the process of capitalist development a certain and continuing tendency either to global equalisation or global disequalisation. In a world in which famine stalks the low-income, high birth- and death-rate lands of sub-Saharan Africa, and where certain NICs contemplate their stilted integration into a 'new international division of labour', we need a theory which measures up to the immense variety of forms and processes which are consistent with the 'logic of capital accumulation'. If I might leave the last words with Lipietz:

> Beware of labels. Beware of the International Division of Labour. Look at how each country 'works', at what it produces and for whom it produces it. Look at how and why specific forms of wage relations and regimes of accumulation developed. And be very careful about "casting a net" over the world in an attempt to grasp relations between regimes of accumulation in different national social formations (Lipietz, 1987, p. 28).

REFERENCES

Adam, G. (1975) Multinational corporations and worldwide sourcing. In H. Radice (ed.), International firms and modern imperialism, Penguin, Harmondsworth

Aglietta, M. (1979) A theory of capitalist regulation: the U.S. experience. New Left Books, London

------ (1982) World capitalism in the eighties New Left Review, 137, 5-41

----- (1985) The creation of international liquidity. In L. Tsoukalis (ed.), The political economy of international money. Sage, London, pp.171-202

Armstrong, W. and McGee, T. (1985) Theatres of accumulation: studies in Asian and Latin American urbanisation. Methuen, London

Auty, R. (1983) MNCs and regional revenue retention in a vertically integrated industry: bauxite/aluminium in the Caribbean. Regional Studies, 17, 3-17

Bacha, E.L. (1978) An interpretation of unequal exchange from Prebisch-Singer to Emmanuel. Journal of Development Economics, 5, 319-30

Balassa, B. (1981) The newly industrialising countries in the world economy. Pergamon, Oxford

Baran, P. (1973) The political economy of growth. Penguin, Harmondsworth

Bath, C. and James, D. (1976) Dependency analysis of Latin America: some criticisms, some suggestions. Latin American Research Review, XI, 3-54

Becker, D. (1984) Development, democracy and dependency in Latin America: a post-imperialist view. Third World Quarterly, 6, 411-31

Beenstock, M. (1983) The gloomy economics of Willy Brandt. Financial Times, March

------ (1984) The world economy in transition. George Allen and Unwin, London

Bell, M. (1986) Contemporary Africa. Longman, London

Benton, J. (1984) The changing position of Aymara women in Bolivia's Lake Titicaca region. In J. Townsend and J. Momsen (eds), Women's role in changing the face of the earth. IBG, London

Bernstein, H. (1982) Industrialisation, development and dependence. In H. Alavi and T. Shanin (eds), Introduction to the Sociology of developing countries. Macmillan, London

Bhagwati, J.N. (1979) Anatomy and consequences of trade control regimes. National Bureau of Economic Research, New York

Binder, L. (1986) The natural history of development theory. Comparative Studies in Sociology and History, 28, 3-33

Booth, D. (1985), Marxism and development sociology: interpreting the impasse. World Development, 13, 761-87

Boyer, R. and Mistral, J. (1983) Accumulation, inflation et crise. Presses Universitaires de France, Paris

Brandt, W. (chairman) (1983) Common crisis, North South. Pan, London

Brenner, R. (1977) The origins of capitalist development: a critique of neo-Smithian Marxism. New Left Review, 104, 25-92

Brett, E.A. (1985), The world economy since the War: the politics of uneven development. Macmillan, London

Brewer, A. (1980) Marxist theories of imperialism. RKP, London

Cardoso, F.H. (1977) The consumption of dependency theory in the U.S. Latin American Research Review, XII, 7-24

Carling, A. (1986) 'Rational choice Marxism'. New Left Review, 160, 24-62

Cheung, S.N.S. (1986) Will China go 'capitalist'?, 2nd edn. IEA, London

Chudnovsky, D. (1979) The challenge by domestic enterprises to the TNC's domination. World Development, 7, 45-58

Cline, W.R. (1983) International debt and the stability of the world economy. Institute for International Economics, Washington, D.C.

Coakley, J. and Harris, L. (1983) The city of capital. Blackwell, Oxford

Cody, J., Hughes, J. and Wall, D. (eds) (1980) Policies for industrial progress in developing countries. OUP, Oxford

Corbridge, S. (1984) Crisis, what crisis? Monetarism, Brandt II and the geopolitics of debt. Political Geography Quarterly, 3, 331-45

------ (1986a) Capitalist world development: a critique of radical development geography. Macmillan, London

------ (1986b), Capitalism, industrialisation and development. Progress in Human Geography, 10, 48-67

------ (1987) The asymmetry of interdependence: the United States and the geopolitics of international financial relations. Studies in Comparative International Development, forthcoming

Cutler, A., Hindess, B., Hirst, P. and Hussain, A. (1977) Marx's Capital and capitalism today: volume 1. RKP, London

Davis, K. (1949) Human society. Macmillan, New York

Dicken, P. (1986) Global shift: industrial change in a turbulent world. Harper and Row, London

Domar, E. (1957) Essays in the theory of economic growth. OUP, London

Dos Santos, T. (1970) The Structure of Dependence. In C.K. Wilber, (ed.), The political economy of development and underdevelopment, Random House, New York

Dutt, R.C. (1901) The economic history of India, 2 vols. Routledge, London

Edwards, C. (1985) The fragmented world: competing perspectives on trade, money and crisis. Methuen, London

Emmanuel, A. (1972) Unequal exchange. Monthly Review Press, London

Evans, P. (1979) Dependent development: the alliance of multinational, state and local capital in Brazil. Princeton University Press, Princeton

Forbes, D. (1984) The geography of underdevelopment. Croom Helm, London

Frank, A.G. (1967) Capitalism and underdevelopment in Latin America. Monthly Review Press, London

------ (1969) Latin America: underdevelopment or revolution. Monthly Review Press, London

------ (1982) Asia's exclusive models. Far Eastern Economic Review, 25 June, 22-3

Friedmann, J. (1966) Regional development policy: a case study of Venezuela. Harvard University Press, Cambridge, Mass.

Frobel, J. (1983) The current development of the world economy: reproduction of labour and accumulation of capital on a world scale. Review, V, 507-55

------ Heinrichs, J. and Kreye, O. (1980) The new international division of labour. CUP, Cambridge

Gould, P. (1970) Tanzania, 1920-1963: the spatial impress of the modernisation process. World Politics, 22, 149-70

Gregory, D.J. (1986) 'Areal differentiation and post-modern human geography', Mimeograph

Griffith-Jones, S. and Lipton, M. (1984) International lenders of last resort: are changes required? Midland Bank International, Occasional Papers on International Trade and Finance

Haberler, G. (1961) A survey of international trade theory. Princeton University Press, Princeton

Hamilton, A. (1986) The financial revolution. Penguin, Harmondsworth

Harris, N. (1983) Of bread and guns: the world in economic crisis. Penguin, Harmondsworth

Harrod, R. (1948) Towards a dynamic economics. Macmillan, London

Harvey, D. (1982) The limits to capital. Blackwell, Oxford

Hayter, T. (1981) The creation of world poverty. Pluto, London

Heckscher, E. (1919) The effect of foreign trade on the distribution of income. Ekonomisk Tidskrift, XXI

Hindess, B. and Hirst, P.Q. (1977) Mode of production and social formation. Macmillan, London

Hoogvelt, A. (1982) The Third World in global development. Macmillan, London

Jalée, P. (1969) The Third World in world economy. Monthly Review Press, London

Jenkins, R. (1984) Divisions over the international division of labour. Capital and Class, 22, 28-57

Kay, G. (1975) Development and underdevelopment. Macmillan, London

Kitching, G. (1982) Development and underdevelopment in historical perspective: populism, nationalism and industrialisation. Methuen, London

Laclau, E. (1979) Politics and ideology in Marxist theory. Verso, London

Lal, D. (1983) The poverty of 'development economics'. IEA, London

Lall, S. (1975) Is dependence a useful concept in analysing underdevelopment? World Development, 3, 799-810

------ (1980) The multinational corporation. Macmillan, London

------ (1984) Transnationals and the Third World: changing perceptions'. National Westminster Bank Quarterly Review, May, 2-16

Lamb, G. (1981) Rapid capitalist development models. In D. Seers (ed.), Dependency theory: a critical reassessment, Pinter, London

Landsberg, M. (1979) Export-led industrialisation in the Third World: manufacturing imperialism. Review of Radical Political Economics, 11

Lenin, V.I. (1970) Imperialism: the highest stage of capitalism. Foreign Languages Press, Peking

Lipietz, A. (1985) The enchanted world. Verso, London

------ (1987) Mirages and miracles: the crisis of global Fordism. Verso, London

Little, I.M.D. (1982) Economic development: theory, policies and international relations. Basic Books, New York

Llewellyn, D. (1985) The role of international banking. In L. Tsoukalis (ed.), The political economy of international money, Sage, London

Luxemburg, R. (1972) The accumulation of capital. Allen Lane, London

McClelland, D. (1961) The achieving society. Von Nostrand, Princeton

McMichael, P., Petras, J. and Rhodes, R. (1978)

Industrialisation in the Third World. In J. Petras (ed.), Critical perspectives on imperialism and social class in the Third World, Monthly Review Press, London

Marx, K. and Engels, F. (1967) The communist manifesto. Penguin, Harmondsworth

Morawetz, D. (1977) Twenty-five years of economic development: 1950-1975. World Bank, Washington, D.C.

Morello, T. (1983) Sweatshops in the sun? Far Eastern Economic Review, 15 September, 88-9

Mouzelis, N. (1978) Modern Greece: facets of under-development. Macmillan, London

Naoroji, D. (1901) Poverty and un-British rule in India. Routledge, London

Nurske, R. (1953) Problems of capital formation in under-developed countries. OUP, Oxford

Ohlin, B. (1967) Interregional and international trade. Harvard University Press, Cambridge, Mass.

Palma, G. (1978) Dependency: a formal theory of under-development or a methodology for the analysis of concrete situations of underdevelopment? World Development, 6, 881-924

Parboni, R. (1981) The dollar and its rivals: recession, inflation and international finance. Verso, London

------ (1986) The dollar weapon: from Nixon to Reagan. New Left Review, 158, 5-18

Parsons, T. (1948) The structure of social action. McGraw Hill, New York

Petras, J. and Brill, H. (1985) The tyranny of globalism. Journal of Contemporary Asia, 15, 403-20

Prebisch, R. (1959) Commercial policy in the under-developed countries. American Economic Review, Papers and Proceedings, XLIX, 251-73

Radice, H. (1984) The national economy: a Keynesian myth? Capital and Class, 22, 11-40

Raj, K.N. (1984) The causes and consequences of world recession. World Development, 12, 151-69

Rey, P.P. (1971) Colonialisme, neo-colonialisme et transition au capitalisme. Maspero, Paris

------ (1973) Les alliances des classes. Maspero, Paris

Ross, R.J. (1983) Facing Leviathan: public policy and global capitalism. Economic Geography, 59, 144-60

Rostow, W.W. (1960) The stages of economic growth: a non-communist manifesto. CUP, London

Scott, A.J. and Storper, M. (eds) (1986) Production, work territory: the geographical anatomy of industrial

capitalism. Allen and Unwin, Boston
Shenoy, S. (1971) India: progress or poverty? IEA, London
Slater, D. (1977) Geography and underdevelopment, II. Antipode, 9, 1-31
Smelser, N.J. (1959) Social change in the Industrial Revolution. Chicago University Press, Chicago
Smith, N. (1984) Uneven development: nature, capital and the production of space. Blackwell, Oxford
Sraffa, P. (1960) Production of commodities by means of commodities. CUP, Cambridge
Stockman, D. (1986) The triumph of politics. Coronet, London
Strange, S. (1986) Casino capitalism. Blackwell, Oxford
Sunkel, O. (1973) Transnational capital and national disintegration in Latin America. Social and Economic Studies, 22, 132-76
Taylor, J. (1979) From modernisation to modes of production. Macmillan, London
Taylor, M. and Thrift, N. (eds) (1982) The geography of multinationals. Croom Helm, London
Taylor, P. (1987) The error of developmentalism in human geography. In D.J. Gregory and R. Walford (eds), Horizons in human geography. Macmillan, London, forthcoming
Teulings, A.W.M. (1984) The internationalisation squeeze. Environment and Planning A, 16, 565-606
Thrift, N. (1986) The geography of international economic disorder. In R.J. Johnston and P. Taylor (eds), A world in crisis? Geographical perspectives. Blackwell, Oxford
Tokman, V.E. (1985) Global monetarism and destruction of industry. Cepal Review, 23, 107-21
Toye, J. (1983) The disparaging of development economics Journal of Development Studies, 20, 87-107
Triffin, R. (1960) Gold and the dollar crisis. Yale University Press, New Haven
United Nations (1983) Transnational corporations in world development: third study. United Nations, New York
UNCTAD (1983) Economic cooperation among developing countries. UNCTAD, Geneva
Vernon, R. (1966) International investment and international trade in the product cycle. Quarterly Journal of Economics, 80, 190-207
De Vroey, M. (1984) A regulation approach interpretation of contemporary crisis. Capital and Class, 23, 45-66
Wallerstein, I. (1974) The modern world system. Academic

Books, New York
------ (1979) The capitalist world economy. CUP, Cambridge
------ (1984) The politics of the world economy. CUP, Cambridge
Warren, B. (1973) Imperialism and capitalist industrialisation. New Left Review, 81, 3-44
------ (1980) Imperialism: pioneer of capitalism. Verso, London
Williamson, J. (1985) The theorists and the real world. In L. Tsoukalis (ed.), The political economy of international money. Sage, London
Wolf, E. (1982) Europe and the people without history. University of California Press, Berkeley
Wolpe, H. (ed.) (1980) The articulation of modes of production. RKP, London
World Bank (1981) Accelerated development in Sub-Saharan Africa. World Bank, Washington, D.C.
------ (1986) World Development Report, 1986. OUP, Oxford

Chapter Two

AGRARIAN STRUCTURE

W.B. Morgan

AGRICULTURE IN THE THIRD WORLD ECONOMIES

It is a widely expressed view that in most Third World countries agriculture has tended either to stagnate or not to grow as rapidly as the other sectors of their economies (see, for example, the discussion of agricultural policies in World Bank, 1986, p. 61 et seq.). In consequence, what is often the largest sector in any given Third World economy has contributed very little to economic growth. This has happened despite evidence that agriculture can be a dynamic sector with very rapid growth, as in some of those countries which have had successful introductions of high-yielding varieties of wheat and rice.

In Africa south of the Sahara, the situation of the agricultural sector is particularly weak and in some African countries agriculture is even in decline. The persistence of severe droughts in the 1970s and 1980s has been admitted as one contributing factor, particularly in the poorer countries, but generally, stagnation has been blamed on policies (and especially on neglect in development planning), on the allocation of the majority of resources to other sectors of the economies, on the failure to introduce modern inputs, and on tendencies to drain agriculture of its labour resources and even, in some cases, of its little capital (see, for example, Todaro, 1985, Chapter 10). In the 36 poorest countries, by 1984 agriculture was still the most important sector, contributing 36 per cent of GDP, although the percentage has been declining (World Bank, 1986, p. 184; Table 2.1 and Figure 2.1). Between 1960 and 1984 the percentage had fallen from 50 to 35 in India, from 46 to 24

Figure 2.1: Share of agriculture in GDP in the Third World in 1984 (low- and middle-income economy countries as listed by the World Bank)

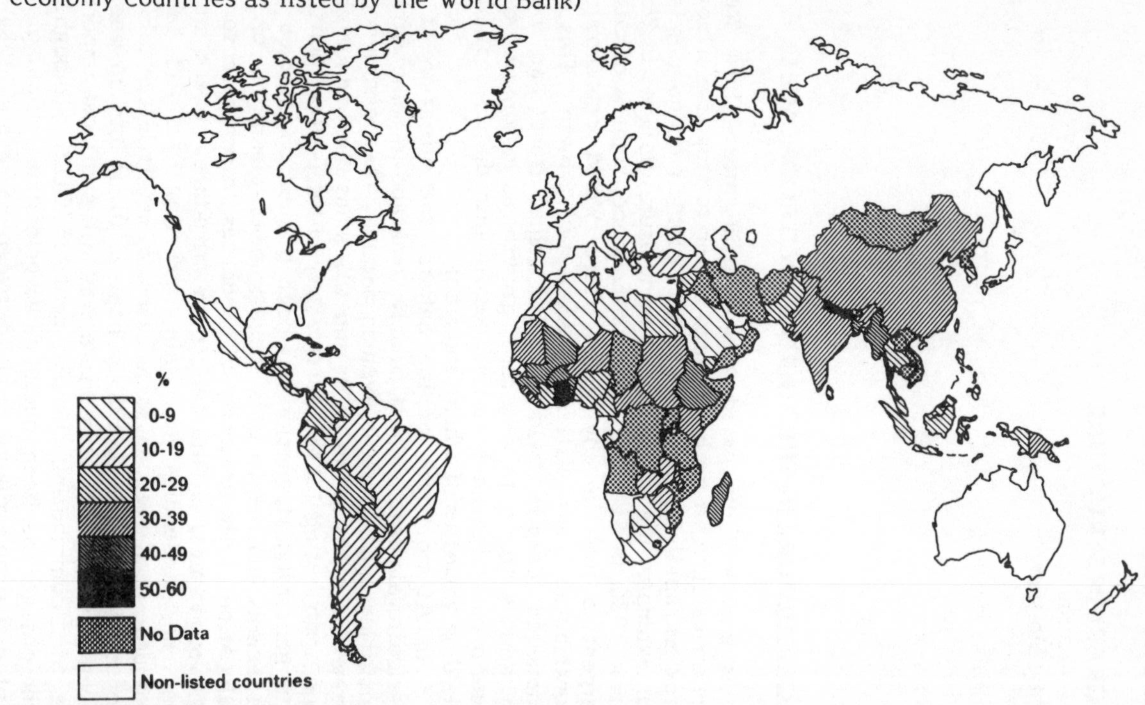

Source: World Bank, World Development Report 1986

Table 2.1: Percentage share of agriculture in GDP: selected regions and countries

	1960	1965	1980	1984
Low-income economies	50	42	36	36
Bangladesh	58	53	54	48
Ethiopia	65	58	61	48
Burkina Faso	62	52	40	43
Tanzania	57	46	54	nd
India	50	47	37	35
Kenya	38	35	34	31
Pakistan	46	40	31	24
Senegal	24	25	15	14
Middle-income economies	24	21	15	14
Nigeria	63	53	20	27
Indonesia	54	59	26	26
Malaysia	37	30	24	21
Thailand	40	35	25	20
Zimbabwe	18	18	12	14
Brazil	16	19	10	13
Argentina	16	17	nd	12
Mexico	16	14	10	9
Jamaica	10	10	8	6

Source: World Bank (1982, 1986).

in Pakistan, from 58 to 48 in Bangladesh, and from 63 to 27 in Nigeria. Generally, agriculture's share of GDP showed a marked positive correlation with poverty - the richer Third World countries exhibiting only small shares for agriculture - e.g. Mexico, 8%; Zimbabwe, 11%; Brazil, 12%; Malaysia, 21%; and Thailand, 23%.

From 1973 to 1984 the annual growth rate in agriculture's contribution to GDP in the low-income economies has been 3.6 per cent, with rather lower rates in the richer middle-income countries (Table 2.2). India did less well despite Green Revolution technology, while Pakistan did rather better. The Philippines, with the International Rice Research Centre at Los Banos, achieved 4.0 per cent, but Malaysia with little use of hybrid grain technology achieved 4.2 per cent, with Brazil also reaching 4.0 per cent. Performances were extremely variable and the best results were apparently achieved in the Middle Eastern oil-

Table 2.2: Average annual percentage growth rate in GDP and agricultural production Third World by selected regions

	GDP		Agriculture	
	1965-73	1973-84	1965-73	1973-84
Low-income economies	5.6	5.3	3.0	3.6
China and India	6.2	5.7	3.2	3.9
Other low-income	3.7	3.5	2.5	2.4
Of which SS Africa	3.7	2.0	2.6	1.4
Middle-income economies	7.4	4.4	3.6	2.7
Oil exporters	7.8	4.6	4.0	2.4
Oil importers	7.1	4.3	3.2	2.9
SS Africa	8.5	1.6	3.0	0.1

Source: World Bank (1986).

exporting countries - overall 6.8 per cent, although admittedly from a small base. In most Third World countries and especially amongst the poorest, the problem was that agricultural growth rates were below total GDP and either equalled or fell below rates of population growth. Per caput increases were extremely poor and in some low-income economies, and especially in Sub-Saharan Africa, have been declining (Table 2.3).

The poorer Third World countries still retain their traditional dependence in world trade on the export of primary products. Recession with falling demand, together with the raising of tariff barriers, have damaged this trade and discouraged farmers. Policies in the countries of North America, the Far East and Western and Eastern Europe have aimed at supporting their own agriculture and at achieving greater levels of self-sufficiency. The EEC, Japan and USA protect against the import of oilseeds, sugar, rice, maize, tobacco, beef and butter. Brazil's attempt to develop 'gasohol' production as a substitute for petrol from sugar cane is hardly surprising, although recent declines in world oil prices may affect that strategy. On the whole the effect of the policies of the industrial countries has been to depress world agricultural product prices and to distort the relative prices of agricultural and manufactured goods. The World Development Report, 1986 (World Bank, 1986) concluded that the developing countries' competitiveness

Table 2.3: Annual growth rates of agricultural and food output, 1960-80: the Third World by regions (A = 1960-70 and B = 1970-80)

	Agricultural output				Food output			
	Total		Per caput		Total		Per caput	
	A	B	A	B	A	B	A	B
Developing countries	2.8	2.7	0.3	0.3	2.9	2.8	0.4	0.4
Low-income	2.5	2.1	0.2	-0.4	2.6	2.2	0.2	-0.3
Middle-income	2.9	3.1	0.4	0.7	3.2	3.3	0.7	0.9
Africa	2.7	1.3	0.2	-1.4	2.6	1.6	0.1	-1.1
S. Asia	2.5	2.2	0.1	0.0	2.6	2.2	0.1	0.0
Middle East	2.5	2.7	0.0	0.0	2.6	2.9	0.1	0.2
Latin America	2.9	3.0	0.1	0.6	3.6	3.3	0.1	0.6
SE Asia	2.9	3.8	0.3	1.4	2.8	3.8	0.3	1.4
Total World	2.6	2.2	0.7	0.4	2.7	2.3	0.8	0.5

Production data weighted by world export unit prices. Decade growth rates based on midpoints of five-year averages, except 1970 which is the average for 1969-71.

Sources: FAO (1981a and 1981b); World Bank (1982, p. 41).

depended less on their own efficiency and more on political decisions in the industrial countries. The EEC's STABEX scheme, established in 1975 under the first Lomé Convention to stabilise the export earnings of certain Third World countries in 48 agricultural products, offers some compensation, but is far from ideal.

The declining dependence on the export of beverages and of agricultural raw materials has been viewed with mixed feelings, partly of anger at or exasperation with the protectionist policies of the richer countries, and partly acceptance as higher priority is given to other sectors of the economies. Declining food production per caput accompanied by rising food imports, however, has very serious implications for Third World economies relating to outbreaks of famine, a failure to provide adequate food supplies for burgeoning urban populations and a diversion of overseas earnings from expenditure on capital goods to expenditure on consumption. The Third World as a whole has been a net importer of food staples since the late 1950s and the volume of food imports has been growing, although the percentage share of merchandise imports devoted to food in the last decade has fallen, partly due to the enormous increases in the cost of oil imports. Amongst the higher percentages of import expenditure devoted to food in 1983 were Egypt, 30%; Senegal, 27% (importing almost one-third of its consumption in energy terms, (Morgan, 1977); Sierra Leone, 27%; Haiti, 26% (1982); Liberia, 25%; Burkina Faso, 23%; Nigeria, 21% (1982); Algeria, 21%; and Bangladesh, 20%. For the Third World as a whole, food and fuel imports combined in 1984 came to almost a third of the value of merchandise imports. Evidence of causes is conflicting, including policies favouring industry or favouring importers with powerful political interests, drought, bad administration and weak agricultural policy (Poleman, 1981). Food aid has achieved a little in some countries, but has failed to touch the core of the problem (Siamwalla and Valdes, 1980). The world food trade also has serious political implications, more especially in the trade in grain (Tarrant, 1981).

The image of dominantly rural societies and economies in the Third World in which peasant, family-based agriculture looks mainly to subsistence as the backbone of the system, is fading. In so many countries, even some of the poorest, expectations have been raised and young people are looking for an escape from agricultural employment.

Higher wages in industry and services - differences of the order of ten times have been recorded - provide an attraction worth even a waiting period of several years of unemployment. The percentage of the work-force in agriculture is falling throughout the world. In the 35 poorest countries in 1981 it was still 73%, but had reached 44% in the middle-income economies, and 30% in the upper middle-income group. India still had 71%, Indonesia, 58%, Nigeria 54%, Malaysia and Egypt 50%, but Brazil had reached 30% (Figure 2.2).

On the whole governments are concerned to increase agricultural productivity, but preferably whilst maintaining investment in industry and the infrastructure. There is evidence of recognition of the need for more expenditure on research, extension services, inputs and marketing. There is also concern to stem the flow of rural out-migration. Where land is still abundant, although peripherally located, resettlement policies are pursued, often reinforcing existing spontaneous settlement, which, until the late 1950s, had been the chief means of increasing agricultural production in the Third World. In Latin America the expansion of the cultivated area is continuing at about 0.6 per cent per annum, but only at about 0.1 per cent per annum in the Third World as a whole. The process appears to be slowing down. In India it has come to a halt, fluctuating at a total area farmed of about 180 million hectares since the early 1970s. As the production frontier disappears, the emphasis in agricultural productivity shifts to higher intensity in order to make agriculture more attractive as an occupation by achieving higher incomes. Policies to this end may hesitate between reinforcing traditional systems and methods (Joy, 1965; Richards, 1979 and 1983) and encouraging a modern inputs strategy combined with improved education (Schultz, 1964) or induced development by 'high-pay-off' inputs (Hayami and Ruttan, 1971). Whatever the general strategy, there are in most cases serious issues with regard to agrarian structure - that is, the implications for increased productivity of the existing state of and trends in land tenure, land holding distribution, fragmentation and subdivision, land reform policies, systems of renting, leasing and share cropping.

Figure 2.2: Percentage of work-force in agriculture in the Third World in 1981 (Low-income and middle-income economy countries as listed by the World Bank)

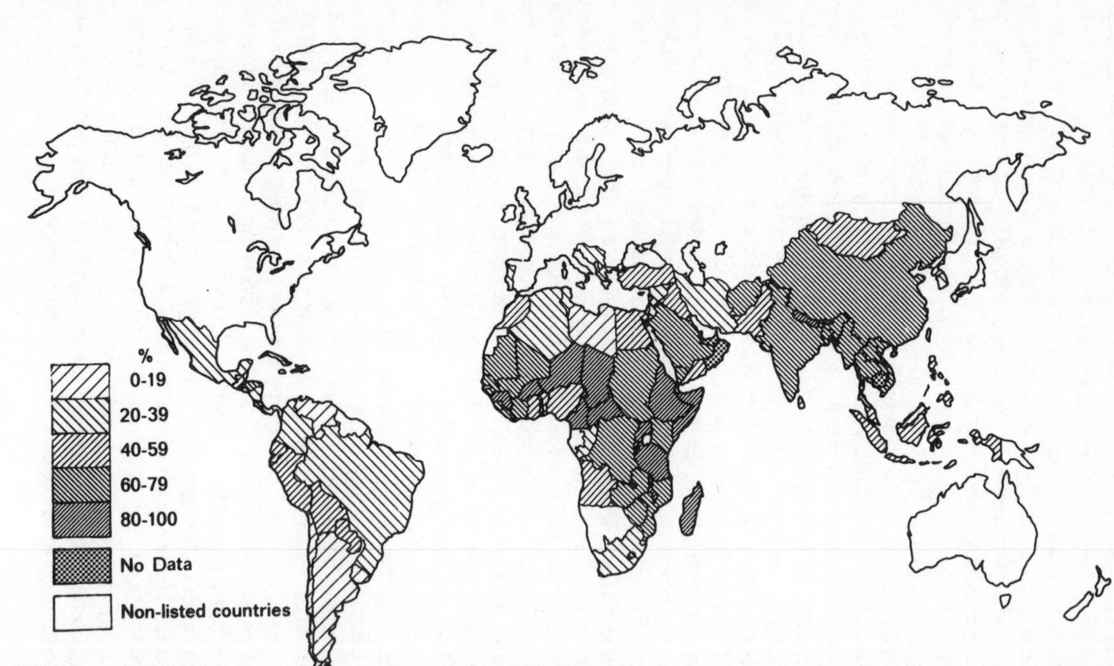

Source: World Bank, World Development Report 1985.

TRADITIONAL AGRICULTURAL SYSTEMS

In a strict sense the distinction between traditional and modern systems of agricultural production in the Third World is very difficult to make. The general picture conveyed by some observers of societies with locally self-sufficient systems of production, still using traditional techniques but invaded by modern modes of production importing a new technology and commercialism from overseas, can give a rather false impression. A great deal of agricultural improvement took place in Third World countries long before contact with Europe was established and has been taking place within the context of so-called traditional systems since (see, for example, Grigg, 1974). Some elements associated with Europe and 'modernisation' were introduced more than three centuries ago - plantations in Brazil and Fernando Po, for example. They are very old-established institutions. How does one classify Chinese commercial estates for nutmeg production in Malaysia at the end of the last century or recently developed commercial food crop production in West Africa which uses the techniques of shifting agriculture? As Richards (1983) has remarked, the contrasts drawn between major crop zones, 'cash' and 'subsistence' sectors and between precolonial and colonial agriculture are too coarse-grained and crude to be of much analytical value, and 'it is worth considering, therefore, the possibility that West African farmers long ago laid the experimental and practical foundation for an improved agriculture ...'. Alternative attempts at classification based on purpose, enterprises and structure have been suggested (Morgan, 1978). However, we can order some of our knowledge by such headings in a way in which it can be related to existing texts.

By 'traditional' we mostly refer simply to farming on peasant small-holdings whether independently or group-managed, or whether part of a large estate. Sizes are mostly in the range of 1 to 5 hectares of land in cultivation. Apart from land, seed and plant stock and a few tools, the main resource of the peasantry is their own labour. Average labour expenditure per worker may be as low as four hours a day, but is constrained by long working days at seasons of peak demand (Clark and Haswell, 1964). There may be a seasonal surplus of labour, often combined with a seasonal shortage, but the impression of a surplus of full-time labour is frequently false. Labour is normally provided by the family and is a fixed input, although in some regions there are seasonal hired labour resources which are old-established. The family labour force is also the chief market for production, since a surplus for sale is normally secondary in quantity and consideration to subsistence production for survival and to the production of the basic raw materials needed for domestic purposes such as gourds, reeds, fibres, dyes and drugs. However, the production of some surplus is necessary in order to acquire tools, weapons, superior clothing, luxuries and even some basic necessities.

In peasant societies market risks are fairly low until commercial production and specialisation become dominant, but weather risks can be high. Rain-dependent systems frequently employ a large number of crops which may maximise the use of both soils and labour and reduce some of the weather risks, besides providing for a range of subsistence needs. Most flood or irrigation systems are more constrained in the range of crops, although they are higher yielding. They may be more dependent on markets or in addition, may produce crops on rain-dependent land. Livestock systems under tropical conditions are even more limited in their product range, although some have additional land for crop production. With a limited market, low incomes and little capital accumulation, risk-taking and innovation are discouraged. Access to land and the possession of children to take over the farm are the chief concerns in the long term. Land tenure, land reform and inheritance are amongst the most important political issues in peasant societies. Despite its management problems, the small-holding may be preferred to group or co-operative farming systems, even though the latter may be supported by financial or operational inducements.

Labour can be saved or its use be more evenly spread over the seasons by various means. One is intercropping or mixed cropping in a given field, and is characteristic of many rain-dependent systems of production. It reduces the labour involved in weeding by concentrating a mixture of crop plants, often with different nutrient demands and able to occupy different positions in ridge-and-furrow or mound systems of tillage because of different moisture needs. Cultivation on level surfaces using hired labour and growing crops for commercial markets discourage intercropping and favour a more specialised crop production. The use of tractors, fertilisers and sprays has had a similar effect.

In peasant agriculture, although increased output from

greater inputs has been achieved, in most instances higher intensity has mainly meant increased applications of labour, even drudgery. Excepting extremely small holdings unable to support a family, one should expect that the smaller the holding the higher the level of intensity in order to achieve a given income. The well-known inverse relationship between input and productivity on the one hand, and size of farm on the other, has provided one argument in favour of land reform programmes. However, Bharadwaj (1974, pp. 11-18, 84), for example, noted in a study using data from five regions in India in the mid-1950s that there was no significant or systematic relation between yield per unit area for individual crops and size of holding, although there was in several, but not all cases, some relation between size of holding and total crop production. Others, such as Palmer (1976, p. 85), have claimed abundant evidence of the inverse relationship between size of holding and a single crop. On very small holdings, off-the-farm work is sought, not only to achieve a sufficient income, but also because of seasonal imbalance in labour use (see Clark and Haswell, 1964, pp. 84-94, 117-18). Population density and local overcrowding have important effects on agricultural practice, tenure and labour migration. The idea that communities respond to increased population density by agricultural innovation has been developed into a major hypothesis concerning agricultural development by Boserup (1965). An alternative thesis of 'running hard in order to stand still' has been proposed by Geertz (1966).

Low-intensity systems, involving long fallow periods and the use of shifting or slash-and-burn techniques, are disappearing or are being pushed outwards to the production frontier, as in 'conuco' farming in Venezuela. However, traditional slash-and-burn areas have persisted until recently in Central Africa (Allan, 1965), highland New Guinea (Brookfield, 1961, 1962, 1964) and in many other locations in West Africa and in South and Southeast Asia (see also Watters (1960) for a general survey of shifting cultivation; and Ruthenberg (1971) for a review of farming system in the tropics). We may also distinguish more intensive variants such as 'rotational bush fallow' where settlement is more or less fixed, field boundaries are nearly permanent, fallows are short, some permanently cropped land is combined with less intensively cultivated land, and use may be made of manures and crop waste.

Most livestock rearing in the Third World is at a low

level of intensity and in relatively dry environments. It is rarely integrated with other land uses, even where families combine rearing with cultivation, although periodically the livestock may graze on stubble or tree fodder and manure may be used. Livestock diseases have been an important constraint in relatively wet tropical regions. Preference for young grass has encouraged seasonal movement across rangelands, accompanied by the burning of old plant growth. Floodlands in areas of relatively low rainfall have the best potential for livestock rearing in the tropics, but are also highly desirable for crop production, leading to competition between graziers and cultivators.

On irrigated lands cash crops may be rather more important where crop variety is limited and where payments have to made for land and water - although sometimes in kind. Tight controls by community or landlord over land and water are frequently exercised. Some crop variety may be achieved where water control makes it possible to combine paddy rice and sugar cane or to follow rice with a succession of other crops such as sweet potatoes, wheat, maize, groundnuts or soya. Labour demands on irrigated land can be very heavy, but yields can be high and the reliability of the production system is generally much greater.

MODERN MARKET-ORIENTED SYSTEMS OF AGRICULTURAL PRODUCTION

Although agricultural production for internal, including urban, demand has existed in some Third World countries for millennia and for export over several centuries, nevertheless the scale of modern agricultural marketing has significantly changed the character of farming in almost every Third World country. The sixteenth century saw the diffusion of several food crops, especially American crops such as maize, manioc, potato, sweet potato and groundnuts to Africa and Asia, and the creation of the first sugar plantations for the export trade. In the eighteenth century, plantations producing sugar, coffee, cocoa, tea and spices spread to many regions and the science of botany was developed, leading to the identification and transference of several hundred plant species, a process which was considerably accelerated in the nineteenth century. The creation of botanical and experimental gardens followed, together with the first model farms, which were intended to remould local farming methods. The imposition of colonial government and the spread of crops for export production, particularly of tree crops, led to new kinds of land tenure and larger units of production.

We may distinguish mainly commercial farms, normally specialised, from mixed commercial and substantial subsistence. Both are associated with major export cropping developments, particularly coffee, cocoa, groundnuts, rubber and oil-palm. Resettlement schemes, notably in Malaysia, Indonesia, Thailand and Amazonia, have been developed to promote small farmer systems of commercial crop production (Uhlig, 1984; Manshard and Morgan, forthcoming). Small farmer commercial systems have been associated particularly with the expansion of the production frontier. In Ghana early railway and road development encouraged the mainly spontaneous creation of the cocoa belt. Similarly coffee planting on small farms spread in Uganda and elsewhere in Central Africa - later in the Ivory Coast where early planting was undertaken mainly by French settlers. Coffee small-holdings or 'sitios' also occur in Brazil, especially in Parana where coffee is combined with food cropping and cattle raising (Ruthenberg, 1971, pp. 224-7). In the early stages of peripheral expansion farming families depended for food, mainly on their own production. Many later maintained food cropping as an economic safeguard or to supply new urban markets, often using seasonally hired labour to supplement the family labour resource. In Senegal the spread of groundnut cultivation on to sandy soils in areas of low rainfall with limited potential for grain production led to a major dependence of farming families on imported rice, wheat and sugar (Morgan, 1977; Morgan and Pugh, 1969, pp. 632-41). Annual crops such as the groundnut can be integrated into an existing food crop system, but tree crops such as cocoa and coffee require different systems of production and long-term tenure. Their spread has frequently been accompanied by marked changes in agrarian structures.

Mainly commercial production may be classified as centrally managed systems, estate systems with rented or share-cropped small-holdings, and smaller owner-farmer holdings employing hired labour. There is a confused use of the terms 'plantation' and 'estate' in tropical Third World countries. They include:

(a) Large-scale, scientifically managed monocultures, with

a paid labour force, usually producing for export and often the product of overseas investment, especially by multinational companies (Jacoby, 1975), although there are several important exceptions. Many plantations of this kind have become state-controlled, e.g. sugar plantations in Cuba and banana plantations in Cameroon. In such cases it is difficult to distinguish the organisation of what is essentially still a plantation from other forms of state farming. Plantations still remain cost-effective systems of plant production for a high-value monoculture in which a high labour input, usually skilled and semi-skilled, is required, as in the cultivation of rubber (for a general discussion, see Courtenay, 1965). Some modern resettlement small-holdings imitate plantation organisation, with a central organisation controlling their operations, as in Malaysia's Federal Land Development Authority (FELDA) schemes for oil-palm and rubber production (Bahrin and Perera, 1977).

(b) Large-scale estates divided into small-holdings with varying degrees of central management and with centralised processing, in which one commercial crop is produced, but in which small-holders are often permitted to grow their own crops in separate fields. Such systems are often of low input, lacking in scientific management, produce a low-quality product, operate on cheap land and try to keep labour costs to a minimum. Estate production of coffee, cocoa and sugar is particularly characteristic of Latin America. Share cropping has been widespread, although there has been some decline with the general depression in export crop production, the increase in small-farm resettlement and the expansion of small- to medium-size commercial food crop farms responding to the growth of urban markets. In Sao Paulo the 'parceria' system consisted of an initial five-year stage of clearance and establishment of coffee, interplanted with food crops, followed by a second stage of either tenant coffee farms, contracted to sell to the owners, or of a centrally managed estate operated by wage labour. Such systems are often associated with absentee landlords and the apparent under-utilisation of land (many use less than a quarter of the area they own), although extremely profitable in their hey-day. They existed as long as cheap land was still accessible and began to decline as land prices rose and the world market prices of their products fell (for a discussion of plantation and estate systems, see Beckford, 1969).

(c) Nucleus estate systems in which a centrally managed estate possessing a processing system, skilled workers and management, modern equipment and access to market, is contractually linked to small-holders who provide a regular supply of a given product for processing and marketing, and receive technical advice and help. Such systems are often the product of an agreement between a government agency and a multinational corporation.

(d) State farms having central management, processing, modern resources and paid labour, sometimes devoted to crops for export, but often producing food staples for which a shortage is feared as urban demand rises and small-holder systems fail to meet it. In many cases such units incur high costs, yet are required to produce for a cheap market and have to be subsidised. Some state-controlled or stage-agency farming schemes involve large estates divided amongst small-holders, e.g. the Inland Niger Delta cotton and rice scheme in Mali and the Gezira cotton and grain scheme in Sudan.

Mainly commercial small-holding schemes are increasing in number. Such farming is associated with the rise of a new, essentially capitalist, richer and more progressive class of farmers, educated and interested in a high-income agriculture using new techniques. Thus, the Green Revolution in South Asia has increased agricultural productivity (for a study of the input/productivity relationship in India, see Dayal, 1984; and for review, see Farmer, 1981), but has also increased income differentials, encouraged an increase in the sizes of farms, increased the numbers of landless labourers in certain areas and laid the foundations of a land-holding revolution (see Singh and Day, 1975: '... mechanization in a labour-surplus environment is something of a paradox that invites further analysis'). The intermediate technology movement, devoted especially to small-scale operations and the interests of the rural poor, has so far tended to produce equipment better suited to the needs of a 'middle class'. In many countries we are witnessing a small-scale agro-industrial revolution, with farmers responding to the needs of food processing companies, whilst small firms are manufacturing cheap diesel engines, pumps, rotavators and other equipment for small to medium sizes of farm (Frievalds, 1973; Tay and Wee, 1973; Child and Kaneda, 1975).

LAND TENURE

Land can be regarded as the most important of the factors of production in peasant agriculture where capital for investment is lacking and family labour is a limited if not fixed resource - increase in family size means an increase in both producers and consumers. The availability of land to allow the production system to expand is essential unless intensification is possible, usually by increased applications of labour with diminishing marginal returns - or alternatively by new resource inputs. Even in societies in which land is still abundant, quality, accessibility and local availability are important constraints. In an age of centripetal population movement, there are few volunteers for the pioneer fringe.

Land ownership is one of the most important forms of wealth, for as population increases and as new uses are found for land, so land increases in value. Land has been the basis of political power. It confers an element of security, and unless planning intervenes, possession of land allows decision making about the location of economic activity and the use of soil. These can be decisions in which the interests of the individual may differ greatly from the interests of society. To work the soil to destruction may be the most profitable short-term individual choice, whatever the results may be for the longer term and society. Few governments are without some kind of land-use policy and some form of planning control, however limited. Land uses are varied and are so unequal by any measures of value that it is difficult to estimate their relative merit. There is, in consequence, widespread doubt about leaving land-use decisions to the market place.

The system of tenure and the relative freedom with which it operates also affect the size of holdings and their freedom to expand or contract. It can constrain the farm business. Some traditional forms of tenure make it impossible to create farm sizes appropriate to new forms of technology or they may create by subdivision very small production units as population increases, together with a landless class. A farm holding may not be limited to one form of tenure, but may combine two or more. Owners may rent or lease additional land. Shared and individual systems may be combined. A holding may become fragmented through repeated subdivision or by expansion through land purchase. Farm structure may be seen as a crucial factor in agricultural improvement, but few governments are willing to adopt policies of large-scale structural change in the interest of higher productivity and farm incomes. They are more likely either to conserve existing structures in the interests of a land-owning class or to adopt land-reform policies for political ends, including social justice, or for the solution of other more pressing problems such as unemployment or urbanward migration.

Six broad classes of land tenure may be distinguished. These are almost all present in every major Third World region, but in particular locations, a given type may predominate, leading to broad regional distinctions. In Latin America large estates, the so-called latifundia, are dominant. Over 60 per cent of the farmed area is divided amongst holdings over 500 hectares which are only 2 per cent of all holdings, whilst holdings less than 5 hectares are over 50 per cent of the total. In Brazil and Venezuela the average size of holdings is about 80 hectares. In Argentina and Uruguay, with ranching dominant, the average size is over 200 hectares. King cites the extreme cases of Ecuador with 705 latifundia (0.17 per cent of land holdings) on 37 per cent of farmland, and Guatemala with 516 holdings (0.15 per cent of the total) on 41 per cent (King, 1977, p. 81). In tropical Africa more communal systems of tenure prevail. Even taking Africa as a whole, over 70 per cent of farmland is in holdings of less than 50 hectares, and about 90 per cent of all holdings are less than 5 hectares. Asian estates and reformed systems have a similar broad distribution with 90 per cent of holdings less than 5 hectares.

The six classes are as follows:

(a) Usufruct or use-right systems of tenure, well developed in tropical Africa, in which the land is held by various kinds of social group, frequently a lineage - that is, a group of people recognising a common ancestor. In a strict sense there is no individual ownership, although there may be individual rights to clear land and plant crops on a temporary basis, so that the land farmed must eventually revert to the group. Traditionally, group-held land cannot be sold, but the intrusion of colonial land law and the spread of systems of commercial farming has led throughout tropical

Africa to the diffusion of individual tenure and the loss of communal rights. Usufructuary land may be thought valueless if unsaleable and useless as collateral, but there is comparative value of a sort in the sense that some locations are more productive than others, more accessible or better provided with resources such as water and building clay. The pressure on easily accessible land has resulted, in some traditional systems, in the recognition of a form of individual or nuclear family tenure over land close to the village. Communal forms of tenure are widespread throughout the Third World. They occur in Mexico as the 'ejido' of Indian tradition, in Iraq as 'dirah', in Syria and Jordan as 'masha', and in many other forms elsewhere, e.g. in Latin America and in the hill lands of Southeast Asia. Most usufructuary systems are seen as obstacles to agricultural improvement, discouraging innovation and investment and in some cases favouring fragmentation by subdividing different qualities of land, and favouring different locations according to accessibility.

(b) <u>Landlord-tenant farmer systems</u> in North Africa, the Middle East and South and Southeast Asia. These systems are very varied in type and origins. In many cases, as in 'miri' land in the Middle East or 'zamindari' land in India, the land was originally held by the state or ruler, but was subsequently transferred to individuals who imposed rents as high as the farm holdings could bear (for some illustration of the complexity of land tenure in India, see Bharadwaj, 1974). Until recent land reforms, estates of tenant farms were the most widespread form of land holding in these regions. In Egypt such tenancy developed late under population pressure, rising from 17 per cent of farmland in 1939 to 75 per cent by 1952 (King, 1977, pp. 380-1). In Syria by the early 1950s tenancy occupied 75 per cent of farmland with high rents and widespread indebtedness (King, 1977, pp. 388-9). In Thailand, tenancy is mainly significant in the Central Plain, where landlords resident in Bangkok, including nobles and officials who received land grants in lieu of salaries, lease holdings mainly to commercial farmers operating market gardens and irrigated rice farms for the urban market (Sternstein, 1967). Colonial rule mostly favoured the development of a land-owning class. In former French Indochina, for example, French landowners existed alongside Chinese and Vietnamese landlords, controlling most of the rice-producing delta lands and the lands suitable for producing rubber, tea and coffee.

(c) Latin American landlord-tenant systems. The latifundia, haciendas or fazendas usually have absentee ownership, depend for their work-force on the 'colono' or labouring tenant, belong to an old-established tradition and frequently depend on antiquated methods. There are also some more intensively operated estates, usually with more efficient management and sometimes linked to foreign investment, chiefly from multinational companies. 'Colono' systems include tenants with short-term holdings or with a traditional usufruct who are required to provide a specified amount of work on the estate on a share cropping or a cash-rent basis. Some share cropping tenants establish tree crops for the estate and grow their own food crops on very small holdings or 'minifundia'. They then move on to another estate, leaving behind a plantation to be maintained by other workers. Colono systems flourished in the days of cheap land and an expanding agricultural frontier, but greater distances, rising transport costs and a growing urban food market have begun to break such systems down and have encouraged more intensive methods of farming. In Latin America it is estimated that over 40 per cent of idle land is in latifundia, although some of it is occupied by squatters.

(d) Socialist forms of tenure. These include land vested in large groups such as collective farms or co-operative forms of farming in which at least some of the operations are performed on a joint basis. They also include communally organised village holdings such as the 'ujamaa' systems of Tanzania. Usually the land is owned by the state, but it can be held by the collective body or the local authority. Collectivisation of farming by combining a number of holdings and eliminating individual or nuclear family ownership has developed for a variety of reasons. The main motive in socialist societies is clearly to eliminate private ownership, profit and the use of land as capital, but other motives include the attempt to achieve economies of scale, particularly in the application of farm machinery, group bulk purchase of inputs, group sales and handling, the amalgamation of scattered holding blocks, the opportunity for a supposed more rational planning of land use, an improved linkage between farmers and extension, advice and information systems, and even the encouragement of crop production for market. In a few cases the attempt to socialise agriculture and land tenure has been linked to an attempt to revive precolonial or precapitalist traditions, related to a belief in the superiority of traditional rural life and morals, or even to the encouragement of self-sufficient communes as a means of survival amidst economic depression or as a means of isolation from capitalist influence (Morgan, 1978, pp. 140-1). State farms may also be included in this category of socialist tenure, although many of them exhibit a character difficult to distinguish from privately owned and operated large-scale units of agricultural production. State-owned or state-agency-owned land also features in planned agricultural resettlement to relieve population pressure (Indonesia), to develop wilderness and occupy frontier lands (Brazil, Colombia, Peru), or to create more productive units of small-scale commercial plantation as in the Malaysian, Federal Land Development Authority scheme (Bahrin and Perera, 1977). Socialist forms of land tenure also occur in land-reform programmes in which former tenants or share croppers acquire tenure on a communal or co-operative basis, relieving them of former rent burdens but creating relatively large productive units.

(e) Tenure in large-scale units: plantations and ranches. We may group plantations and ranches together in a tenurial classification, although the two systems of land management are very different in character. They are both centrally managed, employ wage labour and produce for urban or export markets. Foreign investment or direct ownership is often involved. Some are owned by the state, usually as a result of state acquisition of former private estates, and some are operated by joint state and foreign investment. Both tend to be highly specialised - corned beef production in Argentina, Uruguay and Lesotho, for example, or rubber, tea and palm oil production in Southeast Asia. Traditionally, most livestock in the tropics have been reared on open rangelands such as the Sahelian grasslands of western and central Africa, but more specialised livestock raising for processing plants and overseas markets is being concentrated more and more on extensively operated but enclosed and owned ranges.

(f) Tenure in smaller-scale units: private ownership of land by individual farmers. Owner-farmer tenure includes a considerable range of sizes and types of farm, normally associated with commercial agricultural production and a well-developed marketing and dealer system and transport network. Owner-farmer tenure mostly occurs on larger farms than in usufructuary or estate systems and with higher incomes. In many Third World countries land holdings tend to be at extremes of size from plantations of a thousand hectares or more to peasant farms of as little as a half hectare. Owner-farmer holdings tend to be of intermediate size, frequently in the 10 to 200 hectare range. There are, however, privately owned small-holdings well below 10 hectares, including Latin American 'minifundia' able to support families at 2-5 hectares. Owner-farmer holdings are often associated with innovation, the use of new seed, fertilisers and water-control techniques, and their owners are frequently amongst the better-educated members of the population. As management technology improves they may exhibit size growth and be associated with increasing landlessness and rural unemployment. Economic development in rural areas can bring decline in self-sufficient production and increased commercialism and specialisation, often promoting private land ownership. In Ghana and Nigeria the development of cocoa production brought changes in the tenurial system, including reduction in usufructuary tenure. In India improved seeds and water control technology, combined with fertilisers and tractors, have encouraged some landowners to remove tenants and create commercial grain farms. In Mexico, Argentina, Uruguay and Chile over 50 per cent of the active population in agriculture is landless and the numbers are rising (Morgan, 1978, p. 143).

HOLDING STRUCTURE: FRAGMENTATION AND SUB-DIVISION

Few major texts on agricultural development and policy pay much attention to holding structure, more especially to the problems of holdings with spatial discontinuity. Yet it is a fundamental aspect of agrarian structure which has played a significant role in attempts to raise agricultural productivity. Consolidation of holdings which are poorly organised spatially has been a major objective of reform policy in many countries where large numbers of holdings, although of adequate size to be economically viable, remain difficult to manage because they are fragmented into many small pieces. Taylor (1969) has cited excessive fragmentation in the Fort Hall district of Kenya, as have Spate and Learmonth (1967, p. 263) in the Punjab, and Arulpragasam (1961) in Sri Lanka (quoted in King, 1977, p. 21).

The fragmentation of holdings can arise from traditional systems of land allotment, the desire for equity in distributing different qualities of land, subdivision resulting from inheritance, or the increase in size of farms as a result of purchasing, leasing or renting additional fields or blocks. Fragmentation can thus result from the 'dead hand' of tradition, the desire for both property and equity in a family and economic growth with individual ownership. Parcellisation or fragmentation in traditional usufructuary systems can occur where land is cleared each year by a land-holding group, which then allocates plots to families in sequence, repeating the sequence until all land has been allocated (Morgan, 1955). Such sequences in certain cases distribute differences in soil, aspect and drainage more equitably between families. In other cases land round a village may be zoned and land from different zones allocated to each family, so that there is some equity in the distribution of accessibility to fields (Prothero, 1957). Land purchase in order to increase the size of the farm business leads inevitably to fragmentation, since the desire and ability to purchase will rarely be matched by the availability of adjacent land. The expansion of cocoa and coffee production in Africa and Latin America encouraged the more successful planters to buy land ever further 'into the interior' as it became available, creating 'holdings' of several very widely scattered pieces.

However, most interest in fragmentation has concentrated on the subdivision of holdings in situations where area expansion is impossible and where the increasing size of families has resulted in the further subdivision of land. Subdivision of this kind can be exacerbated by the practice of presenting parcels of land on marriage and by the recognition of equal rights in land for all members of the family. It can create both uneconomic sizes of holding and an unmanageable scatter of tiny properties. In many countries consolidation has been seen as a basic necessity for agricultural improvement, although the results can only be temporary unless the basic causes of subdivision are removed at the same time. The extreme case of the Kikuyu

Highlands of Kenya has already been cited, but there are many others in all the major regions of the Third World. Programmes to combine consolidation with the creation of larger, more economic holdings can only succeed if they not only include changes in the laws of inheritance, but offer work opportunities elsewhere for those dispossessed. Farmer (1960) has rightly stressed that fragmentation and subdivision are the symptoms of a deeper problem of rural employment and job opportunities for a growing population.

LAND REFORM

In its broadest and most literal sense, land reform could refer to any process restructuring the system of land tenure or the right to farm land, together with holding size and distribution and the spatial arrangement of holding portions or blocks. In practice, it normally refers to attempts to redress inequalities in tenure or grievances with regard to rent or leasehold arrangements. More especially, it refers to schemes for the division of large estates into smaller farms or even into small-holdings and to the handing of land or of the rights to farm land to poor peasantry and the landless. Examples range from Egypt's land reform in the 1950s and 1960s, in which over 300,000 hectares were expropriated and handed over to farming families in lots of one hectare per family, to Kenya's complex decolonisation and land reform scheme, also of the 1960s, in which about a half million hectares were handed over to about 40,000 families in different size groups:

(1) a high-density settlement of 35,000 families on 400,000 hectares;
(2) a low-density scheme for 5,000 families with capital on 90,000 hectares; and
(3) a large holdings scheme for a few 'yeoman farmers' (129 transactions only in 1961-2: see summary in King, 1977, p. 340-1).

Many land reform schemes envisage simply the expropriation and redistribution of land or the elimination of exacting rent and share cropping systems. Others involve the creation of new kinds of farm, planned to provide for the needs of progressive farmers in a stable farming system. In Malawi a cheap leasehold system has encouraged the growth of commercial cultivation, especially tobacco production on small private estates, threatening the existence of usufructuary rights (Brown, 1979).

The motives for land reform are rarely simply for the improvement of agricultural productivity or the relief of the economic plight of a mass of poor farmers. In most cases the motives are a blend of political, social and economic considerations, of which the political are normally dominant. Regarding the work of the Land Reform Institute of Venezuela, Kirby (1974) concluded that its real success was not an equitable distribution of farmlands, but the stabilisation of the rural population who would otherwise have inflated the army of urban unemployed. Large estates to be expropriated may consist of centrally managed plantations or of a vast number of small-holdings rented from some absentee owner. In the former case, reform may bring a total change in farm structure; in the latter case, the spatial and size structure may remain, although the tenure system may change. Land reform is first and foremost about land tenure, not farming as such, although a change in tenure may be a necessary condition for an improvement in either farm productivity or in the social and economic status of peasant families. Land reform may be linked to employment policies, the control of migration, agricultural resettlement, attempts to change the social structure and reduce income differences and also attempts to win popular support.

Mexico provides the classic case of land reform as the mainspring of revolutionary movements - in 1910 1 per cent of the population owned 97 per cent of the land, and 92 per cent of the rural population were landless - probably the highest proportion in the world (King, 1977, p. 94). Ideals and policies are one thing, but practice is normally quite another. Lip service is frequently paid to land reform by political parties and governments whose programmes for change often fail to achieve very much. The Egyptian land reform, for example, eliminated the estates over 84 hectares but still preserved inequality in land holding. The redistributed land represented about 10 per cent of Egypt's farmland and benefited 9 per cent of the rural population. The numbers of landless labourers rose and many employees on expropriated estates were displaced (King, 1977, p. 384). In most Third World countries the landowning class has been sufficiently influential politically speaking to reduce or even minimise the effects of land reform, and has often been able to preserve the political <u>status quo</u> despite considerable social pressures. However, this does not mean one should regard all land reform programmes with scepticism. A study of the effects of reform on three Mexican 'ejidos' published in 1960 concluded that the social and economic gains included independence for the farmer, higher real incomes, improved housing, increased literacy and new political strength (Mendieta y Nunez, 1960, quoted in Saco, 1964). In some countries, particularly those of southern Asia, only limited equity can be achieved by land reform because of rural overcrowding. Thus, if in India a maximum holding limit of eight hectares were imposed, then the land made available would be barely sufficient to raise existing very small holdings to a minimum size of two hectares and no land would become available for the landless (World Bank 1974, p. 24).

The case for land reform has been made very fully by Dorner (1972) and by Dovring (1974). In brief summary, the main elements are:

(1) Land reform should be seen as a fundamental part of economic development in Third World countries in that agriculture must be restructured as it contributes to the development process, providing manpower, capital, food and products for industrial processing.
(2) Labour-intensive development is required in agriculture in many Third World countries where there is underemployment in the rural labour force (and growing urban unemployment) and 'is more likely to be achieved on small, family-scale farms under favourable tenure conditions than on large labour-hire operations or under very insecure tenure' (Dovring, 1974, p. 510).
(3) Small farms are to be preferred to large farms which emphasise the use of capital and labour savings. Even in conditions of underemployment it is argued that the institutional wage rate in agriculture may exceed the marginal product of hired labour, which has less incentive to be highly productive than self-employed labour. Moreover, in many forms of agricultural production in the Third World there are claimed to be few economies of scale beyond the size of farm which can fully employ a family and provide them with a better than basic income.
(4) The elimination of depressing rental or share cropping systems should raise real income and provide an incentive for increased productivity. The case for tenancy reform

depends not only on removing rent burdens, but also on the removal of certain forms of landlordism as obstacles to innovation. It should also provide protection against being dispossessed, together with longer leases. Zamindari abolition in India provides a well-known case (Rao, 1963; Metcalf, 1967).

(5) Distributive equity should provide a better basis for co-operation amongst small-scale farmers, extending the possibilities of input supply, marketing and credit to more farms and promoting a greater level of participation in commercial activity - and in consequence, probably in public affairs - amongst the members of rural communities. Non-agricultural people in rural areas may also expect to benefit from the greater economic opportunities offered by the increased number of farmers looking for new inputs and better market and transport facilities.

(6) Increased income equity should be accompanied by improved rural health, education and social services.

(7) The need to mechanise agriculture would be minimised - an advantage in countries with limited capital resources. The emphasis would in any case shift from large- to small-scale implements and power sources for small independent farms, i.e. to an intermediate technology, or to group mechanisation schemes in which co-operative forms of farming are practised.

(8) Land reform may be regarded as an important element in any scheme of large-scale technological change, particularly where it is desired to avoid the creation of a growing class of large-scale farmers, expanding their operations at the expense of others.

(9) Higher yields per hectare may be expected on small family farms than on larger units in countries where little surplus land exists and where it is difficult or expensive to increase non-labour inputs.

Generalisation about land reform and its advantages is, however, not particularly useful when considering actual cases. There is an enormous variety of agricultural and tenurial conditions in the Third World. Secondly, as Saco (1964) points out, it is extremely difficult to prove that land reform has brought about an improvement in rural welfare in any particular country, even in cases where improvement and reform happen to coincide. Possibly, Mendieta y Nunez's study, cited above, may be convincing in that particular case, but in most instances the variety of changing variables is too great to measure the net effect of land reform as such, and the case for land reform in specific instances remains hypothetical rather than proven. Moreover, not all land reform programmes envisage increasing the proportion of small farms nor strengthening the peasant class economically, politically or socially. Some land reform programmes have state ownership of land as their objective and the reorganisation of peasant holdings into collectively operated units. In some cases the benefits of large-scale operation, including economies of scale, have been argued in defence of land reform proposals. Land reform normally involves the creation of very small units of production and may therefore be seen as an obstacle to the future transfer of land to urban and industrial uses or to the creation of larger, more mechanised farms in a future envisaged as having a declining farm population, i.e. it may be regarded as an obstacle to modernisation. Small family farms may exhibit higher yields, but these are normally achieved by higher labour inputs which are subject to diminishing returns. Family farms can be a recipe for low pay and hard work, unless non-labour inputs can be increased (see World Bank, 1974, pp. 22-3). If an improved advisory and extension service, together with other services to agriculture, is thought to be essential, then the costs and problems of providing this will rise as the number of farms is increased. In the poorest countries that may mean that such services can only be available to a few farmers - inevitably those who will be thought most likely to produce the best return for the effort invested, or those who most conform to the system proposed by the government of the day.

Undoubtedly there have been difficulties in developing small-scale farming created by land reform programmes wherever a better-than-basic income is desired, and attempts have been made to introduce new inputs and high-earning market-oriented programmes. Hewitt de Alcantara (1976), in a detailed analysis of modernising Mexican agriculture in Sonora, discusses the growing disadvantage of traditional agriculture and the competitive disadvantage of very small commercial farms. She comments:

> Now, when economic realities dictate the rehabilitation of the land reform sector in order to widen the internal market, most of the best ejido land of commercial farming areas legally belongs to men who have known little but defeat over the past few decades and who

> have been forced to grow accustomed to passivity. The task of mobilizing such long-wasted talent will be great indeed. (ibid., p. 315).

Some of the obstacles to developing effective land reform proposals have already been discussed, including government attitudes and the power of vested land-owning interests. Even in the industrialised West, the landed interest still retains some political power and in the nineteenth century resisted the rising wealth and status of the entrepreneurial class. In the Third World generally land still retains its social and power status and land holders struggle to defend their interests. The great mass of Third World peasantry still suffers from a lack of education, information and organisation, despite some educational progress and the growth of rural improvement programmes. To some extent the peasantry suffers from a constant drain of its more talented and youthful members to the city. Education encourages people to leave agriculture, unless it can provide comparable income levels to those available in industry. Without a marked change in inducements, improvements in rural education seem unlikely to assist peasant farmers. This remains one of the problems affecting the promise of land reform - that unless the result of reform is the creation of farm units with high income potential, the rural areas must lose most of their better-educated population and probably suffer productive decline as their farming population ages. The lack of rural-tenant, share-cropper, farm-labourer and small-owner organisations has been commented on by Dorner (1972, p. 29) as reflecting the intolerance of an opposition to such organisations by those who stand to lose if reforms are implemented. Another obstacle is ineffective legislation, often due to the extreme complexity of the land tenure situation and the great difficulty involved in draughting laws which can cover all aspects and anticipate land owners' reactions. Yet another is the difficulty in enforcing legislation where the leadership in rural areas is against the reforms and where the law-enforcing bodies are relatively powerless or under the control of the local leadership. Finally, the inevitable obstacle in most Third World countries is the poverty of the data base. Few countries have adequate land ownership records or information on land quality and productivity which can be used as a basis for understanding and the formulation of policy.

AGRARIAN STRUCTURE AND AGRICULTURAL DEVELOPMENT POLICY

Agrarian structure is only one element in agricultural development, even though it may be regarded as extremely important. Few programmes for improved productivity, higher incomes and less inequity in income distribution can hope to succeed by reform of the agrarian structure alone. Policy must be based on an understanding of the interrelationship of the variables involved and therefore on an attempt to identify and guide or manage at least the key variables which act as major controls. At the same time it must be appreciated that no scheme of agricultural development can take place in all aspects of production and in all locations simultaneously. Increased productivity, unless created by the expansion of the area farmed, can only be achieved by increased intensity of operation, increased specialisation and increased regional concentration (Bowler, 1986).

Social and economic change in rural areas can only be produced by changes diffused through agricultural space and therefore creating and also increasing distributive inequities until the process is nearly complete. The innovators will lead and may go to success or failure, motivated by the prospect of high rewards. Others will follow until only the laggards remain. To command confidence and support, policy must take the diffusion process into account and must be seen to hold a prospect of long-term equity, despite the apparent inequities in the short term. However, it must be admitted that since diffusion takes time, during which market and technology may change, and since not all farmers will be equally skilled adopters, a degree of inequity in the distribution of income is likely to remain. Moreover, there is a tendency for a given innovation to be followed by another in successive waves, heightening the disparity between the leaders and the laggards, and creating situations in which many farmers have been forced out of business before they have had the opportunity to adopt. Boom is inevitably followed by slump, ruining many businesses, and widening further the gap between the successful and the unsuccessful. Whilst providing incentives to innovate and improve productivity, policy must be seen both to control individual growth in order to prevent abuses and individual wealth and power accumulation, and to provide investment in a number of regions, not just a preferred few. If equal investment in all forms of farming activity and in all regions is a recipe for spreading limited resources in packet lots too small to offer sufficient inducement and into regions too marginally located to provide a satisfactory return, the other extreme of concentrating on very few preferred regions will be seen as promoting the interests of the few and be opposed by the many. Policy has to adopt a practical investment strategy falling between the extremes.

Agricultural development policy must be concerned not only with land but with labour and its relative lack or abundance both in total and seasonally, since an apparent surplus of labour in average terms may conceal an under-provision of labour at certain seasons of peak demand. It has often been argued that the Third World has abundant rural labour associated with rapid population growth, yet it is apparent in several countries that there is a drain on rural labour resources created by urban demand and high wage attraction which has resulted in falling agricultural productivity. Labour may have been 'under-employed' overall, but in such countries it has been impossible to reallocate labour resources in a way which would adjust for the loss. The idea of labour surplus has helped to promote policies favouring higher yields on the grounds that land is the scarce resource, particularly in the overcrowded countries of southern Asia. If total productivity is to be increased, such policies must be pursued unless the farmed area can be extended. However, yield includes costs and is not the same as net income or profit. Cost minimisation is as important as output maximisation in a policy to raise farm incomes relative to those of industrial workers. It may be sensible to produce less and profit more or to seek new low-cost techniques for raising yields. The price of labour is a vital factor in farm production and there has to be a limit to the extent to which family labour in a peasant system may be seen as a 'free' or cheap good. The seasonality of farm operations must mean that the hiring of short-term wage labour is essential to expand the productive system and raise income levels or that mechanisation of certain key operations is required. Often the case for the mechanisation of agriculture in the Third World rests more on the problem of relieving seasonal bottlenecks than on any argument concerning the relative costs of labour and machinery.

Agricultural development policy must also be concerned with the availability and use of capital, both turnover capital, the immediate 'life-blood' of the young farmer trying to expand the productive system and needing to pay for seed, fertilisers and temporary labour, and fixed or long-term capital to provide for larger investments in land, buildings or machinery. Usufructuary systems in which land cannot be sold, and share cropping and estate tenancies, have no collateral to borrow from banks or other agencies concerned with the credit-worthiness of their clients. It is hardly surprising that dealers and money-lenders, exacting extremely high rates of interest as a reward for the risks taken, have had an important role in agricultural development in the Third World. For farmers to take the risk of moving towards more specialised systems of crop production and greater dependence on the market, there needs to be a high reward prospect as an inducement to innovate. The range of crops which are sufficiently attractive is very small unless the risks are removed. The development process itself holds a major problem in the provision of investment capital for agriculture, in that, in the poorest countries most dependent on farming, capital availability is low and the greatest pressure exists to concentrate investment into industry, infrastructure and services. On the whole, it is the industrially advanced countries which have shown the greatest propensity to invest in agriculture.

Arguably, management of the farm business is the key element in agricultural improvement. The quality of a farmer's education and experience is a major element in decisions to innovate and willingness to take risks. The development of high quality management may be seen as important in all aspects of production, marketing and services. The rural areas generally have poorer education facilities and lack the social services and contacts which would attract the more able managers to live there. In agriculture it is not only a problem of farmers' education but also the education and training of those who provide farming services, work in agriculture processing and maintain the fabric of rural society.

A policy for agricultural improvement must therefore deal with much more than agriculture to be successful and must offer a plan for change over a wide range of social, economic and political activities. The case for integrated rural development has been made by Kotter (1974). Agriculture may be seen as an economic and ecological activity within a social and political system having a number of social development goals. For any programme of land reform to work, it must be accompanied by the technical education of farmers, the provision of credit, the organisation of inputs supply and marketing, and other appropriate measures (Saco, 1964). One example of a package programme is the Indian Intensive Agricultural Districts Programme (IADP), which contained a package of educational and supporting services, including extension, credit, technical supply and assistance to farmers, together with an agricultural improvement package of new seed, fertiliser, plant protection and water management. In part the IADP strategy did not succeed because it did not go far enough. It over-emphasised supply and there were failures in the provision of technical assistance for irrigation and in the help given to farmers to change their patterns of cropping (Desai, 1969; Malone, 1970; Brown, 1971). The Kenya programme of consolidation and holding-size reform in the Kikuyu Highlands was also based on an integrated package of farm planning, new inputs and service provision related to different levels of farm investment and to programmes of soil conservation and transport improvement (Clayton, 1964, 1970; Taylor, 1969; Barber, 1971).

The idea of integrated agricultural development or of what has been called 'integral reform' owes much to the thinking of American agricultural economists and rural sociologists. In some cases it has been criticised as attempting to transplant institutions developed in the more industrial countries and which are unsuited to Third world development (King, 1977, p. 48), and as 'social welfare' to help 'pockets of agrarian poverty and backwardness'. It may also involve the creation of an expensive bureaucracy - the fate of most development programmes. Nevertheless, there can be no question that rural change, which is attempted solely through the agrarian structure and which does not deal with the consequences of related social and economic activities, is likely not only to fail and waste the resources invested, but to worsen the conditions of rural life. The problem is not the question of integration in itself, but the form it must take and the objectives it must serve.

Here, it must be admitted that the policies of many Third World governments and of many of the international and bilateral assistance agencies with regard to agricultural development have been far from satisfactory. As the World Bank's World Development Report, 1986 (World Bank, 1986, p. 61) observes, government interventions at all stages of production, consumption and marketing of agricultural products and inputs, although undertaken to improve market efficiency, have frequently done the opposite and lowered output and incomes. Thus:

> Paradoxically, many countries which have been stressing the importance of agricultural development have established a complex set of policies that is strongly biased against agriculture ... Many subsidize consumers to help the poor, but end up reducing the incomes of farmers who are much poorer than many of the urban consumers who actually benefit from the subsidies. Most developing countries pronounce self-sufficiency as an important objective, but follow policies that tax farmers, subsidize consumers, and increase dependence upon imported food. (ibid.)

REFERENCES

Allan, W. (1965) The African husbandman. Oliver and Boyd, Edinburgh

Arulpragasam, L.C. (1961) A consideration of the problems arising from the size and subdivision of paddy holdings in Ceylon, and the principles and provisions of the Paddy Lands Act pertaining to them. Ceylon Journ. Hist. and Soc. Stud., 4, (1), 59-70

Bahrin, Tunku S. and Perera, P.D.A. (1977) FELDA, 21 years of land development, Federal Land Development Authority. Kuala Lumpur

Barber, W.J. (1971) Land reform and economic change among African farmers in Kenya. Econ. Dev. and Cult. Change, 19 (1), 6-24

Beckford, G.L. (1969) The economics of agricultural resource use and development in plantation economies. Soc. and Econ. Stud. (Jamaica), 18, 321-47. Reprinted in H. Bernstein (ed.) (1973) Underdevelopment and development. Penguin, Harmondsworth, 115-51

Bharadwaj, K. (1974) Production conditions in Indian agriculture: a study based on farm management surveys. Univ. Cambridge, Dep. of Appl. Econ., Occasional Paper, 33, (Cambridge University Press, London)

Boserup, E. (1965) The conditions of agricultural growth: the economics of agrarian change under population

pressure. Allen and Unwin, London
Bowler, I.R. (1986) Intensification, concentration and specialization in agriculture: the case of the European Community. Geogr., 71 (1), 14-24
Brookfield, H.C. (1961) The highland peoples of New Guinea: a study of distribution and localization. Geogr. Journ., 127 (4), 436-48
------ (1962) Local study and comparative method: an example from central New Guinea. Ann. Assoc. Amer. Geogr., 52, (3), 242-54
------ (1964) The ecology of highland settlement: some suggestions. Amer. Anthrop., 66, (4), 20-38, 309-22
Brown, D.D. (1971) Agricultural development in India's districts. Harvard University Press, Cambridge, Mass.
Brown, P. (1979) Land tenure creates problems in Malawi. Chart. Surveyor, Aug.-Sept., 26-8
Chaudri, D.P. (1979) Education, innovation and agricultural development. Croom Helm, London
Child, F.C. and Kaneda, H. (1975) Links to the Green Revolution: a study of small-scale, agriculturally related industry in the Pakistan Punjab. Econ. Dev. and Cult. Change, 23 (2), 249-75
Clark, C. and Haswell, M.R. (1964) The economics of subsistence agriculture. Macmillan, London
Clayton, E.S. (1964) Agrarian development in peasant economies: some lessons from Kenya. Pergamon, Oxford
------ (1970) Agrarian reform, agricultural planning and employment in Kenya. Internat. Lab. Rev,, 102 (5), 431-56
Courtenay, P.P. (1965) Plantation agriculture. Bell, London
Dayal, E. (1984) Agricultural productivity in India: a spatial analysis. Ann. Assoc. Amer. Geogr., 74 (1), 98-123
Desai, D.K. (1969) Intensive Agricultural Districts Programme. Econ. and Polit. Weekly (Bombay), 4, (26), 83-90
Dorner, P. (1972) Land reform and economic development. Penguin, Harmondsworth
Dovring, F. (1974) Land reform: a key to change in agriculture. In N. Islam (ed.), Agricultural policy in developing countries. Macmillan, London, 509-33
Farmer, B.H. (1960) On not controlling subdivision in paddy lands. Trans. Inst. Brit. Geogr., 28, 225-35
------ (1981) The 'Green Revolution' in southern Asia. Geogr., 66, 202-7

Food and Agriculture Organization (1981a) State of food and agriculture. FAO, Rome

------ (1981b) Production yearbook. FAO, Rome

Frankel, F.R. (1971) India's Green Revolution; economic gains and political costs. Princeton University Press, Princeton, NJ

Freivalds, J. (1973) Agro-industry in Africa. World Crops, 25, (3), 124-6

Geertz, C. (1966) Agricultural involution: the process of ecological change in Indonesia. University of California Press, Berkeley, CA

Grigg, D.B. (1974) The agricultural systems of the world: an evolutionary approach. Cambridge University Press, London

Hayami, Y. and Ruttan, V.W. (1971) Agricultural development: an international perspective. Johns Hopkins University Press, Baltimore

Hewitt de Alcantara, C. (1976) Modernizing Mexican agriculture: socioeconomic implications of technological change 1940-1970. UN Res. Inst. for Soc. Dev., Geneva

Issawi, C.P. (1963) Egypt in revolution. An economic analysis. Oxford University Press, London

Jacoby, E.H. (1975) Transnational corporations and Third World agriculture. Econ. Dev. and Cult. Change, 6, (3), 90-7

Joy, L. (1965) Transforming traditional agriculture: a research agenda, some new evidence. A/D/C Seminar on Subsistence and Peasant Economies, Hawaii

King, R. (1977) Land reform: a world survey. Bell, London

Kirby, J.M. (1974) Venezuela: land reform and agricultural development. World Crops, 26 (3), 118-21

Kotter, H.R. (1974) Some observations on the basic principles and general strategy underlying integrated rural development. FAO Monthly Bull. of Agric. Econ. and Stat., 23 (4), 1-12

Malone, C.C. (1970) The Intensive Agricultural Districts Programme ('package' programme) India. In A.H. Bunting (ed.), Change in agriculture. Duckworth, London, 371-80

Manshard, W. and Morgan, W.B. (eds), (forthcoming) Resource use of frontiers and pioneer settlements. UN University Workshop in Kuala Lumpur, Sept. 1985, UN University, Tokyo

Mendieta y Nunez, L. (1960) Efectos sociales de la reforma

agraria en tres comunidades ejidales de la Republica Mexicana. Universidad Nacional Autonoma de Mexico, Mexico

Metcalf, T.R. (1967) Landlords without land: the U.P. zamindars today. Pacific Affairs, 40, (1-2), 5-18

Morgan, W.B. (1955) Farming practice, settlement pattern and population density in South-Eastern Nigeria. Geogr. Journ., 121, (3), 320-33

------ (1977) Food supply and staple imports of tropical Africa. Afr. Affairs, 76 (303), 167-76

------ (1978) Agriculture in the Third World: a spatial analysis. Bell, London

------ and Pugh, J.C. (1969) West Africa. Methuen, London

Palmer, I. (1976) The new rice in Asia: conclusions from four country studies. UN Res. Inst. for Soc. Dev., Geneva

Poleman, T.T. (1981) A reappraisal of the extent of world hunger. Food Policy, 6, (4), 236-52

Prothero, R.M. (1957) Land use at Soba, Zaria Province, Northern Nigeria. Econ. Geogr., 33, 72-86

Rao, B.S. (1963) Economic and social effects of Zamindari abolition in Andhra. Government of India Press, Delhi

Richards, P. (1979) Community environmental knowledge and African rural development. Inst. of Dev. Stud. (IDS) Bull., 10 (2), 28-36

------ (1983) Farming systems and agrarian change in West Africa. Progress in Human Geography, 7 (1), 1-39

Ruthenberg, H. (1971) Farming systems in the tropics. Oxford University Press, London

Saco, A.M. (1964) Land reform as an instrument of change: with special reference to Latin America. FAO Monthly Bull. of Agric. Econ. and Stat., 13 (12), 1-9

Schultz, T.W. (1964) Transforming traditional agriculture. Yale University Press, New Haven and London

Siamwalla, A. and Valdes, A. (1980) Food insecurity in developing countries. Food Policy, 5 (4), 258-72

Singh, I. and Day, R.H. (1975) A microeconometric chronicle of the Green Revolution. Econ. Dev. and Cult. Change, 23 (4), 661-86

Spate, O.H.K. and Learmonth, A.T.A. (1967) India and Pakistan. Methuen, London

Sternstein, L. (1967) Aspects of agricultural land tenure in Thailand. J. of Trop. Geogr., 24, 22-9

Tarrant, J.R. (1981) Food as a weapon? The embargo on grain trade between USA and USSR. Appl. Geogr., 1, 273-86

Tay, T.H. and Wee, Y.C. (1973) Success of a Malaysian agro-industry. World Crops, 25 (2), 84-6

Taylor, D.R.F. (1969) Agricultural change in Kikuyuland. In M.F. Thomas and G.W. Whittington (eds) Environment and land use in Africa. Methuen, London, 463-93

Todaro, M.P. (1985) Economic development in the Third World, 3rd edn. Longman, New York

Uhlig, H. (ed.) (1984) Spontaneous and planned settlement in Southeast Asia. Inst. of Asian Affairs, Hamburg and Giessener Geographische Schriften 58, Hamburg

Watters, R.F. (1960) The nature of shifting cultivation. Pacific Viewpoint, 1 (1), 59-99

World Bank (1974) Land reform. WB Paper, Rural Dev. Ser., Washington, DC

------ (1982) World Development Report, 1982, Oxford University Press, New York

------ (1985) World Development Report, 1985, Oxford University Press, New York

------ (1986) World Development Report, 1986. Oxford University Press, New York

Chapter Three

URBAN ECONOMY AND EMPLOYMENT

R. Bromley and C. Birkbeck

Studies of economic activity in Third World cities have been bedevilled by confusions resulting from the application of inappropriate conceptual frameworks and terminology derived from the industrial experience of advanced capitalist economies and from the classificatory zeal of international organisations anxious to cast the whole world in the same statistical mould. Geographers and other social scientists have frequently used terms such as 'poverty', 'work', 'employment', 'self-employment', 'underemployment', 'labour force', 'formal', 'productive', 'firm', 'proletariat' and 'working class' without prior definition, and sweeping generalisations have often been made without adequate substantiation. Even when definitions are made, many analyses turn out to be tautologous or contradictory because ill-defined categories such as 'the informal sector' are duly studied and shown to have the very characteristics which were used in their initial definition, or because case studies of specific activities and places are often used to characterise vast, hetereogeneous aggregates which have never been enumerated.

Since 1976 we have been involved in research on urban poverty and the labour process in Third World cities. (1) In this essay we present a summary and revision of our conceptual framework in the hope that it will prove useful to other scholars. The framework can be applied successfully in any Third World urban areas which are characterised by high levels of socio-economic inequality; by strongly materialistic, individualistic and competitive ethics; by the decline of 'traditional', 'tribal' and 'peasant' social organisation; by high levels of monetisation, in the sense that most goods and services have a widely recognised market price; and by the predominance of exchange-oriented economic activities over subsistence activities. Most of the ideas and terms that we develop can also be applied both in rural areas of the Third World and in the First and Second Worlds, but we do not feel able within the constraints of length and experience to present a universal framework. Hopefully increased opportunities will arise to interrelate First, Second and Third World urban and rural research, to seek more common bases of analysis, and to identify the special and unique features of the Third World urban context. In the meantime our concern is to facilitate the analysis of poverty and employment in the Third World cities as integral parts of urban, regional, national and international systems characterised by mass poverty, high levels of inequality, and the conspicuous wealth and consumption of a minority. In the concluding section of this essay, we discuss how our conceptual framework and terminology can be applied to the analysis of such topics as the 'putting-out system', franchising, international sub-contracting, sweat-shops and the 'underground economy'.

INEQUALITY AND POVERTY

Most social formations are characterised by marked inequalities between individuals in terms of five major variables: income, wealth, control over the productive process, influence over the state (political power), and access to information. Such inequalities are particularly marked in the peripheral (dependent) capitalist economies of the world system - the so-called Third World. In general, there is a positive correlation between the five variables, so that those who have most of any one of these usually have a substantial amount of all the others, and those who have least of any one usually have relatively little of all the others.

Our approach focuses on the vertical interrelationships between the relatively large numbers of worse-off people towards the bottom of the economic, political and social continuum, and the comparatively reduced numbers of better-off people towards the top. We emphasise both the inequalities of the socio-economic system and also the links between different levels, and we are concerned to explore the mechanisms which articulate and perpetuate inequalities within social formations. We concentrate mainly on individuals and how they relate to one another in situations of inequality, rather than on the behaviour of such social aggregates as classes or ethnic groups. This is not to deny the significance of social aggregates forming all or part of a particular horizontal stratum of the population, but simply to suggest that the vertical interrelationships are initially more striking to the observer, and ultimately, perhaps, more significant in perpetuating and even accentuating inequalities. We proceed from the analysis of relationships between individuals in a particular social formation to the identification of groups, classes, and relationships between such aggregates, rather than from supposed class structures to the stereotyping of individuals. In essence, we are following E.P. Thompson's lead in trying to 'think seriously and precisely about actual social relationships' (Bechhofer and Elliott, 1981, 182), and not simply assigning class terminology to aggregate groups which may well have no functional pattern, channels or sense of identity. Thus, Thompson (1968, p. 9) argues that:

> I do not see class as a 'structure' nor even as a 'category' but as something which in fact happens ... in human relationships ... And class happens when some men, as a result of common experiences (inherited or shared), feel and articulate the identity of their interests as between themselves, and as against other men, whose interests are different from (and usually opposed to) theirs.

Thus, what we are primarily concerned with are 'the social relations of production' - the ways in which capital, labour power and technology are linked together in the productive process. Through our focus on the high levels of inequality characteristic of dependent capitalist economies, our attention is drawn to the ways in which the few individuals with substantial amounts of capital interrelate with others with relatively modest amounts of capital, and with the great majority of individuals who have little more to offer in the productive process than their own labour power or that of their immediate family. Understanding such interrelationships can lead us towards a broader comprehension of the economy as a whole - the interlocking processes of production and consumption of goods and services, articulated through the medium of money. In turn, such a comprehension can contribute to our understanding of how labour power is exploited, or in other words, how surplus value is appropriated, in particular economies, and how such exploitation contributes to the accumulation of wealth and the perpetuation or accentuation of inequality.

Though much has been written about poverty and the poor in many different parts of the world, there is little agreement as to how the two terms should be defined or to how those characterised by poverty might be described and enumerated. As we have reviewed some of the theories and literature on poverty in other works (Bromley and Gerry, 1979, pp. 11-14, Birkbeck, 1980, pp. 85-9), and as the issues involved are somewhat tangential to our general line of analysis, it is sufficient here for us to summarise our position. We are concerned to examine poverty as relative deprivation, but we do not impose strict empirical limits on the use of the terms by defining our own 'poverty line', an upper limit beyond which members of the population do not fall into the category of 'the poor'. Though we focus our attention on relatively poor individuals and occupational groups, and on the economic, social and political processes affecting those groups, the processes that we are studying recur at, and have implications for, all levels of society. We therefore reject the imposition of a sharp division between 'the poor' and 'the rest' of society, as expressed through the concept of a poverty line, and we instead focus on interrelationships in situations of high inequality. This approach stems directly from our conviction that poverty can only be interpreted in terms of the historical development of complete social formations, whether local, regional, national or global, and in terms of the current social relations of production in those formations. Not surprisingly, given this conviction, we reject interpretations of poverty which are derived solely from the description of the poor or from notions of a 'culture' or 'subculture' of poverty (see, for example, Lewis, 1970; Wade, 1973).

THE DEMISE OF DUALISM AND THE AMBIGUITIES OF UNDEREMPLOYMENT

Much of the research which has been conducted on Third World economies and urban labour markets has been cast within a dualist framework. In such a framework the economy or labour market is divided into two contrasting sectors, most commonly described as 'modern' and 'traditional', and these two sectors are presumed to have considerable internal coherence and homogeneity, but to contrast very strongly with one another. The modern sector, which is based on large-scale state and corporate enterprises, is generally thought to be superior to the traditional sector, which is based on small-scale family and individual enterprises. Modern-sector firms are assumed to have relatively high levels of income, wealth, control over the productive process, influence over the state and access to information, while traditional sector enterprises score low on most or all of these variables.

Our preference is for a strict avoidance of dualist frameworks. This not only reflects a greater enthusiasm for studying relationships and interactions than for defining and studying specific categories, sectors and segments of the population, but also a strong belief that dualistic (two-sector) models have outlived their utility for academic analysis and policy-making. We recognise that all models and theories abstract from the complexity of the real world to present a simplified view, emphasising divisions and relationships which are considered to be particularly important. Our criticism of dualistic models, therefore, is not so much that they simplify, as that they emphasise what to us seem the wrong divisions and relationships, consequently underplaying the significance of the relationships which to us seem most important. Viewing the contrast between the relatively few large enterprises and the many small ones, we feel that a 'horizontal' division between modern and traditional, formal and informal, upper circuit and lower circuit, or what ever dualistic terminology is preferred, is being made precisely where the dominant focus of attention should be the vertical interlinkages between large and small. The study of these interlinkages could reveal the patterns of exploitation and accumulation maintaining and even accentuating socioeconomic inequalities, while the study of the individual sectors may do little more than confirm the characteristics which were assigned in their original definition.

Our views on the deficiencies of the dualistic models, and particularly on the division between the informal and formal sectors, have already been published in various papers that we have authored or edited (see Bromley, 1978-9; Bromley and Gerry, 1979; Bromley, 1985), and there is little need to repeat most such views here. Suffice it to say that those two-sector models which assume either that there is no relationship between sectors, or that relationships are essentially benign, seem to us to be entirely inappropriate, perpetuating misconceptions and inevitably leading to misguided policy formulations. In contrast, those two-sector models which assume that one sector dominates and subordinates the other, and that interrelationships are both important and fundamentally exploitative, present credible and relatively useful images of Third World economies (e.g. Tokman, 1978; Santos, 1979; Berger and Piore, 1981). Although dualistic models emphasising domination, subordination and exploitation have considerable didactic value for relatively simplistic explanations of how national and urban economies function in the Third World, however, the substantial analytical problems of 'structural overlap' between the two sectors (see Breman, 1976; Harriss, 1978) make such models more of an impediment than a help for advanced and detailed analyses.

Just as we believe that analysis is facilitated by abandoning dualistic formulations because they are potentially misleading, ambiguous and over-simplistic, we see great advantages in dropping the concept of 'underemployment'. While 'unemployment' is a fairly simple and useful concept, indicating the total absence of work, 'underemployment' is a more complex and relative concept, indicating that there is somehow less work than there should be according to some norm. The implicit assumption behind the term underemployment is that workers to whom the term is applied are doing something, but not much, and that they are in some sense 'less employed' and 'less productive' than those workers who are considered simply as 'employed'. Often, the term underemployment, contrasted with employment, simply forms the basis for a new dualistic division of labour markets into a 'modern sector' in which workers are fully employed, and a 'traditional sector' in which workers are underemployed, having a low, zero, or even negative marginal productivity.

Specific occupations tend to be type-cast as classic examples of underemployment, particularly peasant farming, artisan manufacturing and repair, petty trading, small-scale transport, and personal and domestic services, and these occupations are then viewed as 'inferior' or 'substandard' relative to work in government, public services, factories or capitalist agriculture. Building on such assumptions, it is easy to assume that the underemployed work less than the fully employed, or that their work is simply of less value or functional utility than that of the fully employed, and to go on to the obvious conclusion that development will take place by a transfer of workers from 'the underemployed sector' to 'the fully employed sector'. This deterministic formulation is effectively a restatement of Lewis's classic dual-economy model. (Lewis, 1954; see also that of Henning, 1976, pp. 1-11). Clothed in underemployment terminology, however, its sociopolitical implications are rather more evident. The term underemployment is applied by bureaucratic, intellectual and managerial elites to describe the work of most of the poor, yet in many senses the whole socioeconomic system relies upon the hyperexploitation of the labour power of the poor. In other words, the term underemployment is often being applied to those who work hardest for least rewards, and is serving to reinforce social status differences by undervaluing the contribution of the poor and implicitly blaming them for their own poverty and the deficiencies of the socioeconomic system as a whole. For local elites, the concept of underemployment can be a convenient euphemism avoiding more specific reference to the exploitation of labour power, the prevalence of gross inequality and mass poverty, and the exclusion of the poor from many opportunities and services, and even from access to the means of production. For the poor, however, underemployment is generally an unknown or unused concept. In most countries the poor do not complain about being underemployed, but rather about being poor, underpaid, overworked, exploited, swindled, excluded, forgotten or marginalised. The bureaucratic niceties of 'underemployment' are simply irrelevant in most poor persons' descriptions of their situations, but in sharp contrast the word 'unemployment' is used frequently. The lack of a job is usually an ever-present threat, and sometimes an overbearing reality; little wonder, therefore, that it is a frequent topic of conversation among the poor.

In those elite circles in which the term underemployment is frequently used, four main approaches are taken, individually or in combinations, to its definition:

(1) that the workers work less hours than they wish to, or ought to according to some elite-defined norm;
(2) that the workers have very low productivity in terms of

the amount of work completed per unit time;

(3) that the workers are inadequately remunerated for their labour; and

(4) that the workers do not have a 'normal relationship' which an employer - or in other words, that they do not have a steady wage-working job.

Given this variety of possible definitions, and the fact that the four sets of variables are not necessarily mutually correlated, it is hardly surprising that confusion and ambiguity are often the end-products of discussions and analyses of underemployment. The population that wishes to work longer hours is not necessarily the same as the population that governments or intellectual elites feel should work longer hours, nor as the population that has a very low productivity per unit of time worked, nor as the population with low earned incomes, nor as the population of casual workers. Each social group and each definitional criterion is clearly distinct, and in our view it is far easier to deal with the underlying processes and issues by tackling them individually than by agglomerating them together under some nebulous concept of underemployment.

PRODUCTION, WORK AND EMPLOYMENT

The terms 'production', 'work' and 'employment' are so commonly used in our everyday language that many might consider them almost universally understood. We have already used all three terms somewhat loosely in the preceding discussion, but it is necessary to define them more precisely before we can fully explain our analytical framework.

In this chapter we take a deliberately inclusive and wide-ranging approach to the definition of both production and work, placing them firmly within the context of monetary values and the market economy. This approach reflects a powerful disillusionment with the moralistic and political implications of restricting the terms 'productive' and 'work' to specific activities which particular ruling or intellectual elites define as 'good' or 'desirable'. For us, production is simply the creation or making available of goods and services intended for use or consumption by someone, somewhere. It can be broadly divided into 'subsistence production', in which the goods and services are used or consumed by those who create them or make them available; and 'production for exchange', in which they are intended to be exchanged for money or for other goods and services, so that others can use or consume them either in their present form and location, or in some changed form and/or location. In this approach, production is effectively half of economic activity, the other half being consumption, and there is no marked distinction between the production of goods and the production of services, both of which can be treated as commodities because they can be bought and sold through monetised exchange or barter.

To take an example: the coal miner produces by extracting coal from the ground and transporting it to the surface in appropriately sized lumps, the transporter produces by shifting that coal from the mine to places of sale, and the retailer produces by making that coal available to consumers at convenient times and locations, and in convenient quantities and forms of packaging. Each activity adds value to the coal by bringing it closer to its ultimate objective of consumption through burning to generate needed energy, and hence each activity is productive. Whether or not the miners or the company for which they work actually legally own the coal they are mining, they are considered to produce coal if they make it available to transporters, retailers and consumers, or if they use it themselves for their own energy needs. Similarly, whether a transporter has a driving licence or not, or whether his or her truck is legally owned, stolen, or home-made from scrap parts, s/he can be considered to produce a service and to give himself or herself a potential for gaining an income by shifting coal from the mine to retailers or consumers. If some of the coal falls off the back of the lorry into a deep gorge where it cannot be recovered, this is simply an operating loss in the transporter's production process, but if some of the coal is stolen by a person who can then sell it or use it for personal subsistence, this is an operating loss for the transporter but a productive activity for the thief because it generates an income or avoids an otherwise-necessary expenditure. Even if the market price for stolen coal is lower than that for legally acquired coal because of some social stigma attached to theft, or because of the danger of prosecution for receivers of stolen goods, the thief has added value to his or her own total property by adding a quantity of coal which can subsequently be sold or used for subsistence. Similarly, if a coal retailer gives a few lumps of coal to old people who come begging for coal to heat their houses, this is merely an operating loss for the retailer, but it is actually a subsistence production activity for the beggars, creating a potential for coal use which was not previously available to them.

The characterisation of illegal activities like theft and parasitic activities like begging as productive for their participants, and as production in the economic system as a whole, may seem curious to many readers. It has the advantages, however, of avoiding confusion between economic activities and legal or moral constraints, and of treating the so-called 'black', 'clandestine', 'hidden' or 'counter'-economy (see, for example, Smith, 1976, pp. 53-123; Junguito and Caballero, 1978; De Grazia, 1980; Pahl, 1980) as an integral part of the economy as whole rather than as some autonomous system. National-accounts statistics, of course, cannot include all economic activities officially deemed to be illegal or immoral, as many such activities are clandestine and difficult to quantify. This is no reason, however, why such activities should not be taken into account in localised field research projects, and in reality many of the enterprises and individuals which contravene legal or moral sanctions are recorded in government statistics and received the official accolade of being considered 'productive'. Thus, for example, a factory is not normally considered unproductive or left unrecorded simply because it generates as a by-product more air pollution than is permitted in some government decree, nor a mine simply because its owners bribed government officials to get the mining concession. Similarly, we would not call economic activity by state enterprises unproductive or leave them undocumented after a military coup has overthrown elected civilian rulers simply because the state itself has lost legitimacy according to democratic norms. In effect, there is always a degree of recognition accorded to a substantial number of economic activities deemed illegal by the state apparatus or deemed immoral by the dominant value system, and we are simply taking that process to its logical conclusion by attempting to take account of all exchange-oriented economic activity.

Our approach to defining 'work' stems directly from our definition of production. Work is considered as any activity in which time and effort are expended in the pursuit of financial gain, or of material gain derived from other persons in exchange for the worker's labour or the products of such labour. Anyone who works is described as 'a worker', any financial or material gain resulting from work is 'earned income', and any opportunity to achieve such a gain is an 'income opportunity'. Thus, work is the labour involved in production for exchange, and it is an economic activity in the sense that it is 'income-generating'.

Our definition of what work is can be further clarified by defining two major areas which are not considered as work. Firstly, there is a variety of forms of income which do not derive directly from the work of the recipient: payments from government or private pension funds, and from social welfare and charitable institutions; payments derived from investments, moneylending and renting (except where the recipients devote so much time to supervising their assets that they can be described as 'professional investors, moneylenders or landlords'); and windfall gains such as legacies, lottery wins and fortuitously finding buried treasure. Secondly, there are possibilities of holding expenditures down so that less income is required to maintain a balanced budget. Thus, some individuals devote considerable time, effort and skill to such activities as building, maintaining and cleaning their own homes, cooking their own food and looking after their own children, so as to avoid paying others to perform the same tasks, and to walking or cycling to places of work or recreation, so as to avoid paying transport fares. Though they are immensely important, particularly for the poor, these expenditure-reducing activities are not described here as work, but rather as subsistence-labour. They are, in effect, forms of subsistence production analogous to the peasant farmer growing his or her own food, rather than production for exchange. It should be recognised, however, that the effort and skills applied to subsistence labour can easily be transferred to work, and vice versa, as when a good domestic cook opens a restaurant, or when bricklayers or carpenters working in the construction industry uses their skills to build or extend their own houses at weekends.

The social relations of production for exchange should be examined within the broader context of the economy as a whole, with its complex processes of accumulation, regulation and cross-subsidisation, and with its interchanges and long-term shifts between subsistence and exchange. At a more specific level, it is important to examine the ways in which individuals insert themselves, or are inserted, into the labour market (the market for income opportunities), and the possibilities that they have to substitute subsistence labour for income and vice versa. This approach, in turn, can be broadened slightly by a consideration of the division of labour in terms of age, sex, experience and skills, and of the extent to which there are 'household strategies' as well as 'individual strategies' for insertion into the labour market. An obvious example of household strategies, for example, is the idea of the 'multi-occupational household' (Smith, 1980, p. 359) or 'domestic economic archipelago', whereby each member of the household engages in different productive activities so as to provide opportunities for interchange of products and cushioning against possible economic depressions in specific activities.

Any form of work which is regularly performed by a given person may be described as an 'occupation', and most specific occupations (e.g. bricklayer, seamstress, washerwoman, and advertising executive) have characteristic working relationships, condition and regimes. In turn, when an individual performs an occupation in a specific location with specific working relationships and conditions, and with a definite working regime, he or she can be said to have a 'job' (e.g. shopkeeper operating his or her own premises at no. 1407, 18th Street, or washerwoman for the Diaz-Munoz household). Working relationships are the interpersonal relations between fellow workers (including managers and other workers with supervisory functions) in the same occupation or type of economic activity. They are one of the three main aspects of the social relations of production, the other two being the investment/accumulation process, and the technology of production. When working relationships are of approximate equality they can range from bitter competition to partnership or joint membership of a co-operative enterprise. When they are unequal they can range from the unremunerated use of family labour through apprenticeship and patron-client links, to simple wage-working and the links between superiors (e.g. foremen, managers and bosses) and inferiors (e.g. shop-floor workers) in the same enterprises. Working conditions relate to the physical conditions of work in terms of such variables as noise, temperature, exposure to wind, rain or sunshine, and the likelihood and types of possible accidents. Lastly, working regimes relate to the spatial and temporal dimensions of work - where and when work takes place, including the degree of regimentation of location and timing, and the duration of particular tasks or projects.

Wherever an unequal working relationship exists in which someone works for someone else or for a specific company or institution, there is a verbal or written contract which specifies the mutual obligations and expectations of the two parties. Similar contracts are made whenever a loan of equipment, raw material, premises or merchandise is made from one enterprise to another, or when one enterprise commissions another enterprise to perform a specific task. Under certain circumstances, contracts may be legally binding and supported by the forces of law and order and the regulatory mechanisms of the state. At least as often, however, they are merely informal agreements on working relationships, enforceable by nothing more than the withdrawal of assistance, collaboration, materials or facilities; or the threat or use of violence, sabotage, acts of protest or invocation. It is to such informal agreements that our attention is mainly drawn in any study of the economic activities of the urban poor in the Third World.

In our analysis, the term 'work' has a different meaning to the term 'employment'. 'Employment' is used to denote a relationship between two parties, an 'employer' and an 'employee', the former paying the latter to work on the former's behalf for a significant period of time (at least a working day) or for lesser periods on a regular basis. When there is a direct two-tier employer-employee relationship based on some form of contract, there are two main forms of working relationship: 'on-premises working', when the employee works at a site owned, rented, or operated by the employer; and, 'outworking', when the employee works away from the employer, usually in his/her own home, in the streets, or in some door-to-door operation. An employee may be paid wages per unit of time worked, per unit of 'output', or by some combination of the two. When payment is partly or entirely per unit of 'output', the employee is effectively paid a commission, and this mode of remuneration is known as piece-work. When work is remunerated wholly or partly by the unit of time worked, whether as 'on-premises working' or 'outworking', it is generally recognised as a form of wage-work. When it is remunerated solely per unit of 'output', however, it is usually only viewed as wage-work if it is conducted 'on premises'. When conducted off premises, piece-work is conventionally viewed as a form of 'self-employment' or 'own-account working' and such a conception is embodied in most labour legislation. The distinction that is implied is that between 'contracting' labour, whereby an employer hires an employee to do such work as the employer deems necessary according to the working regime and conditions specified in the contract; and 'subcontracting' for the performance of specific tasks, whereby all that is required is the performance of the tasks by a specified deadline date, without insistence on a specific working regime or conditions.

For most of the world's workers, the prime motive for work is undoubtedly the simple need to ensure their own means of survival and that of any dependents such as children or elderly relatives that they choose or feel obliged to support. Their income is either made up of, or provides the means to purchase, food, drink, shelter and clothing, and even when all these necessities are provided, individuals may choose to work so as to be able to purchase other desired goods and services. As well as income, however, work has the potential to give fulfilment, status, security, and social interrelationships, and for many better-off workers these factors are at least as important as income in determining their motivation to work and choice of job. In situations of high socio-economic inequality, the rich can often treat work as a pleasant luxury to be indulged in when it seems attractive. Meanwhile, the poor must view it as a basic necessity, usually taking occupations which the rich would consider degrading, and frequently being forced to change occupation and/or employer because of cyclical downturns in specific economic activities, the whims of key entrepreneurs, long-term shifts in the terms of trade for particular goods and services, or such domestic disasters as illness, theft of personal property, and flooding or fire in the home.

THE SOCIAL RELATIONS OF PRODUCTION

The social relations of production assume particular significance in the analysis of high levels of socioeconomic inequality and mass poverty. They are intimately related to inequalities, the accumulation of wealth and the perpetuation of poverty, manifesting these disparities and processes in the central functioning of the economy, and contributing to the maintenance of the existing social and economic order. In dependent capitalist social formations, in which such institutions as slavery and feudalism have largely or entirely broken down, and in which co-operative and collective enterprises are of relatively little significance, our attention tends to be heavily drawn towards employer-employee relationships and the significance of self-employment. (2) Indeed, these two categories are frequently viewed as the only basic working relationships, so that individuals seeking work are faced with the simply three-way choice of responding to someone else's initiative by taking up offers of employment, taking an entrepreneurial initiative and establishing their own self-employment, or at a higher level, establishing their own enterprises and employing others to work for them. Whole social formations can apparently be broken down into 'the bourgeoisie', who employ; 'the petite bourgeoisie', who engage in self-employment, sometimes assisted by unpaid family labour; 'the proletariat', who do wage-work for the bourgeoisie; and 'the lumpen-proletariat' of criminals, prostitutes, beggars and others surviving on the margins of society and effectively being parasitic on other social groups. Such a stratification, however, embodies a remarkably simplistic conception of the social relations of production, and our own research and that of many others focusing on unequal relationships rather than on pure social stratification (e.g. Broadbridge, 1966; Breman, 1976; MacEwen Scott, 1979) presents a more complex and intricate picture.

Rather than simply dividing workers' employment into two basic categories, 'wage-work' in which employees work for employers, and 'self-employment', in which individuals work independently on their own account, we believe that there is a continuum of employment relationships which can be subdivided in terms of two major variables, 'relative stability and security of work opportunities', and 'relative autonomy and flexibility of working regimes and conditions' (Figure 3.1). (3) The two extremes of the continuum have relatively high levels of stability and security, and are described as 'career wage-work' and 'career self-employment', while the various intermediate relationships have relatively low levels of stability and security and are described collectively as 'casual work'. Our approach is intended to form the basis for a focus on relationships between large and small enterprises, between enterprises and workers, and between the state and the labour process. Our categories may not be well-known, and ambiguous cases will arise as in all classifications, but they do provide a

Figure 3.1: The continuum of employment relationships

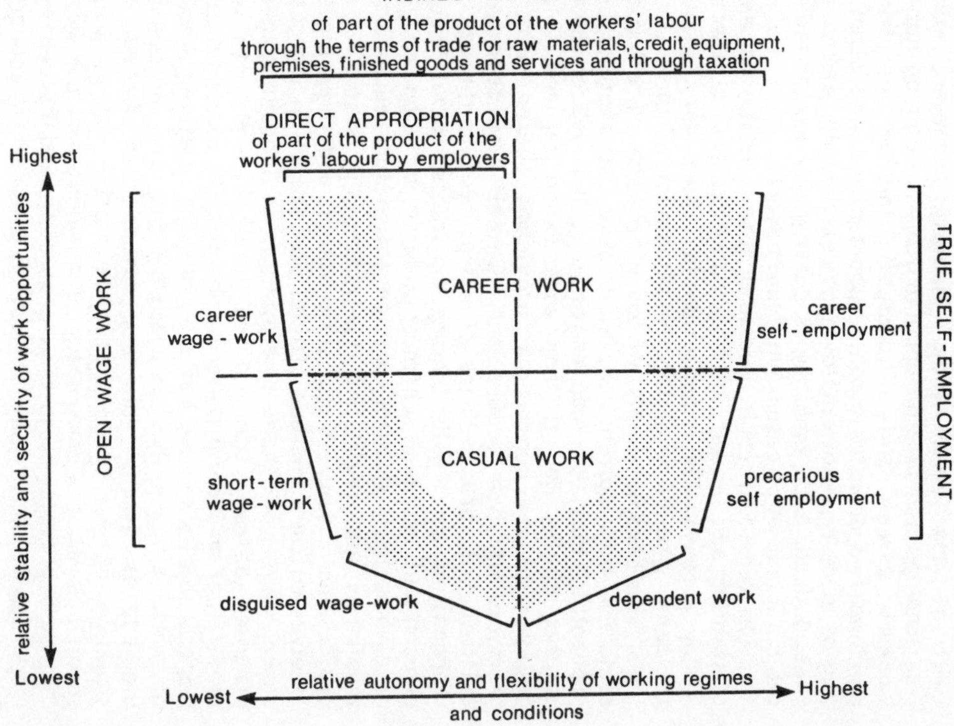

much richer and more appropriate focus for studies of contractual relationships, economic linkages and changes in employment structure than the more conventional 'dualist' distinction between wage-work and self-employment.

As with all employment classifications, our classification does not necessarily allow individuals to be placed in one category for the whole of their lifetimes, or even at any given time. Mobility between working relationships and income opportunities is common during a worker's career, and a substantial number of workers combine two or more income opportunities, and often also two or more types of employment relationship, at the same time. Indeed, such mobility is particularly associated with the combination of casual work and poverty.

The term 'career' is used to denote a substantial degree of stability and security of work opportunities, with workers having a reasonable expectation that they can continue in the same type of work for the remainder of their working lives. This does not imply, of course, that they are forced to remain in their current job, but merely that there is a high probability that they can continue in that job or move to another similar job if they wish to do so, and if they conform to the contractual provisions and/or customary practices of their occupations. 'Career' also denotes the existence of various forms of job protection through such mechanisms as: legally binding contracts; social security, pension and insurance schemes; powerful trade unions, guilds or associations; statutory compensation for redundancy or invalidity; and legislation protecting the status of such professions as lawyers and accountants, and requiring their involvement in numerous official procedures. In the case of career wage-work there is normally a medium- to long-term contract (between one and ten years) with expectation of renewal, or an indefinite-term contract. Of course no career work is totally secure, but the risks are _relatively_ low, loss of work is _normally_ compensated for, and occupational mobility is mainly voluntary rather than forced. In a situation of war, _coup d'état_, revolution, or at a more specific level, the bankruptcy of an employer, the workers may lose protection, just as protection and career wage-work opportunities may be lost if a worker is thought to have committed a crime against the employer or a workmate. Such risks, however, are small in relation to those faced by most casual workers.

Though casual work is characterised by relative

instability and insecurity of work opportunities, this does not mean that all these workers change occupation, or even change job, with great frequency. The term 'casual' simply implies that the workers are continually faced with considerable risks, which may force them to change occupation in order to survive. Thus, for us the term 'casual work' not only encompasses such well-known precarious occupations as unskilled wage-labouring on construction projects or in fruit-picking during the harvest season, but also a wide range of occupations characterised by considerable risk, poverty and instability which are not usually described as 'casual' in common parlance. As an extreme example of apparent stability which actually embodies considerable risk and instability, it is worth considering the self-employed poor male peasant farmer. The farmer may farm for the whole of a career, but throughout that career he or she is subject to numerous natural hazards (flood, drought, pests, etc.) and to fluctuations in market conditions. Such a small enterprise barely maintaining itself on the margins of subsistence is for ever subject to the risk that a poor harvest will lead to the mortgaging of land or the next crop, increasing dependence upon suppliers, buyers or moneylenders, and sometimes leading on to the total loss of livelihood. Continual exposure to serious risk of loss of livelihood implies that work is 'precarious' or casual, just as much as oscillation between different income opportunities. While the poor peasant farmer has no more than a precarious self-employment, a form of casual work, the richer peasant farmer has more land and accumulated savings to act as a cushion against risks, so that the former can genuinely be considered as being engaged in career self-employment. Indeed, in times of crisis, such as drought or widespread crop disease, the farmer may buy out neighbouring poor peasant farmers altogether, confirming both his or her own career status and the others' casual status.

Casual work can be crudely divided into four broad, and occasionally overlapping, ideal-type categories, which, if put in order of increasing dissimilarity from career wage-work, are: 'short-term wage-work', 'disguised wage-work', 'dependent work', and 'precarious self-employment' (which grades into 'career self-employment', the other extreme of the continuum). These four categories are alternative relations of production, affecting the individual worker. In turn, they have ramifications at the aggregate level in terms of the types and structure of enterprises, and the composition and nature of social classes. The four categories can be described as follows:

(1) Short-term wage-work. This is paid and contracted by the day, week, month or season (so-called casual labour), or paid and contracted for fixed terms or tasks, with no assurance of continuity of employment, for example 'on probation', 'for a specific one-off job', or as 'peak period assistance'. Short-term wage-work is recognised as 'wage-work' in law, and so it can be grouped with career wage-work in a broader category of 'open wage-work', but it carries relatively few of the benefits to the worker which are associated with career wage-work. The employer normally provides short-term wage-workers with most or all of their equipment, raw materials and other inputs, and in most cases short-term wage-workers work on the employer's premises at specific times, and for specific periods, as defined in a verbal or written contract of employment.

(2) Disguised wage-work. In this case, an enterprise or group of enterprises regularly and directly appropriates a fixed proportion of the product of a person's work without that person legally being an employee of the enterprise or group of enterprises. Thus, for example, many enterprises utilise 'outworkers' for manufacturing and repair operations; workers who work in their own homes performing specific functions in return for a payment predetermined by the enterprise for each task or piece - a form of off-premises work which is not defined as 'employment' in most countries, but rather as a type of subcontracting. Similarly, many manufacturing firms, wholesalers and insurance companies retail through commission sellers, vendors who receive an agreed sum as their commission for each sale, and who only sell the products of one firm or of a few related firms. Both 'outworkers' and 'commission sellers' can select their own working hours as long as they bear in mind the need to obtain a subsistence income and to maintain a good relationship with 'the firm'. Often, an enterprise using disguised wage-workers also supplies these workers with equipment, credit and raw materials, and sometimes even with premises, as a means to increase their production and to tie (subordinate) them more securely to the enterprise.

(3) Dependent work. In this case, the worker is not in open or disguised wage-work, as there is no wage payment or

fixed commission, but is dependent upon one or more larger enterprises for credit, and/or the rental of premises or equipment. The credit and/or rental arrangement provides a mechanism for the appropriation of part of the product of the worker's labour, and the opportunity to work is dependent upon its continued availability. Typical examples are the carpenter who works in rented premises and buys the necessary equipment with a loan, and the taxi driver who operates someone else's vehicle, keeping any earnings over and above a fixed rental and running costs. A further example is the street trader who normally sells products bought from a specific wholesaler because only that wholesaler will allow credit on the purchase of merchandise. The key differences between dependent work and disguised wage-work concern the relationship between appropriation and production. In dependent work, appropriation is through the payment of a fixed sum of rent and/or interest by the worker, agreed prior to the onset of production, rather than through the operation of a fixed margin of commission. Thus, in dependent work the worker is not usually constrained by fixed prices and proportional deductions, and there is no fixed relationship between the amount of appropriation and the amount of production.

(4) Precarious self-employment. In this case the worker has two major characteristics, considerable instability and insecurity of work opportunities (i.e. the work is casual), and self-employment (i.e. working independently, doing a job without engaging in open or disguised wage-work, or in dependent work). Most self-employment is clearly precarious, both because of the sharp fluctuations in environmental and economic circumstances which characterise the work opportunities involved, and because of the worker's lack of protection against loss of work opportunities through such natural and human-made disasters as drought, fire, accidents at work, and arbitrary closure by order of local officials.

The minority of self-employed workers who are career workers rather than casual workers have attained their career status through significant capital investment and savings, through collective solidarity with others in the same occupation so as to establish an oligopoly, through official patronage and protection, through insurance schemes, and through the possession of scarce skills or other resources which are likely to be heavily in demand for many

Table 3.1: Ideal-type characteristics of the six major categories in the continuum of employment relationships

	career wage-work	short-term wage-work	disguised wage-work	dependent work	precarious self-employment	career self-employment
Reliance of worker on capitalist or state for the opportunity to work	yes	yes	yes	yes	no	no
Direct appropriation of surplus value from worker by capitalist or state	yes	yes	yes	no	no	no
Subordination of worker's labour to capitalist or state in terms of working regime and conditions	yes	yes	no	no	no	no

Table 3.1: continued

Legally recognised employment relationship (and work normally on employer's premises)	yes	yes	no	no	no	no
'Full' available protection for worker through labour legislation and the social security system	yes	no	no	no	no	no
Protection for worker through relative prosperity, and/or statutory guarantees of professional status, and/or guild-like occupational associations, and/or the possession of scarce skills, and/or insurance	no	no	no	no	no	yes

Worker normally remunerated by unit of output and not by time worked	no	no	yes	yes	yes	yes
Worker selects what price to charge for the products of his/her labour (though constrained by market conditions)	no	no	no	yes	yes	yes
Worker owns his/her means of production	no	no	varies	varies	yes	yes
Worker is able to obtain equipment and raw materials at going market prices	no	no	partial	partial	yes	yes
Worker is able to choose both suppliers and customers	no	no	no	varies	yes	yes

years. The category of career self-employment is made up of such groups as lawyers, doctors, well-equipped mechanics and well-off peasants, many of whom may also be small-scale employers of wage-labour. In contrast, the category of precarious self-employment is made up of much larger numbers of petty traders, transporters, artisans and other groups without significant capital, savings, protection, insurance, or scarce and highly demanded skills.

Together, career, self-employment and precarious self-employment make up a broader category which we call 'true self-employment' so as to emphasise the point that many of those who are conventionally described as self-employed are in fact disguised wage-workers or dependent workers, and are therefore not truly self-employed. 'Truly self-employed workers' must, of course, rely on inputs provided by others, on the receipt of outputs by others, and on a system of payment in goods, services or money. However, the keys to their self-employment are that they have a considerable and relatively free choice of suppliers and outlets, and also that they own their means of production. 'Truly self-employed workers' are affected by general economic and social conditions, and particularly by the supply and demand situation for their products (the goods or services that they provide), but they are not reliant upon specific enterprises for the means to obtain their livelihoods. In this sense they are genuinely 'independent workers'.

The ideal-type characteristics of each category in the continuum from 'career wage-work' through to 'career self-employment' are summarised in Table 3.1. Most occupations and individuals can be placed fairly easily within this typology, but there are inevitably anomalous cases and definitional problems, most notably with individuals who have more than one occupation, who are highly mobile between occupations, or who alternate between two different employment relationships within the same occupation.

So far, our consideration of the social relations of production has focused on the employment relationships of individual workers, and on the links between workers and those who appropriate part of the product of their labour. In order to establish the bases for a more comprehensive understanding, it is necessary to go on to consider enterprises and enterprise structure. This enables us to examine the numerous ways in which some workers combine their labour with the roles of supervisors and/or employers, and also the multiple and interlocking forms of economic organisation which encompass both employment relationships and subcontracting.

We define an enterprise as 'an economic unit by means of which the productive process develops', and the fundamental elements of its activity as 'the productive factors which it utilises, the product which is the result of such utilisation, and the surplus - that is, the difference between the value of the product (revenue) and of the productive factors (cost)'. Under this deliberately broad and general definition, enterprises may range from ephemeral to highly stable ventures, and they may or may not be registered and functioning according to the legal norms established by the local and national authorities. They may be very small, often having only one worker - who is 'independent' and 'truly self-employed' - or they may involve two or more workers and incorporate relationships of employment, supervision or partnership. Single-individual enterprises (I-enterprises) are too simple to have an enterprise structure, but all multi-person enterprises (M-enterprises) are considered to have such a structure. This comprises the distribution of powers, resources, status and responsibilities between the various workers involved, and also the patterns of relationships between workers, of flows of goods and services, and of capital accumulation. The workers of M-enterprises may be characterised as 'enterprise workers', so as to distinguish them from the 'independent workers' of I-enterprises.

While we employ the term 'enterprise' very broadly, we limit the terms 'firm' or 'company' to the description of those M-enterprises which have legal recognition of their status as 'economic institutions' undertaking productive activities. Such legal recognition specifies both the identification, central location and character of the firm, and also its statutory obligations towards workers, suppliers and customers. Such recognition does not imply that a firm is necessarily wholly law-abiding and open in its operations, but it does clearly indicate the legal frameworks (company law, labour legislation, contractual procedures, etc.) within which the firm's legality can be assessed.

Our distinction between enterprises and firms is intended to avoid confusion between functional and formal, and between economic and legal/administrative relationships. All firms are enterprises, but many enterprises are not firms and some enterprises never evolve into firms. Those enterprises which are not also firms may be smaller, the same size, or larger than most firms. Thus, each independent worker constitutes an I-enterprise, but he or she does not qualify for the description 'firm'. Similarly, highly ephemeral, illegal or clandestine M-enterprises are not registered as 'firms', even though some of them generate large profits for their 'owner-organisers'. Several 'firms' may even be interrelated by a single owner so as to maximise that owner's profits, constituting a single large enterprise even though there is no legal recognition of that enterprise so as to give it the status of a 'firm'. Thus, enterprises may exist as numerous levels, with some enterprises encompassing or overlapping with others, and with only a minority assuming the character of 'firms'.

Two key characteristics normally define the enterprise as an economic institution: firstly, the capacity to generate sufficient product to cover the minimum cost of the reproduction of the labour power involved; and secondly, the suppression of the price mechanism within the enterprise. The first criterion can be relaxed for a time if there is a subsidy for the enterprise from accumulated wealth, transfers from related enterprises, or state patronage, but the life of an enterprise which is unable to generate sufficient product to satisfy the minimum food, drink, clothing and housing requirements of its worker(s) is likely to be relatively short. Indeed, in most M-enterprises with wage-workers, the expectations of the owners of the means of production are considerably more ambitious - to achieve a substantial profit for themselves over and above any wages paid to their wage-workers.

The suppression of the price mechanism refers to the fact that within an enterprise the factors of production (land, labour, raw materials, premises, equipment, merchandise, money, etc.) are not normally valued and transacted at their current market prices based on the interplay of supply and demand (see Coase, 1937). These factors are devoted to the objectives of production and accumulation, but there is little or no effort to conduct an internal process of bargaining, contracting and profit-loss accounting between the different components of the enterprise. Such procedures are generally considered to be excessively time-consuming and costly for intra-enterprise transactions, and it is often argued that their introduction will lead to the eventual break-up of the enterprise and the loss of any sense of common objective. The suppression of the price mechanism within the enterprise ensures the primacy of external economic relations over internal ones, facilitating the enterprise's image and functional identity as a coherent unit. Though it is more notable and significant in M-enterprises, it is also evident in I-enterprises, particularly in the tendency of independent workers not to assign a clear value to their time or to keep accounts of their expenditures and use of materials. In M-enterprises, the suppression of the price mechanism may also serve to permit and disguise exploitation which, for the purposes of this discussion, may be defined as 'the difference between remuneration and productivity which accrues to the owners of the means of production, constituting an appropriation of surplus value from workers because they obtain less than the equivalent of the value which they actually produce'.

M-enterprises often start off as relatively precarious and unstable entities, some of which dissolve within relatively short periods of time, while others gradually consolidate. In some cases, consolidation may be carried through to a legal formalisation of the enterprise as a 'firm', while in others it may remain almost indefinitely as a 'partial consolidation' permitting the continued functioning of the enterprise but retaining the possibility of its prompt dissolution. Partially consolidated large enterprises may encompass various smaller enterprises through the use of subcontracting mechanisms which arrange for outworking, collecting or distributing functions to be performed on a piece-work basis. Such subordination processes may be reinforced through 'support' to small enterprises based on the provision of credit or the rental of premises or equipment, converting small enterprises into 'dependents' of a large enterprise or group of larger enterprises. In extreme cases, large enterprises may establish a whole cluster of relationships with smaller enterprises, so that a hierarchy of indirect relations of production is established forming an archipelago of interrelated economic activities centred on a particular large enterprise.

THE BROADER SIGNIFICANCE OF CASUAL WORK AND SUBCONTRACTING

It is in the context of partially decentralised production that the full significance of casual work and subcontracting can be appreciated as means of increasing the diversity and profitability of production, and of bringing into existence levels and types of production which would otherwise not be profitable for the owners of capital. Thus, casual workers lack the security of employment and stability of income which is given to employees with long-term (indefinite) contracts. Career wage-working normally includes provision for a variety of forms of job security, including some or all of the following: minimum wages, regularised working hours, fixed overtime payments, 'minimum notice requirements' for both employer and employee, paid holidays, sickness benefit, redundancy pay, life insurance, and even access to subsidised consumer purchasing, mortgage, and public housing arrangements. Although the range of benefits available to the employee with such a contract varies from firm to firm, between types of employees, and between economic sectors and countries, there is a widespread tendency for the job security of the regular employee to be strengthened, imposing greater obligations upon the state. In any economy in which career wage-working opportunites are scarce and job security provisions are selectively imposed, a situation tends to arise whereby a minority of workers are within the system of 'regulated job security', while a majority are outside this system, relying upon different forms of casual work.

The tendency of governments to respond to pressure from trade unions, associations of civil servants, the armed forces, the police, and other organised groups of workers with a degree of job security, and the pressures exerted upon governments by international organisations (and particularly the International Labour Office), leads to an increasing provision for regulated job security. At times, provision may be extended to new groups of society, but the stronger tendency is for provision to remain concentrated upon a minority of workers, and to be improved for them, further differentiating this group from the casual workers. In many cases there is a stabilisation of work opportunities and remunerations for a minority, at the cost of the destabilisation (casualisation) of work opportunities for the majority of the labour force. This destabilisation process has four major causes (see Bromley and Gerry, 1979, pp. 15-19): firstly, the tendency of employers of wage-labour to find loopholes in the legislation intended to stabilise employment relationships, and hence to seek means of overtly or covertly employing casual workers, or of indirectly appropriating surplus value; secondly, the accelerating pace of worldwide technological change; thirdly, the increasing economic instability in most national economies and the widespread recession which began in the early to mid-seventies; and finally, the breakdown of the traditional rural economies in most parts of the Third World.

Some organised groups of career wage-workers with considerable amounts of regulated job security are clearly more concerned with the defence and increase of their relative privilege than with the expression of solidarity towards casual workers. The minority of relatively well-off and organised workers who may ally with the interests of capital and the state against the interests of less privileged groups is usually described as the 'labour aristocracy' (see Hobsbawm, 1964, pp. 272-315; Arrighi, 1970; Saul, 1975). We feel that this description has been applied too liberally by some authors, and that the term 'aristocracy' is hardly appropriate for individuals whose incomes are considerably less than those of most professionals and managers, and far less than those of most large-scale landlords, speculators and investors. Despite the inappropriateness of the name, however, there can be no denying the relative privilege and prosperity of a small minority of career wage-workers in many countries, or the degree to which their class interests may be affected by their situation of relative advantage. Governments and large companies, for example, frequently use the concession of regulated job security and relatively high incomes to potentially powerful groups of workers as a means of 'buying their support' and giving them a vested interest in the continuity of the system.

Creating and rewarding a 'labour aristocracy' is costly, and part of the price is usually the avoidance of significant provision or protection for casual workers. Under such circumstances, employers may seek to utilise casual workers through short-term wage-employment, sub-contracting to smaller enterprises, and the harnessing of formerly self-employed workers into their supply and distribution system as disguised wage-workers or dependent workers (see, for example, Forbes, 1981, pp. 133-40 and 159-64), rather than take on more career wage-workers with regulated job security. In a sense, therefore, the casual workers are used as a 'reserve army of labour', and as an alternative to more regularised forms of contracting and the expansion of the 'labour aristocracy' (see Williams and Tumusiime-Mutebile, 1978; Oliveira, 1985). They are used to carry much of the burden of risk in unstable and insecure situations, being incorporated into the economic system when extra labour and production is advantageous, and being excluded from the same system when they are no longer needed.

At times, employers can use casual labour as a bargaining counter in negotiating with their regularised labour force, threatening to pass jobs over to casual workers, or even to use casual workers as a strike-breaking force if the regularised labour force demands major improvements in their situations. At other times, employers may operate substantial enterprises with little or no regularised labour force, relying upon casual workers, and hence avoiding many of the costs imposed by government job security legislation. By using outworkers and/or outside commission sellers, they avoid the need for substantial premises, and they may even pass the responsibility for the acquisition and maintenance of equipment on to their casual workers and subcontractors (see Broadbridge, 1966; HKRP, 1974; Ozorio de Almeida, 1977).

The general picture that we have painted of casual work concentrates heavily on exploitation by larger-scale enterprises, and on the appropriation of surplus value from casual workers. This view is intended to counter the naive optimism of many observers who ignore the social relations of production and simply focus on the casual worker's apparent freedom, flexibility and entrepreneurial potential. In situations of considerable insecurity and instability, freedom may represent little more than the opportunity to become destitute, and with minute amounts of capital available for investment the aspiring entrepreneur may find it a hard job even to make enough to support a family at the barest level of subsistence. Despite the realities of continuing poverty facing most casual workers, however, Horatio Alger cases of seemingly miraculous success occasionally occur, and such cases are frequently recounted and exaggerated. Though casual work carries obvious risks of economic disaster, it also offers the remote possibility of windfall gains and the more tangible possibility to conduct clandestine operations which might not be permitted in regularised wage-work. In most cases, however, workers have no real choice between career wage-work and casual work, as the former is simply not available to them. The privileges of the so-called 'labour aristocracy' or of career self-employed professionals are not even a remote possibility for the majority of workers, and the real choices for most of the poor are simply between alternative forms of drudgery.

NOTES

1. This paper emanates from the reading, discussions and fieldwork associated with a research project on the socioeconomic organisation of the street occupations in the city of Cali, Colombia. We are grateful to the UK Overseas Development Administration for financial support, but the views here are our own and do not necessarily represent those of the ODA or of any other institution or individual who provided assistance. Examples of work already published include the chapters by Bromley, Birkbeck and Gerry in Bromley (1978-9, 1985) and Bromley and Gerry (1979), and also Gerry and Birkbeck (1981), Birkbeck (1982), Bromley (1982) and Bromley and Birkbeck (1984).

2. An interesting attempt to produce a comprehensive typology of social relations of production applicable to all types of social formations is available in Cox (1971), Cox, Harrod et al. (1972, pp. 5-11), and Harrod (1980). An alternative but related approach deals with 'the articulation of different modes and forms of production' (see, for example, Hindess and Hirst, 1975, 1977; Wolpe, 1980).

3. This section presents an improved version of a typology which was originally proposed in Bromley (1978-9, 1982), and Bromley and Gerry (1979, pp. 5-8).

REFERENCES

Arrighi, G. (1970) International corporations, labour aristocracies and economic development in Tropical Africa. Journal of Modern African Studies, VI, 141-69

Bechhofer, F. and Elliott, B. (1981) Petty property: the survival of a moral economy. In F. Bechhofer and B. Elliott (eds), The petite bourgeoisie: comparative studies of the uneasy stratum, Macmillan, London, pp. 182-200

Berger, S. and Piore, M. J. (1981) Dualism and discontinuity in industrial societies. Cambridge University Press, Cambridge

Birkbeck. C. (1980) 'Property crime and urban poverty: a case study of Cali, Colombia'. Unpublished PhD dissertation, University of Wales, UK, 2 vols

------ (1982) Property crime and the poor: some evidence from Cali, Colombia. In C. Sumner (ed.), Crime, justice and underdevelopment, Heinemann, London, pp. 162-91

Breman, J. (1976) A dualistic labour system? A critique of the informal sector concept. Economic and Political Weekly, XI, (no. 48), 1870-6; no. 49, 1905-8, no. 50, 1939-44

Broadbridge, S. (1966) Industrial dualism in Japan. Frank Cass, London

Bromley, R. (ed.) (1978-9) The urban informal sector: critical perspectives on employment and housing policies (special double issue of World Development, VI (no. 9-10), later published as a book). Pergamon, Oxford

------ (1982) Working in the streets: survival strategy, necessity, or unavoidable evil? In A. Gilbert, J. Hardoy and R. Ramirez (eds.), Urbanization in contemporary Latin America, John Wiley, Chichester, pp. 59-77

------ (ed.) (1985) Planning for small enterprises in Third World cities. Pergamon, Oxford

------ and Gerry, C. (eds), (1979) Casual work and poverty in Third World cities. John Wiley, Chichester

------ and Birkbeck, C. (1984) Researching the street occupations of Cali: the rationale and methods of what many would call an 'informal sector study'. Regional Development Dialogue, V, (no. 2), 184-203

Coase, R.W. (1937) The nature of the firm. Economica, NS, IV, 386-405

Cox, R.W. (1971) Approaches to a futurology of industrial relations. International Institute for Labour Studies Bulletin. 8, 139-64

------, Harrod, J. et al. (1972) Future industrial relations: an interim report. International Institute for Labour Studies, Geneva

De Grazia, R. (1980) Clandestine employment: a problem of our times. International Labour Review, CXIX, 549-63

Forbes, D.K. (1981) Petty commodity production and underdevelopment: the case of pedlars and trishaw riders in Ujung Pandang, Indonesia. Pergamon, Oxford; Progress in Planning, vol. 16, part 2

Gerry, C. and Birkbeck, C. (1981) The petty commodity producer in Third World cities: petit bourgeois or disguised proletarian? In F. Bechhofer and B. Elliott (eds.), The petite bourgeoisie: comparative studies of the uneasy stratum, Macmillan, London, pp. 121-54

Harriss, B. (1978) Quasi-formal employment structures and behaviour in the unorganized urban economy, and the reverse: some evidence from South India. World Development, VI, 1077-86

Harrod, J. (1980) Informal sector and urban masses: a social relations of production approach. Institute of Social Studies, Discussion Papers (The Hague), March

Henning, P. H. (1976) 'The popular economy: an approach to the urban employment problem in developing countries'. Unpublished PhD dissertation, University of Michigan, Ann Arbor (UM order no. 76-19, 154)

Hindess, B. and Hirst, P.Q. (1975) Pre-capitalist modes of production. Routledge and Kegan Paul, London

------ (1977) Mode of production and social formation. Macmillan, London

HKRP (Hong Kong Research Project) (1974) Hong Kong: a case to answer. Spokesman Books, Nottingham

Hobsbawm, E.J. (1964) Labouring men: studies in the history of labour. Weidenfeld and Nicolson, London

Junguito, R. and Caballero, C. (1978) La otra economia. Coyuntura Economica (Bogota), VIII (no. 4), 103-39

Lewis, O. (1970) The culture of poverty. In O. Lewis, Anthropological essays. Random House, New York, pp. 67-80

Lewis, W.A. (1954) Economic development with unlimited supplies of labour. Manchester School of Economic and Social Studies, XXII, 139-91

MacEwen Scott, A. (1979) Who are the self-employed? In R. Bromley and C. Gerry (eds.), Casual work and poverty in Third World cities, John Wiley, Chichester, pp. 105-29

Oliveira, F. de (1985) A critique of dualist reason: the Brazilian economy since 1930. In. R Bromley (ed.), Planning for small enterprises in Third World cities. Pergamon, Oxford, 65-95

Ozorio de Almeida, A.L. (1977) 'Industrial subcontracting of low-skill service workers in Brazil'. Unpublished PhD dissertation, Stanford University (UM order no. 77-12, 680)

Pahl, R.E. (1980) Employment, work and the domestic division of labour. International Journal of Urban and Regional Research, IV, 1-20

Santos, M. (1979) The shared space: the two circuits of the urban economy in underdeveloped countries, Methuen, London

Saul, J. S. (1975) The 'labour aristocracy' thesis reconsidered. In R. Sandbrook and R. Cohen (eds.), The development of an African working class, Longman, London

Smith, G.A. (1980) Huasicanchino livelihoods: a study of extended domestic enterprises in rural and urban Peru. Canadian Review of Sociology and Anthropology, XVII, 357-66

Smith, H. (1976) The Russians. Times Books, London

Thompson, E.P. (1968) The making of the English working class. Pelican, Harmondsworth

Tokman, V.E. (1978) An exploration into the nature of informal-formal sector relationships. World Development, VI, 1065-75

Wade, R. (1973) A culture of poverty? IDS Bulletin, V (no. 2-3), 4-30

Williams, G. and Tumusiime-Mutebile, E. (1978) Capitalist and petty commodity production in Nigeria - a note. World Development, VI, 1103-4

Wolpe, H. (ed.) (1980) The articulation of modes of production: essays from economy and society. Routledge and Kegan Paul, London

Chapter Four

HOUSING

D. Drakakis-Smith

INTRODUCTION

Of all the basic needs that have increasingly attracted the attention of development strategists over recent years, housing has probably formed the major focus. Although in most Third World countries the majority of people still live in rural areas, the attention given to housing has overwhelmingly been urban in its orientation (Hardoy and Satterthwaite, 1981). In some ways this is understandable since the main shifts of population have been to the towns and cities, but the standard of most rural accommodation would probably fall below any minimum levels of habitability set for urban dwellers throughout the Third World.

However, it is not only the migrational pressure upon the city which has focused attention upon urban housing provision: it is also the result of the city itself metaphorically occupying a central place in development analysis since the 1950s. As the physically dominant proportion of the built environment, residential space has held a strong place within the literature in both a direct (housing _per se_) and indirect (its inhabitants) sense. However, it is only within the last ten years or so that housing has assumed an important role in theories seeking to explain the process of urban development in the Third World. Indeed, many of the studies of the 1950s and 1960s were primarily concerned with establishing the differences between the cities of the developed and developing countries, as evidence of the need for modern, Western programmes of development. Even the more sympathetic observers, such as Abrams (1964), were still preoccupied more with documentation of housing conditions than with the exploration of their evolution as part of a specific built environment.

Over the last 20 years, however, perceptions on the role of housing in development, particularly low-cost housing, have changed in parallel with major conceptual shifts in development studies. Thus, with the growth of sector or dualist concepts from the late 1960s onwards (much of the research for which was carried out in squatter settlements), there was a widespread recognition of the scale, permanence and positive nature of the self-built housing of the urban poor. John Turner (1967, 1968, 1969) personified this approach. In the broader and more ambitious purview of dependency and world-system theory, housing provision has received scant attention, largely because such theories have been highly econometric in nature and concerned more with uneven development as a consequence of unequal trading relationships. Indeed, the implicit notion of the poor of the Third World as passive victims of such exploitation did not fit easily with the evidence being amassed by continued work in low-income settlements.

From the mid-1970s, the role of housing in the analytical critiques developed by the neo-Marxist school has steadily assumed greater importance. For many this grew out of a reaction to the way in which liberal capitalists had sought to use the findings of those such as Turner to expand incrementalist programmes of asssistance through aided self-help (ASH) schemes which had relatively limited impact on the basic inequalities in access to housing resources. More recent analysts have sought to investigate the overall role of the built environment, including housing, in underpinning continued capitalistic growth in the Third World. These arguments will be discussed at length in later sections of this chapter, together with the crucial question of how the global recession of the 1980s has or is likely to affect urban housing provision in the Third World city.

Whilst the focus for much of this burgeoning literature has (rightly) been the provision of housing for the urban poor, it is clear that there are other housing markets in the Third World city. Indeed, the existence of conflicting groups competing for limited resources of land, labour, materials and capital is crucial to an understanding and appreciation of the nature of the low-cost housing market which functions within a complex network of economic, social and political relationships. Such arguments imply that in any single location, the low-cost housing market is subject both to local forces (positive and negative) and to broader national, or even global, patterns of development. It will be the task of this chapter to tread the difficult path of emphasising the common ground within this hierarchical process.

The next section examines such influences and their impact on two distinct levels within the housing supply system. The first is the broad realm of housing policy, the second is the more specific and subordinate level of housing programmes. Following this overview, a typology of the principal forms of low-cost housing and their interrelationships is presented. The basic components of all types of low-cost housing, viz. land, finance, infrastructure, materials and design are then examined in a little more detail. The next section analyses how all these factors integrate into a process of change in terms of policies, programme and projects in relation to aided self-help housing, the expansion of which has been so rapid over the last decade. The final section returns to the opening theme of the links between development in general and low-cost housing in particular, to discuss the ways in which the current global recession might affect the housing of the urban poor.

HOUSING POLICIES AND HOUSING PROGRAMMES

Payne (1984) has drawn a distinction within the analysis of housing systems between what he terms political and economic factors, administrative and institutional considerations, and design and technical aspects: these he relates broadly to questions of policy, programmes and projects. Whilst this categorisation is useful in identifying different concentrations of decision making, it is also true that the broader influences penetrate down the full range of components. Put in another way, projects are shaped by programmes and programmes are shaped by policies, so that influences of policy are all-pervading.

It is perhaps useful in this context to distinguish between scales of influence in relation to housing policy. Clearly there are policy decisions of the broadest kind, outlined in the previous section, which affect policy attitudes <u>towards</u> housing as one of many different aspects of development. These measures impinge directly or indirectly on housing systems and, for want of a better term, may be called macro-decisions. Figure 4.1 is an amended version of a diagrammatic representation of the structure of these macro-influences which has been discussed in detail elsewhere (Drakakis-Smith, 1981). In brief, governmental views on the role of housing within development planning can be affected by:

(1) Very broad political philosophies influencing the attitude of the state towards the goals of equality and efficiency. This could affect the commitment to state investment in housing, tolerance of squatters, etc.
(2) Policies on urbanisation and the extent to which the city is favoured over the countryside in development priorities.
(3) The comparative claims of social, political and economic objectives within national urbanisation policies. Most governments view housing programmes as social overheads and rarely invest in low-cost housing provision unless important economic or political objectives dictate this. Singapore, for example, initiated its huge public housing programme shortly after becoming a republic in 1965, partly to give its potentially unstable immigrant, ethnically mixed population a stake in political stability. Furthermore much of the most recent expansion of its housing programme has been undertaken to sustain economic momentum during the global recession.
(4) Finally, within the framework of social investment, housing policies may reflect priorities placed on other basic needs, or may be shaped by a whole range of influences not acting primarily in the interest of the poor. A good example would be the adoption of high-rise blocks simply because of their apparent 'success' elsewhere.

It would be wrong to examine housing in the Third World solely in econometric terms of supply and demand. Strong social and political forces shape both sides of what is seldom a simple equation: for example, the demand for housing cannot be equated with housing needs, even within the same city. The former can be articulated through market forces and perhaps even elicit some sort of governmental response; but housing needs are far more

Figure 4.1: Hierarchical influences on macro-policies towards housing

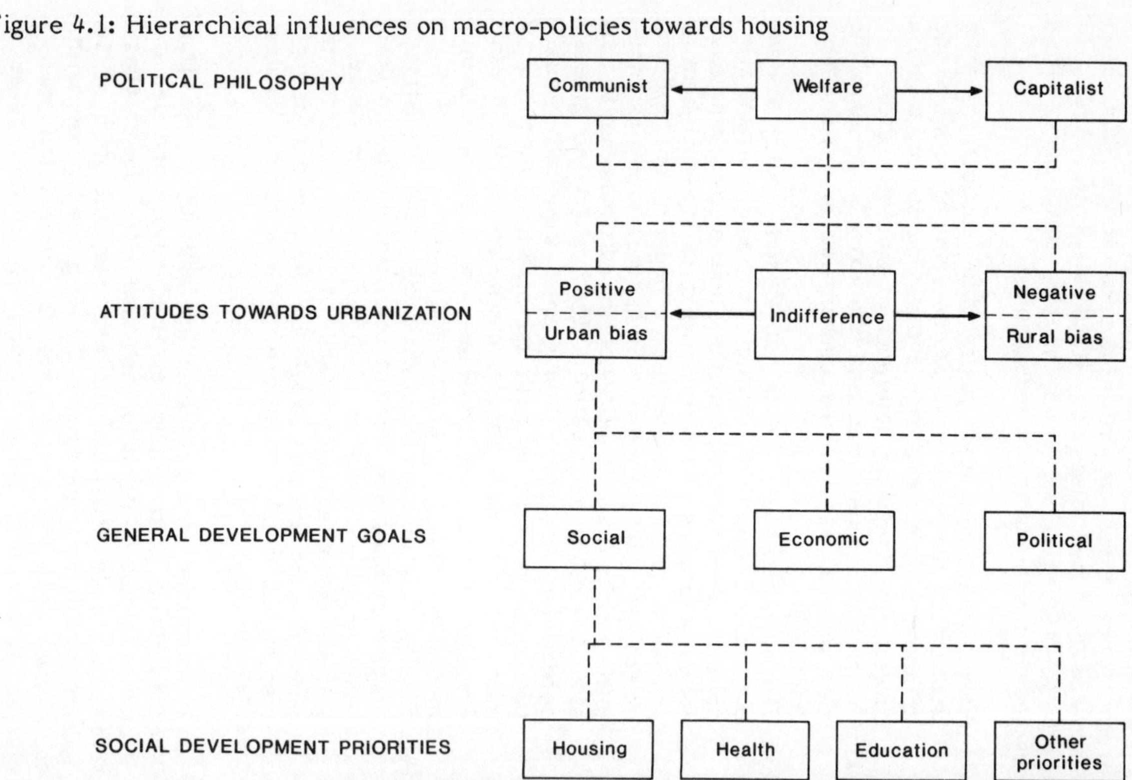

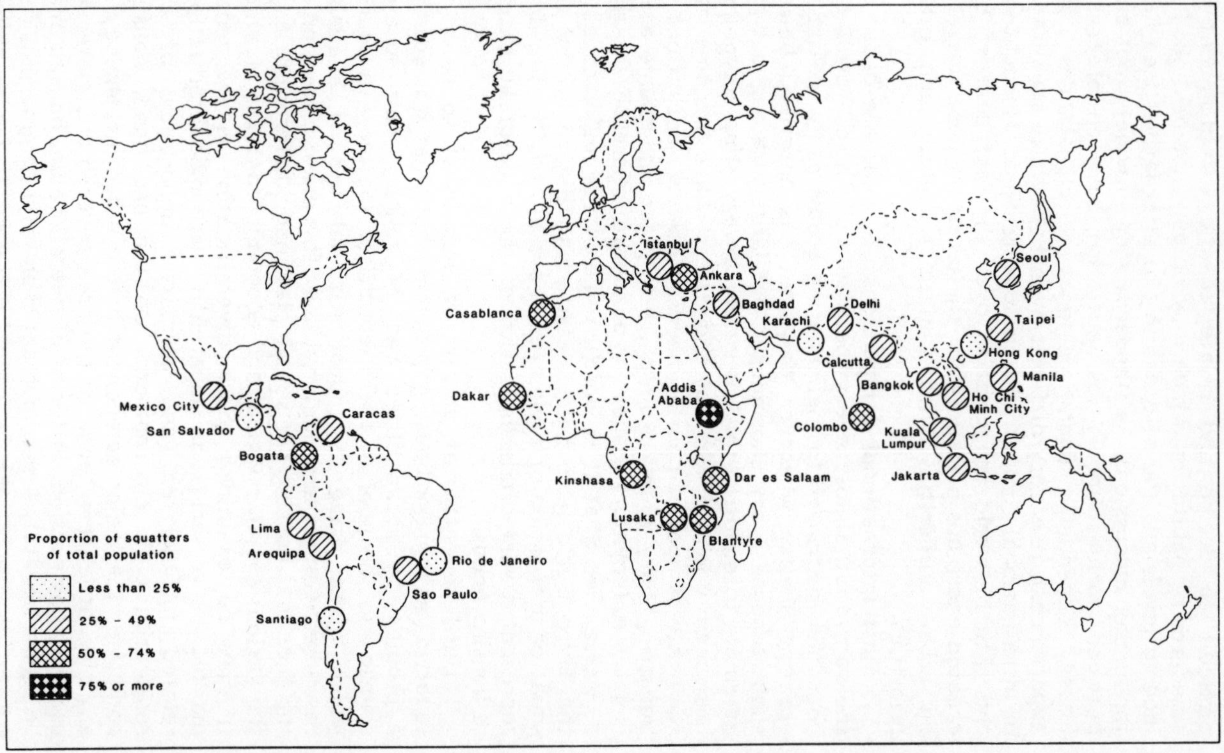

Figure 4.2: Squatter proportions of total population in selected cities

difficult to define and relate to the establishment of standards of habitation which may or may not be linked to income levels and the affordability of alternatives (see ibid. for further discussion). As standards vary so widely, not only between nations and cities, but also between observers (often depending upon personal circumstances and experiences), it is very difficult even to begin to seek inter- or intra-national comparisons on housing needs. Undoubtedly the rate of world urbanisation over the last decades has created enormous pressures for shelter from poor migrants on urban authorities with limited resources and other priorities.

Most international comparisons of needs are based on the fact that almost everywhere in the Third World, the poor have responded to their housing problems by seeking or creating shelter wherever they can. As described below, such responses vary enormously, but largely throughout the attempts to quantify this situation, a very simple typology has been adopted whereby 'squatters' have become the surrogate symbol of housing needs. However, definitions of 'squatting' vary widely from place to place and enumeration is often rudimentary at best or deliberate misrepresentation at worse. For what it is worth, some of the major cities of the Third World have been classified in Figure 4.2 by the squatter proportion of their overall population; but this does represent a very crude indication of the widespread extent of housing needs.

Until recently, of course, it was usual to refer to squatting as a major housing 'problem', but what was seen as a problem by urban elites was, of course, some sort of solution to the poor in their search for shelter - a search which often made the most rational use of available resources (Gilbert and Gugler, 1982). This draws attention to the fact that housing per se can play many different roles in the various social, political and economic contexts in which it may be examined. Such diversification is frequently marked by a series of apparently paradoxical situations related to the real and perceived importance of the residential built environment. On the one hand, housing comprises the major component of the urban morphology of any city, at least in terms of the area covered and materials used in construction. In terms of overall value, housing may be proportionately less important than (say) commercial real estate because of the locational differential. Nevertheless, the overall value of residential areas can be substantial,

Table 4.1: The contribution of construction to the Third World economies

	Average *per capita* income (US$)			
	Under $350	$350-700	$700-2000	Over $2000
Value added as % GDP	3.6	5.2	5.4	7.3
Capital formation as % GDP	8.9	10.6	13.6	13.5
Employment as % EAP (a)	3.1	3.4	6.6	8.1
Rate of growth (% pa) of construction	5.9	5.2	8.6	3.6
Rate of growth of GDP	3.5	4.4	5.9	5.0
Rate of growth of manufacturing	5.5	6.4	7.7	6.4

Note: (a) EAP - Economically active population

Source: Wells (1986)

particularly if squatter settlements and other low-income housing are properly taken into account. In addition, state housing programmes can take up a substantial part of overall expenditure even when they meet only a fraction of total needs.

Yet despite this ostensible importance, the contribution of housing construction to GDP and employment in the Third World is quite limited. Recent calculations by Wells (1986) have revealed the relatively restricted scale of this contribution for construction as a whole (Table 4.1). Formal or commercially built housing usually accounts for only one-third of this total. Thus, despite its sheer physical scale, housing is not so proportionally prominent in the investment profile of the Third World (for various political and economic reasons discussed below). What is perhaps even more paradoxical, however, is that despite the assumed importance of shelter as the most fundamental basic need for the individual household, there is clear evidence that for many of the poorest families in the Third World, there are other higher priorities (Table 4.2). This is particularly true for households recently arrived in the city or located in the periphery of a city experiencing very rapid land use change. For these families, the subsistence mode of production

Table 4.2: Bangladesh: percentage distribution of household expenditure

	Urban		Rural	
	1966-7	1976-7	1966-7	1976-7
All food	73.8	75.9	62.5	63.3
Fuel/energy	6.1	9.2	6.1	8.2
Clothing/footwear	5.3	5.1	6.3	6.7
Others (inc. housing)	14.8	9.8	25.1	21.8

Source: Islam (1982)

which provided basic needs in food and energy is being rapidly eroded as cultivable plots and fuel wood sources are overwhelmed by speculative property markets. When forced to purchase these very basic requirements of food and fuel with limited and erratic cash incomes, many families choose to downgrade investment in housing in terms of priorities, even to the stage where no investment is made at all and the household is reduced to sleeping on the pavements (see Mukherjee and Singh, 1981).

Notwithstanding the above arguments, the role of housing in the reproduction of labour is also an important function of shelter, one which has been emphasised consistently since the growth of neo-Marxist writings on the subject over the last decade. Nevertheless, the function of housing in this context should not be overstressed: it is only the persistence and continued supply of cheap housing that enables labour to be reproduced at low cost (eventually) to capitalism. The role of women, the changing structure of the household, links with other families and with rural branches of the family (through remittances or circular migration), the practice of sub-letting, etc. - all contribute to the lowered costs of the reproduction of labour.

Clearly, housing and its role in the urban political economy of the Third World is an extremely complex phenomenon, one that has links to all levels of development (and all modes of production) from the household to the global, each of which is filtered through a wide range of social and political screens. It is at this point in the discussion that the overlap between policy and programme formulation begins to emerge. Although it may seem 'natural' for national concerns to be with issues of policy and local concerns to revolve around programmes, the situation is by no means as clear-cut as this. Indeed, this is a very indistinct interface, one which has received limited direct attention from housers, to use Abu-Lughod's (1981) useful term. In essence, there has been a change of focus from policies towards housing (as one of many outlets for investment), to policies for housing, i.e. the micro-level policies that determine attitudes on existing supply systems and any changes to be made to them. Obviously, the latter will be strongly influenced by macro-policy decisions and, indeed, one of the most important issues at the macro-policy level in recent years has been the way in which the state has committed itself to aided self-help schemes. This forms the focus of the penultimate section of this chapter.

Within this indistinct blending of macro- and micro-level policy structures and determinants, one crucial factor which has emerged as worthy of further exploration by housers is the role of central and local state relationships in policy formulation. This is a factor which often features indirectly in empirical studies which attempt to assess the effectiveness of particular projects (see Skinner and Rodell, 1983; Payne, 1984). Usually it is policy decisions by the central state that receive the bulk of the criticism, particularly if the empirical focus lies outside the capital cities (for which rational policies are frequently structured). However, the local state, and all its component parts, can be equally important in determining policy as well as programmes in a wide variety of ways. In Zimbabwe, for example, the urban authorities are now merely the operational extension of central state policy, but in Ankara local politicians frequently use housing policy as a device for building up electoral popularity (Drakakis-Smith, 1981; Tokman, 1983).

There has been a growing body of literature on the interrelationships between the national and local state, but few have concentrated directly on the formulation of housing policy. Alan Gilbert and Peter Ward, working in Latin America, have been notable exceptions to this rule (Gilbert, 1986, 1987; Gilbert and Ward, 1985). Clearly, however, more direct attention must be paid to such a vital aspect of policy formulation, and with the growing interest in the role of the built environment in capitalist expansion, there may be some spillover to the periphery and semi-periphery (Drakakis-Smith, 1987a). One note of caution which perhaps needs to be expressed at this point, with regard to housing policy, is that central and local state relationships within the Third World are increasingly being overlain and infused by a third factor, viz. links with international agencies and institutions, ranging from individual consultancy firms to the World Bank.

Reviews concentrating specifically on the housing role of such agencies are relatively few in number (see Walton, 1984; Williams, 1984; Young and Belisle, 1986) and tend to be written from an establishment perspective, although not always so. Certainly, there is an emphasis on 'learning by doing' and essential background research tends to be the thinnest area of activity (Young and Belisle, 1986). It would be foolish to deny that international aid has helped to provide more and better housing for many people in Third World cities, but it is also pertinent to enquire whether the loans are too restrictive in impact, too inflexible in structure and intended to have effects other than improve the housing of the poor. Many of the more general critics of World Bank activity would not be very positive on these points (see, for example, Bello et al., 1982). Much of the work undertaken by consultancy firms, often in conjunction with international aid programmes, also tends to be subject to the same criticism (Walton, 1983).

In order to appreciate the impact of these various forces and influences on housing provision, it might be helpful to examine in detail the nature and evolution of one aspect of the housing supply system in the Third World. The penultimate section undertakes such a case study in relation to the most widespread of contemporary programmes, that of aided self-help housing. However, before this can be done it is necessary to summarise two important background features which will help to set later discussions in their proper context. These are, firstly, the place of aided self-help and housing within the present range of programme and project responses, and secondly, the nature of the basic components of any housing programme.

A TYPOLOGY OF HOUSING SUPPLY

There have been many attempts to construct a simple model which illustrates the principal types of urban housing in the Third World, particularly since the revolution of sector theory in the late 1960s and early 1970s 'legitimised' the initial classification into formal and informal categories.

Many models are haphazard in construction, merely reflecting the conscious experiences of a particular observer and are of little use in attempting to draw commonalities from the urban housing situation across the Third World (see Leeds, 1981, for example). Descriptive typologies admittedly have limited uses but much illustrative and comparative value can be gleaned from them if a systematic approach is adopted.

Most models have been based on patterns of housing consumption, i.e. they are based on the nature of the commodity in which the consumer lives (see Drakakis-Smith, 1978; Johnstone, 1979), but others have been constructed on the basis of alleged production criteria (Burgess, 1979). Both approaches are clearly drastic simplifications of complex, real-world situations and under-emphasise the complexity of factors influencing the dynamics of formation. In this context and within such caveats, Table 4.3 perhaps takes such a model as far as it can go in its fusion of consumer and producer categorisations, and its attempt to link these with programme and project objectives.

Figure 4.3, in contrast, is intended simply to bring together most of the main housing types identified in the literature in terms of their position in relation to the three pivotal sectors of the supply system initially identified by Turner (1967, 1969) - viz. the public, private and popular sectors. The principal use of such a diagram lies in its function not as an end-product in itself, but as a point of departure for the exploration and analysis of the evolution of the contemporary situation and the changes (if any) which are affecting it. Even if such a model merely stresses the importance of the much-neglected private sector in low-cost housing provision (existing and future), it will have served its purpose.

However, it is not my intention to discuss in detail the role of the private sector: in a brief overview of urban housing provision in the Third World, it would not be appropriate to describe in detail the nature of each of the sub-sections in Figure 4.3. It would be far more useful to review the constituent components of all of these categories, because this will reveal the nature of the processes at work in manipulating the individual parts of what can then be reconstituted in a variety of ways i.e. the housing supply system.

Table 4.3: Urban housing sources and their relationship to self-help strategies

Urban housing sources: Production criteria	Urban housing sources: Consumption criteria		Goals	Strategies	Instruments
Industrial	Conventional	Public	to house low-income migrants to house urban poor to remove slums	public works sub-contracting to private sectors aided self-help	sites and services standards and codes subsidies public financial institutions provided housing
		Private	commercial profit (construction) commercial profit (rent) home ownership	contract building own build (self-help)	corporate finance housing associations residential community action rental housing
Manufactured	Hybrid		combined private and popular goals	combined strategies including self-help	combined instruments
Artisanal	Non-conventional	Popular	acquisition of shelter	self-help	squatting street sleeping private finance (money lending) rooms for rent residential community action

Source: Lea (1983)

Figure 4.3: A typology of low-cost housing types

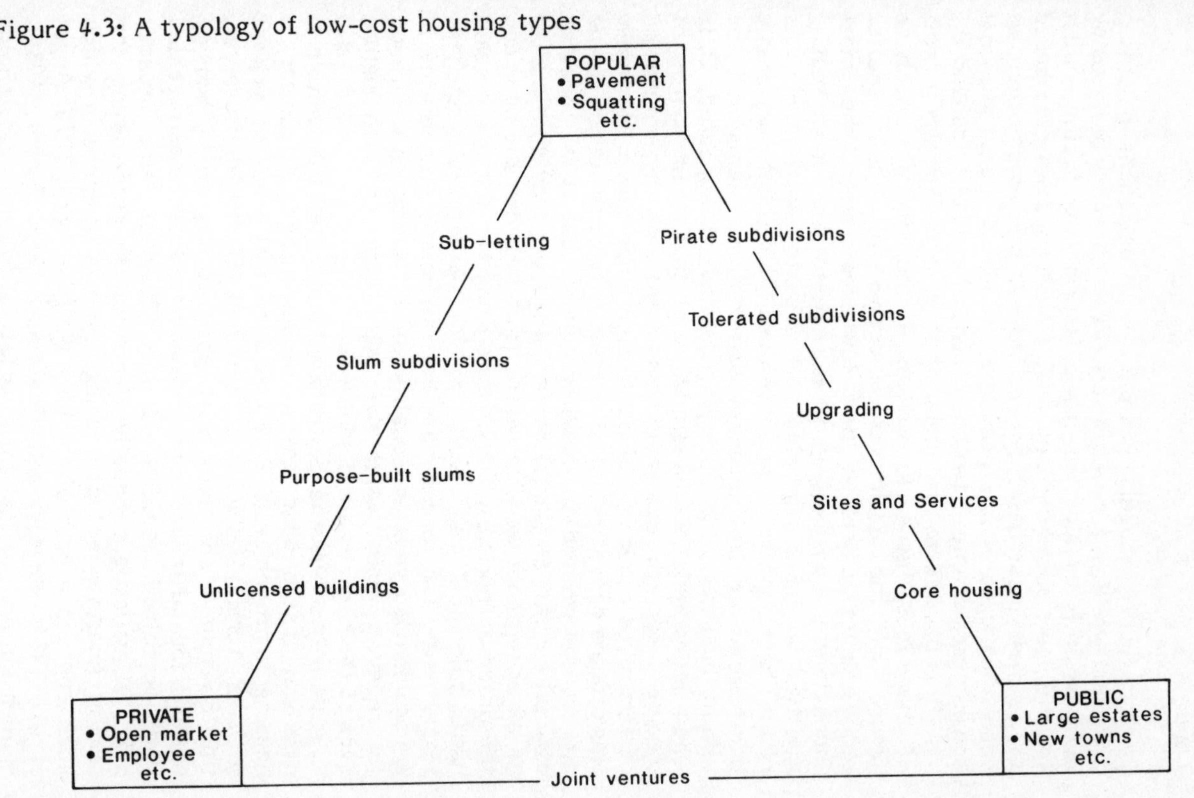

THE COMPONENTS OF HOUSING PRODUCTION

Land

Land is widely recognised as the major component in housing provision whether for rich or poor. For conventional housing markets it is usually the single largest cost factor, whilst for the poor, obtaining land tenure has long been considered to be their most basic desire. Indeed, this assumption has been the fundamental principle on which many low-cost housing programmes have been founded and have foundered.

It is as well to remember that even in Turner's models of the 1960s, his early arrivals in the city did not have tenure as the principle objective of their shelter requirements. Location and low cost were the most fundamental considerations and these are most easily satisfied in inner-city slums or squatter settlements. As cities have grown and developed more internationally commercial central business districts, so slums and the vacant land available for squatting have disappeared. Only when individuals or households feel that their source of income is secure, regular and sufficient does the acquisition of tenure title become of paramount importance. However, this process does vary and where security of residence in the city is not related to dwelling tenure, but to some other official permit, an illegal or rental system may continue to be an acceptable working relationship between the poor and the urban authorities. Thus, Hong Kong has for many years had 'tolerated' as well as 'non-tolerated' squatter settlements. The former have been officially enumerated in government surveys and qualify for rehousing on clearance for redevelopment, while the latter have no such rights, even if their homes are demolished.

Many observers have strongly emphasised that tenure is essential if the poor are to be encouraged to invest in housing improvements. This was a central tenet of writers such as Abrams (1964) and Turner (1967, 1969). However, this downplays the fact that poor do invest in housing without formal tenure, and subsequent research has revealed that perceived security, family finances and broader political issues are equally important in determining the rate of improvement of housing, even in schemes in which land tenure is a part of the package (see Ward, 1978; Angel et al., 1983; Wegelin and Chantana, 1983, Zetter, 1984). Momentum may be lost for a variety of reasons - households may not be able to sustain sufficient income, communities may not mobilise, or authorities may not provide back-up services. Even successful consolidation and development may be self-defeating since it can lead to rising rents and land values (that accrue to the landowner within a leasehold system), and this in turn can result in 'invasions' from middle-income households for whom tenure is a strong attraction. In short, government intervention can often merely formalise and consolidate the commercialisation of the land market which in itself reduces opportunities for informal/illegal access to land by the poor. Little wonder that some governments seek to separate building and land tenure in aided self-help schemes, granting the latter only after satisfactory progress on construction has been made.

Moreover, the granting of title is not in itself a simple matter, particularly in societies with a palimpsest of pre-colonial, colonial and post-colonial tenure systems. Many cities are infused and surrounded by customary land which is owned not by individuals but by community groups. The incursions of migrants into these peri-urban areas has frequently led to informal rented subdivisions, recognised by the traditional owners but not by the local authorities (see Oram, 1976; Lea, 1983, on Papua New Guinea). In circumstances such as these, ownership of the house is clearly distinguished from ownership of the land, at least in the minds of the residents. So strong and persistent is traditional land ownership that it has survived the colonial overlay in many parts of the world, even when legally ignored. Thus, for example, traditional Aboriginal land rights around Alice Springs have always had clear demarcation to the tribes concerned and fringe camp residents of differing groups have always sought the permission of the traditional owners before establishing their settlement (see Heppell, 1979).

Most land in the cities of the Third World is, however, part of a commercial land market. Although investment is usually by individuals, and extensive government ownership is rare (except for the old crown colonies such as Hong Kong, Fiji or Singapore), institutional ownership has been and continues to be very important in some countries. Islamic *waqf* organisations and Chinese clans, for example, own considerable areas in Middle Eastern and Southeast Asian cities, whilst former plantation or mining land often fringes some of the most rapidly growing cities, such as Kuala Lumpur, where squatter settlements may be found amongst the old tin tailings.

As might be imagined, the poor are incorporated into this complex land market in many different ways, depending upon local circumstances. Baross (1983) has categorised the social articulation of the informal land market into three main categories:

(1) Non-commercial articulation, in which those who build on the land either do not pay or else make a 'voluntary gift' to comply with social custom. Occupation of land can therefore be through settlement on customary land, or illegal occupation of nominally controlled government land, abandoned private land (for example, after internal disorder), or physically marginal land. In the last of these cases, of course, occupation by the poor often consolidates poor building land to the position where it is usable for conventional construction, thus creating capital for the state.

(2) Commercial articulation, in which a whole range of rent, lease or sale mechanisms release small plots of serviced or unserviced land to the poor on a fully commercial basis. The landowners may be individuals but in some cities are more likely to be companies specialising in this type of activity. Moreover, such companies are often working in conjunction with institutional, government or political agents or agencies who oversee and protect their activities for their own profit. Extension of these types of dealings into the buildings themselves gives a further range of what might loosely be called slums. These quasi-legal commercial arrangements relating to land are particularly complex and well-researched in parts of Latin America (see Harthe-Deneke, 1981; Gilbert and Ward, 1982; Ward, 1984; Gilbert, 1986).

(3) Administrative articulation, which covers the release of land to the poor through formal housing programmes. In recent years this has largely been through the medium of sites and service projects. Administrative articulation may also refer to the legitimisation of *de facto* occupation of government land referred to in (1). These are both forms of forced or technical 'release' of housing in which access by the poor remains controlled, they do not free the land market for the poor in ways which will enable the poor at an early stage to acquire secure tenure and be encouraged to consolidate in partnership with the state.

As the above classification clearly reveals there is considerably more to the land component in low-cost housing than the technological and juridical nature of its provision. In technical and professional terms, even where the government is willing to legalise and broaden land ownership to incorporate the poor, the organisational problems may be immense and lead to long delays in the transfer of title. In Malaysia, for example, on average land is held for six years whilst conversion, subdivision and distribution of legal title occurs. During this time the land is subject to a holding charge that is in turn passed on to the consumer. Such delays are also prompted by and provide opportunities for speculative holding. The use of land taxation to influence the behaviour of owners, rather than resuming land, is an alternative tactic (Zetter, 1984), but is of little impact where ownership is unclear or if the land lies outside the effective administrative area of the authorities. Moreover, as such taxes are often the fiscal basis of service provision, most local authorities are reluctant to forgo them.

One underlying difficulty in all programmes involving land is the fact that despite - or perhaps because of - its fundamental importance in low-cost housing supply, policy decisions are seldom co-ordinated into a single authority. Different government agencies will be responsible for different aspects of management and the end-result is internal conflict of interests rather than co-ordination. In the short run, chaotic management would seem to favour the poor in that it enables informal access to land to persist, but it also permits ambitious and unscrupulous land agents, entrepreneurs, developers and speculators to make considerable profit from various forms of commercial articulation with the poor, without necessarily bringing long-term security.

Zetter (1984, p. 222-3) has noted that achieving fairer access to land depends 'in the last analysis, on political feasibility ... extensive public ownership will generally only bring advantages if there is a clear relationship between ideological objectives and the detailed operational needs of housing policies'. As Hardoy and Satterthwaite (1981) illustrate, this is all too rare, but in any case such ideological commitment cannot prevent the betterment in land values which low-income communities help to create from falling into the lap of the land owners. In this context, Zetter comments that 'to a large extent squatting is ... the risk that the landowning elite accepts in order to maintain ownership and land values'. Eventually, 'legislation' within the framework of either private or public ownership allows the surplus value to be capitalised by the owner, either in compensation from the government or through higher rents on upgraded and titled land. 'Giving title, far from (being) a radical process, is an eminently rational (if reluctant) reaction of capitalist landowners' (Zetter, 1984, p. 222). Seen in this light, as Gilbert (1986, p. 185) has observed, the World Bank approach 'is very much premised on neo-classical economic orthodoxy (viz) that the unfettered operation of the land market is a positive way of encouraging the efficient use of land'.

If the provision of land for low-cost shelter seems to be a complex issue, it must also be remembered that this is literally only the starting point for housing programmes. The house itself needs to be constructed and linked to the land by site preparation and infrastructural provision. Each needs materials, labour, capital and organisation and the following sub-sections briefly outline some of the major points to be considered in relation to these.

Infrastructure and utility services

It is only in recent years that the design and provision of roads, footpaths, water, sewerage disposal, electricity and such, have received the attention they deserve from housers. The catalyst for this increased attention was the realisation that the existing standards which were the legacies of colonialism and neo-colonialism were hopelessly unrelated to the needs of contemporary programmes. Kirke (1984), for example, has noted that the *per capita* water provision levels in the Cairo master plan are higher than those of London and are much greater than the realistic demands likely to come from small dwellings with very limited 'wet' areas. Similarly, in Malaysia, the statutory minimum width for road reserves was until recently 40.5 feet, and together with educational and recreational space took up between 50 and 60 per cent of the land in most housing developments (Drakakis-Smith, 1977). Such standards are not only wasteful of limited resources but are often incongruous and unworkable in the wider urban context - for example, the reliance of a solid waste disposal system on daily collection facilities that do not exist.

There are many other instances of inappropriate policies and programmes too detailed to discuss here, but one or two problems that are common to other components of the supply system may be worthwhile emphasising. One of the most important of these is the extent to which the beneficiaries of any scheme are expected to pay the costs. Too often this appears to be related more to the demands of the financial source than the capacity of those being rehoused. This is a criticism that is often levied at World Bank involvement in low-cost housing schemes, but it has been refuted by Linn (1983), who argues that cost-recovery for service and infrastructure provision must be 'realistic' because the alternative of high subsidies simply reduces the overall funds available to the urban authorities to extend such provision, thus preparing even more land for incorporation into the housing market (with a concomitant rise in value). Linn's suggestion that the poor prefer user charges because these help to extend utilities and services to more people is an altruistic interpretation of events that I suspect would be very difficult to support in reality, particularly when such cost-recovery charges often include services provided 'free' by the state to middle-income public and private housing developments. The harsh reality of life for the poor means that their fiscal generosity must be quite limited. More acceptable is Gilbert's assertion (1986) that such services would be welcome only when and if their charges are lower than those paid for inferior private, illegal or informal alternatives. Cost-recovery charges related to infrastructural facilities must also be related to the flexibility of the project, with initially low but affordable standards being capable of later upgrading as and when family circumstances permit. This suggests a need for careful consultation between the poor, the engineers and the agencies involved in planning the project - before, during and after its execution. As we will see below, however, such consultations are either absent or superficial.

Building materials and house design

This section covers a very wide range of factors, from the choice of the overall nature of programmes, i.e. high-rise estates or sites and services, to the design of individual units within these schemes and the nature of the materials used in the construction process. The selection of programmes is perhaps too broad an issue to discuss in this section and relates more to the interface between macro- and micro-policy decisions; but housing design and construction do experience problems more akin to those affecting land and infrastructural provision. Housing design, in particular, has often been hamstrung by inappropriate building regulations or by notions of standards and suitability introduced by 'experts' from outside the affected community. Unfortunately the desirability of the Western style of residence as a goal for most urban families, planners and administrators has also been enshrined as part of the general Westernisation of values and culture that has occurred over the last two decades in the Third World.

In parallel with this process, there has occurred a widespread adoption of Western technology, epitomised as far as housing is concerned by the prefabrication of high-rise blocks. The fact that such designs and technology were being 'pushed' by firms in the Third World after their discreditation in Europe and North America is indicative not only of the Eurocentric bias of education and professional training but also of the extensive role of foreign consultants in maintaining profitability for multinational construction firms. The institutionalisation of self-help schemes has also helped to formalise the materials used by the poor, to the profit of the suppliers.

By the mid 1970s liberal ideas on development had reached the construction industry which began in the wake of informal sector research to pay ostensible attention to intermediate and small-scale technology. Whilst there was undoubtedly much genuine desire to devise a technology more appropriate to the new aided self-help programmes, a great deal was mere lip-service. Some of the consequences were farcical: in the Sudan, for example, one low-cost housing scheme reproduced in concrete and brick the traditional round mud house and conical thatched roof; the resultant units were unsuited either to the climate or the finances of the poor (Cain et al., 1976).

Although most designers and planners have undoubtedly moved towards a better appreciation of real community needs, technology (whatever its scale) is still an over-dominant element within discussions of the construction process at the expense of the social and political issues involved (see Sheriden, 1979; Parry, 1984). Thus, the high cost of building materials, frequently comprising up to two-thirds of total construction costs (Hardoy and Satterthwaite,

1981, p. 247), is not unconnected to the fact that most materials are still imported. To some extent this reflects neo-colonial links but it also reveals the limited development of local resources. The need to exploit indigenous building materials has often been noted (see Murison, 1979) and has formed an integral part of the arguments for the new intermediate technologies (Table 4.4). It is also true that most of the urban poor are currently housed in dwellings which use local (or recycled) materials in the petty-commodity sector. However, Wells (1986) cautions against anticipating an easy or broad expansion of local sources, particularly as they are prone to bottlenecks in production or transportation, usually because of undercapitalisation and/or a lack of skilled labour and management experience. Other factors also curb the growth of local material production, particularly the influence of vested interests (individual or corporate) in maintaining imports of building materials or exports of local resources. Hardwood timber, for example, is one of the most valuable of local building materials and yet is also a major export of many Third World countries when commodity prices are high, causing the domestic market to suffer shortages (see Makanas 1974-5 re the Philippines).

Clearly, standards for individual buildings and projects as a whole need to be broad-based and flexible. In this context Mabogunje et al., (1978) claim that six conditions must be satisfied, viz. cultural compatability, social responsiveness, economic feasibility, technological suitability, physical and biological harmony, and temporal relevance. However, the continual evaluation and monitoring essential to achieve these are unlikely to be realised in practice in situations in which local professional input is limited or is Western-oriented. The standards must be established and reviewed by indigenous planners and builders, trained to appreciate local values fully.

Labour

Much of the labour input for low-income housing construction in the Third World is, of course, in the form of petty-commodity production by the individual, household or community. However, it would be incorrect to assume that this means that all 'popularly' constructed houses are self-built. Many of the 'artisanal' (Burgess, 1982) dwellings within

Table 4.4: Choice of building materials and the quality of consumer income, source of basic material and labour intensity

Material	Quality of building	Income class (a) of consumer	Source of basic material (b)		Labour intensity (c)		
			Place	Economic sector	Material production	Processing	Building
Mud and wattle	low	subsistence	local	informal	high	high	high
Sun-dried clay blocks	low to medium	slightly above subsistence (rural) low (urban)	local	informal	high	high	high
Murram-enforced blocks			local	informal formal (d)	high	high-medium	high
Black cotton bricks			local	informal informal	high	high	high
Stones	high	mainly high	local	informal formal formal	high	high	high
Timber	medium	medium and high	local	informal	(e)	(e)	(f)
Pre-cast concrete panels	medium	medium and high	local	formal	medium	medium	medium
Pre-cast concrete panels made with a foaming agent	medium	medium and high	local and imported	formal	medium	medium	low-medium
Cement blocks made with a chemical additive	high	medium and high	local and imported	formal	low	low	low-medium
Cement blocks	medium	medium and high	local	formal	medium	medium	medium

Table 4.4: continued

Notes:

a. Low: rural subsistence and most small-scale market production, urban unemployed, underemployed - including most informal sector activities; Medium: unskilled and semi-skilled formal sector (mostly urban) workers; High: skilled and professional. The use of this classification means that the majority of the population should be classified as low-income.
b. This refers both to geographical location (local or imported) and to sector of the local economy, which is regarded as being divided into an informal and a formal sector. Broadly, the informal sector is more labour-using, while formal sector production involves more equipment, often imported.
c. To classify production methods according to labour intensity implies that each material may be identified with a single technique; as the survey of cement-block making showed, that is not the case, and considerable variation may be possible.
d. Cement.
e. Intensive use of skilled labour and natural resources.
f. Intensive use of skilled labour and some use of machinery.

Source: Burch (1979)

the Third World, including squatter housing, are constructed on a commercial basis within a system of peripheral capitalism. Nor can community assistance be assumed to be constantly available within building programmes in low-income settlements. It is true that such communal effort is well documented in a wide range of Third World countries (see Johnstone, 1978, for Malaysia; Lloyd, 1980, for Peru), but there is also evidence of it being much less common in settlements with traditional ethnic or tribal rivalries or suspicion (see Teedon, 1987).

As far as the private and public sectors are concerned, the abundant urban labour supplies reduce the labour component of construction costs to about one-third, compared with a half in advanced capitalist societies. However, severe manpower shortages are considered to exist in terms of skilled manual, technical, managerial and professional fields. Such shortages exist for two reasons: firstly, there is the rapid rate at which Western technological 'solutions' to the housing programme have been introduced; and secondly, the limited training facilities for indigenous workers within their own countries. Those who go abroad for their training tend to perpetuate the Western values of the policy-makers - what Henry (1979) has termed 'colonial karma'.

Given widespread criticisms of the technological Westernised programmes that prevailed in the 1960s and 1970s, it must be stated that shortages in skilled manpower (whether blue- or white-collar) are not the most critical problems facing labour within the construction industry. The most pressing problem is clearly to train the abundant unskilled, and often under- or unemployed labour, in the basic techniques needed within the intermediate technology of aided self-help programmes. As both Henry (1979) and Thiedeke (1979) note, this is the most lamentable failure of labour organisation in relation to housing. Nor should such involvement be limited to construction *per se*: there are many labour inputs in planning, supervision, maintenance and administration which indigenous personnel can make, given appropriate training. Indeed, such input is more likely to be of the long-term kind needed in gradually expanding aided self-help programmes - the type of labour input least offered by the short-term contracts of expatriates.

Table 4.5: Sources of funds for squatter settlement improvements (per cent)

	Bangkok (a) (principal source)	Karachi (b) (all sources)
Savings:		55
personal	39.5	
pooled with relatives	28.1	
Loans:		48
moneylenders	3.2	
friends	4.3	
relatives	7.0	
employers welfare fund	4.3	
Community save/share	7.6	32
Other	6.0	19

Sources:
a. Wegelin and Chantana (1983)
b. Van der Harst (1983)

Consumer finance

The formal credit schemes related to housing are largely restricted to those families with incomes which are sufficient to meet regular repayment demands. Few of the urban poor, those really in need of assistance, could contemplate taking out such loans even if they were permitted by the institutions themselves. Indeed, there is considerable evidence from some countries that whilst credit institutions are willing to take in the savings of all low-income households, they are very selective about those to whom they give loans (see Bell, 1974; Batley, 1977).

However, the above restrictions do not mean that the poor are not involved in some form of financial scheme or indebtedness related to their housing. Many are simply in arrears with rents, either for slum or squatter dwellings; and others seek credit in order to improve or repair their existing home. Very few detailed studies have been undertaken specifically of the housing finance and credit schemes that operate within peripheral capitalism, but it seems to be the case that few special sources of finance exist for housing per se, i.e. longer-term credit. Most finance, if not from personal or family savings, is through short-term high interest loans (Table 4.5). However, whilst interest rates may be very high, repayment expectations are often flexible enough to recognise alternative emergency pressures upon what are low and irregular incomes. Social controls of various kinds still ensure that debts are eventually paid.

Clearly, the financial needs of low-income housing consumers in the Third World must, like the design and construction elements of the dwellings themselves, extend beyond technical considerations and incorporate in addition relevant social and political factors. Thus, for example, greater emphasis could be given to waiving collateral requirements, or accepting de facto tenure as security, and to accepting fluctuations in repayments. Unfortunately, the increasing involvement of international agencies in the financing of low-cost housing schemes has firmed the management of such projects and forced out many of those originally rehoused in favour of others who are more able to meet repayment requirements. As evidenced below, this is particularly true of aided self-help projects.

AIDED SELF-HELP HOUSING

The emergence of aided self-help (ASH) housing programmes over the last decade or so is unusual in a field not particularly noted for its innovation or sympathy for the plight of the poor. The roots of this change lay in the reactions of development strategists to the general failure of modernisation along Western lines to bring demonstrable improvements to the living conditions of the urban poor. In the wake of the criticisms of dependency and world-system theorists, and following the revelations of sector theorists about the positive values of the informal sector, liberal capitalists began to reshape development theory to take into account the necessity to meet some of the 'basic needs' of the poor. Explicit to the basic-needs approach was the utilisation of the positive energies of the poor in small-scale gradual incrementalist projects of aided self-help (see Richards and Thompson, 1983; Drakakis-Smith, 1987b). The role of housing was central to this approach. Squatter settlements, in particular, were at once the manifestation of inequality and the personification of the energies of the poor. It was therefore but a short step to assume that a combination of public and popular effort/investment would produce a more extensive but still acceptable response to housing needs than had existed theretofor.

As Rodell and Skinner (1983) have noted, aided self-help is not new. It has always been the normal way of housing people in rural areas, with community involvement in land selection, preparation, material assembly, design and construction. When migrants arrived in the city, however, they found that their access to many of these elements was limited or controlled by their poverty, by shortages or by regulations. The only uncontrolled element remaining was their labour. Early recommendations for aided self-help programmes (Abrams, 1964) envisaged labour as the only contribution by squatters, but by the early 1970s researchers such as Turner and Mangin had convincingly argued that the poor had much more to offer. Given the scale of the squatting problem by this time, many international agencies, if not Third World governments, were ready to consider expanded finances for schemes of this nature (Table 4.6).

There are many different types of ASH programme, but three broad categories may be usefully distinguished:

(1) Upgrading schemes: this is the simplest and in many ways the most effective approach, leaving the squatter settlement *in situ* and providing better infrastructure or utilities and/or supplying materials for improving the dwelling itself. However, if plot sizes are also expanded, some residents may be displaced (Figure 4.4) The main advantage is that upgrading leaves the residents where they chose to live, usually because of the proximity of work. The main disadvantage is that upgrading seldom involves a transfer of tenure, particularly if the settlement is located in a potentially valuable central area. However, *de facto* or perceived security is enhanced by upgrading schemes which often serve to encourage further investment by the residents. Nevertheless, as infrastructural improvements to sewerage and water supplies can be expensive, upgrading schemes can degenerate to mere cosmetic improvements of dwellings.

(2) Sites and services: this is the provision of a plot of prepared land and basic services to a household which constructs its dwelling as and when it can so do through self-building or contractors. Most site and service schemes involve transfer of tenure, but also tend to be on the urban

Table 4.6: Comparative concepts of urban self-help housing in the 1950s and 1970s

1950s	1970s
Basic idea Self-help = unpaid family labour	Basic ideas Self-help = families deciding about investments Self-help = investment inputs supplied by families, either inputs purchased with cash savings, or unpaid labour, or both
Examples Shacks and shanties in squatter neighbourhoods	Examples Covers the range from shacks and shanties to standard neighbourhoods
Construction process	Construction process
1 Squatted land 2 Scrap and waste materials 3 Unpaid labour on weekends and at nights 4 Construction starts from occupancy, and ends a day to a year after this date	1 Various and unpredictable Construction can last one day to five, ten or fifteen years 2 Shacks and shanties can be a final product or an intermediate stage within the process
The end-product is constrained by families' income during one year, materials costs and illegality of tenure	The end-product is constrained by families' income over a decade or more, materials costs and perceived security of tenure

Table 4.6: continued

Who uses self-help? Recent migrants from distressed rural areas	Who uses self-help? Anyone can; usually, families who start with self-help are low- to middle-income at the date they start
Policy implications 1 The basic theory: self-help reduces construction costs of a given type of housing	Policy implications 1 The basic theory: under certain conditions, self-help increases investment in housing because (a) it adds unpaid labour to the resources used in housing (b) it adds inputs purchased by families to resources used in housing
2 Conditions for success: self-help reduces costs under all conditions	2 Conditions for success: self-help increases investment when (a) families have to wait more than five or six years for conventional rental or ownership housing (b) families find currently available conventional neighbourhoods do not meet their definition of desirable housing
3 Problems to overcome: (a) lack of land	3 Problems to overcome: (a) lack of secure, good, well-located building sites

Table 4.6: continued

(b) lack of credit to buy standard materials for a complete house at the date of occupancy	(b) building and land use regulations make a number of self-help options illegal
(c) lack of planning and construction skills needed to build conventionally	(c) lack of public services
	(d) lack of small construction loans
4 Solutions: plan house construction to use as much unpaid labour as possible, with appropriate assistance	4 Solutions: plan new neighbourhoods in which families have secure, good, well-located sites, infrastructure and access to technical assistance, and in which families can invest as much or as little as they wish. Upgrade old neighbourhoods
5 Minimum role of government: (a) develop land and infrastructure (b) design houses (c) buy materials (d) organise and supervise construction	5 Minimum role of government: (a) secure land and develop infrastructure
6 Ideal project: aided self-help, complete housing scheme	6 Ideal project: varies with local conditions

Table 4.6: continued

7 Result: high standard housing accessible to low-income families because it is inexpensive	7 Result: a large number of units varying in cost, and accessible to low-income families because (a) there are enough units (b) the least costly have secure land, infrastructure and affordable house standards

Source: Rodell and Skinner (1983)

Figure 4.4: Colombo, Sri Lanka: The impact of upgrading on land holding patterns

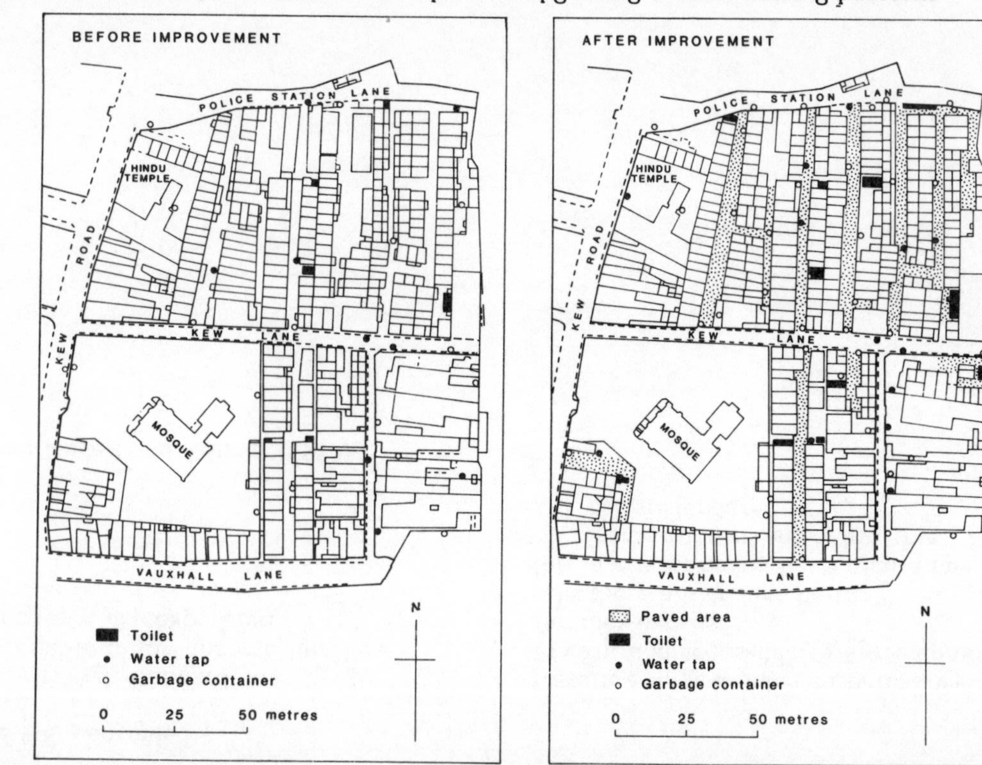

periphery because this is where suitable land is usually located. Many schemes also have stipulated dwelling completion rates and, when combined with their location and regular repayment demands, such projects can prove to be difficult for the poor to resist, being bought out by more financially secure families or by speculators. In some Karachi Metrovilles, for example, only 20 per cent of the site and service plots have been occupied long after allocation, and most of those who have not moved have incomes in excess of the limits and are involved in speculation (Siddiqi, 1983).

(3) Core housing: this is essentially a more sophisticated and expensive version of site and services which includes the provision of a basic core-dwelling unit on the plot itself. The self-improvement programmes of such schemes are often encouraged by the provision of on-site construction or material-production facilities. However, few of these relatively expensive programmes have been in existence long enough to determine whether on-site facilities accelerate or cheapen production, or even last as long as they may be needed. Again, the principal problem is the upward filtering of the new units, either officially or unofficially, to less poor households. This process is often accelerated where credit is advanced for the purchase of materials by the agencies involved in the scheme, an action which also shifts the control of development from the builder to the lender. As a result there is abundant evidence that site and service schemes become infiltrated by the less needy and even by bourgeois households (Atman, 1975; Guhr, 1983; Rodell, 1983). Ward (1984), for example, estimates that as much as half of the site and service residents in Mexico earned more than the permitted financial maximum, and that salaried workers were often given preference in the allocation process because they were more likely to meet repayments.

It is evidence such as this that constitutes one of the most basic criticisms of ASH programmes - viz. that they do not benefit the poorest households (Ward, 1982). Many believe that such situations exist because recovery costs are too high, incorporating infrastructural elements that the state normally bears in private or public housing projects (see earlier discussion); but there are also other factors involved. Several of these relate to the fact that economic circumstances do not stay constant and, indeed, on a global

Table 4.7: Tanzania: actors involved in urban housing development

	MINISTRY OF LANDS HOUSING AND URBAN DEVELOPMENT							PRIME MINISTER'S OFFICE			OTHER MINISTRIES					PARASTATAL ORGANISATIONS				OTHERS			
	1	2	3	4	5	6	7	8	9	10	11	12	13	14	15	16	17	18	19	20	21	22	23
Activity	CABINET	HOUSING DEV.	SURVEYS	LANDS	SEWERAGE AND DRAINAGE	URBAN PLANNING	BUILDING RESEARCH	HEADQUARTERS	REGIONAL ADMIN.	URBAN COUNCILS	PLANNING	FINANCE	INDUSTRY	WORKS	OTHERS	HOUSING BANK	BOARD OF INTERNAL TRADE	TANESCO (POWDER)	OTHERS	CONSULTANTS	CONTRACTORS	INDIVIDUALS	LOCAL COMMUNITY ORGANISATION
1. Policy formulation	o	o									o	o											
2. Land preparation																							
(a) Designation of land for residential development				x		o				o													
(b) Acquisition of land for residential development		x		o					x	o													
(c) Site planning (layouts)		x				o			x	o													
(d) Plot definition (land surveying)		x	o						o	o													
(e) Plot allocation and tenure mechanisms				o					o	o													
(f) Provision of land services (infrastructure)																							
(i) Designs		o			o				x	o				x				o	x	x			
(ii) Budgeting	o	o			o			o	x	o	o	o		x				o					
(iii) Construction		o			o				x	o				x				o			o		

3. Dwelling-unit construction																		
(a) House designs	o	x		x	x	o				x						x	o	o
(b) Housing finance							o	o				o			x		o	o
(c) Construction standards				x	x	o												o
(d) Building permits					o	o												
(e) Building materials	o						o	o	o		x	x	o		x		o	o
(f) Construction mechanism		x		x		x						x					o	
4. Maintenance of residential																		
environment					x	o											o	o
5. Revenue structures																		
(a) Establishing rates for land	o		o															
rent and service charges	o		o			x	x	o										
(b) Valuation of properties			o		x	x												
(c) Revenue collection																		
(i) Service charges					o	o		o										
(ii) Utilities					o	o								o	o			
6. Sales and transfers			o		x	x												x

Key: o — Primary direct responsibility; x — Secondary responsibility
Source: Mghweno (1984)

Table 4.8: Tanzania: comparative construction processes in squatter and site and service settlements

Actual process: squatter settlement	Hypothetical process: sites and services area
1 Time needed to obtain building land before the project could be started	
Two days	Four to twelve days
2 Cost of land	
One month's minimum salary	More than two months' minimum salary
3 Choice of location	
No restriction	Restricted to two areas under development
4 Procedure for obtaining finance	
First using the funds available, i.e. savings plus loan from employer (one week's delay during the negotiations with employer)	First using the funds available for building land and loan approval (seven to eight months for bank to approve application), then regular disbursement of loan instalments until completion on the basis of site inspections by bank and city engineer
Moving in	Moving in
Using funds whenever they are available for completion of projects	

Table 4.8: continued

5 Choice of materials and standards	
No restriction	Following the bank's conditions and the official building regulations
6 Choice of labour force	
No restriction	Craftsmen with the ability to comply with the bank's conditions and extant building regulations
7 Construction costs	
Equivalent to 13 months' minimum salaries	Equivalent to 46 months' minimum salaries
8 Monthly obligation and duration	
50 shillings for less than three years	154 shillings for 20 years

Source: Guhr (1983)

level have worsened considerably over the last ten years. One macro-level impact of this recession has been a tightening of international finance, including that provided by the World Bank and similar agencies for housing projects. Stringent repayment of such loans has in most cases been passed on to the consumer in the form of higher repayments and punitive controls on defaulting. However, the world recession has also filtered down to the urban poor in other more direct ways; by reducing real wages and job opportunities, so as far as housing is concerned the urban poor have been doubly affected by economic problems. Little wonder, in such circumstances, that Ward (1984) can report that in some site and service schemes in Mexico, only half of the plots were occupied several years after they were made available.

Other problems associated with site and service projects may be attributed more firmly to the administrative and institutional frameworks in which they are planned and delivered. A particularly common problem, one which has constantly bedevilled housing provision of all types in the Third World, is the complex, cumbersome, costly and ineffective combination of local, national and international agencies involved in the whole process. Table 4.7 indicates only the domestic section of this complexity in Tanzania (Mghweno, 1984). Such confusion not only involves overlapping functions, inducing inter- and intra-agency rivalry, but also confronts the poor household with a labyrinth of regulations and tasks through which it must pass in order to obtain benefits. Such processes are often too time-consuming and therefore costly for poor families (Table 4.8).

Professional involvement in such schemes can also be criticised for the conventional way in which their role is conceived as just an initial input of planning, design and construction. The result is often a two-stage process in which all the planning is undertaken in advance by experts and expatriates with limited understanding of the local situation. The execution of the scheme is left to local administrators who appreciate its limitations and avoid responsibilities wherever possible. What is most needed within ASH programmes is a long-term commitment to assistance and advice over the period during which a community consolidates the original project. This reinforces the suggestion that what is required is the training of indigenous semi-professionals or residents who will remain available for consultation beyond the initial phases of the programme.

The involvement of local personnel leads directly to one of the major criticisms of aided self-help housing which is that the urban poor who are involved in the programme itself seldom contribute in any sense other than that of their labour power (Martin, 1983). Despite the theoretical advances of the 1970s, participation by the poor in decision-making, planning and allocation processes related to housing provision is minimal (see Skinner and Rodell, 1983, for an excellent discussion of this). What many programmes comprise, therefore, is not aided self-help but institutionalised self-help. Under such circumstances it is not surprising if the poor choose to exhibit little enthusiasm for schemes foisted upon them in the name of community improvement.

There are, of course, many real problems associated with trying to incorporate the poor more comprehensively into low-cost housing policy and programme formulation, although the argument that they are 'not accustomed to being consulted' (Payne 1984, p. 7) is not one of them. Potter (1985) has examined in some detail the difficulties involved in taking into account 'the perceptions, cognitions and aspirations' of the urban poor and concludes that greater empathy is desperately needed. However, whilst such behavioural studies may well reveal valuable information, there is a danger that 'participation (becomes) a substitute for democracy' (Rieser, in ibid., p. 240) and Potter (ibid., p. 239) himself argues carefully that 'participation must be optimized, not maximised in purely qualitative terms'. The crucial question is who determines what is optimal and why.

The political reality of developing countries, in common with most of the rest of the world, is that those presently in control of policy or programme formulation and project administration are not likely to give up voluntarily their political, economic and social power with respect to housing provision. In this context, Payne (1984) has observed that planning and bureaucratic procedures in socialist countries can be just as restrictive as those in capitalist countries. Only when the urban poor are perceived as a real threat to the ruling elite, whatever their political leaning, do opportunities occur for some progress.

Neo-Marxist critics of the present situation would disagree with this, arguing that capitalism has a vested interest in maintaining the present contradictions within the low-cost housing supply system. One of their arguments is that by introducing projects that benefit selective groups, particularly the 'rising poor', capitalist governments can fragment the solidarity of the poor by incorporating some, potentially the most politically dangerous, into the capitalist system through the property market, thus giving them a stake in maintaining political stability. Others, such as Burgess (1979, 1985) would argue that even this incorporation is secondary and that ASH schemes, in particular, seek simply to placate the poor whilst keeping them in essentially the same exploited position. They remain at the bottom of the urban housing hierarchy, deprived of access to the standards of the bourgeoisie, whilst at the same time, because of continued low housing costs, still constituting the pool of cheap labour upon which domestic and international capital continue to rely heavily. In short, aided self-help housing assists in the reproduction of cheap labour.

In addition to ASH programmes indirectly aiding capital, there is also increasing evidence that capitalist interests have realised that because of the vast numbers of the urban poor and the funds currently being made available for housing by domestic and international agencies, there are large profits to be made in low-cost housing provision. This is perhaps nowhere more vividly evident than in South Africa. White capitalist firms have always made handsome profits out of the construction of black townships, but recently this has been extended by the encouragement given to the expansion of consumerism via home improvements and the emergence of middle-income housing for a black bourgeoisie (Mather and Parnell, 1987; Hendler, 1986). This deeper penetration of the black housing market has partly occurred because of the downturn in demand for white housing due to the prevailing political and economic situation in South Africa.

In some ways events such as these present contradictory theoretical positions (Gilbert, 1986). On the one hand, capitalism seeks to conserve cheap housing because it lowers wage demands; on the other, it seeks to destroy petty-commodity production by formalising the land, material, and finance markets in order to achieve greater profits. This conflict is explainable by there being different fractions of capital with different interests with regard to housing for the poor, for example, industrialists want cheap labour, property/construction firms want an expanded conventional housing market. Which ever is dominant will seek to influence housing policies accordingly; but the situation is not as clear-cut as this, and even within the same city can be very complex, with conservation and dissolution of pre-capitalist and petty-commodity sections of the housing market occurring at the same time.

Gilbert (1986, p. 181) claims that the state acts as an arbiter in situations such as these, in mediating between major areas of capital whilst trying to respond to the worsening needs of the poor. However, as the concluding section of this chapter will indicate, much of the alleged competition between fractions of capital within the housing market is fictitious, and the state is not so much responding to the needs of the poor as moderating the political consequences of their growing demands. In short, it is seeking to extend its control over unauthorised and uncontrolled communities within the city and is willing to make concessions in basic needs in order to achieve this.

It is clear, therefore, that policies, programmes and projects related to low-cost housing provision must be placed in an appropriate context before their impact and import can be fully appreciated. It is undoubtedly true that many of the urban poor have benefited from recent changes in housing policy in favour of aided self-help schemes, and this welfare goal has been a principal objective for many of the planners and administrators involved. It is not difficult, therefore, to produce examples to refute Burgess's (1985) extremist claim that aided self-help leads to a worsening of housing conditions (see Gilbert, 1986, on Colombia). However, it is also true that the poorest urban households have yet to be reached by such programmes, that objectives other than social welfare have influenced their adoption, and that the great majority of countries have failed to make aided self-help a central objective of their urban housing policies and have failed to use the opportunities it provides to restructure access to the component resources of land, finance, materials and infrastructure (Hardoy and Satterthwaite, 1981; Ward, 1982; Skinner and Rodell, 1983).

Above all it must be recognised that housing improvement is not and must not be seen, in Victorian terms, as an end in itself. It is but part of the seamless web of basic needs, such as education, health care and nutrition, which all revolve around the central spindle of employment and income. Real and lasting improvement in any basic need is heavily dependent on increasing the access of the poor to employment with fair and adequate remuneration. That such access is dependent on structural changes is axiomatic.

CONCLUSION: WORLD RECESSION AND HOUSING PROVISION

Much of the previous discussion suggests that urban low-cost housing provision in developing countries is linked (through policy formulation, programme management and project inputs) to the broader national and international economy. If this is so, then the downturn in the world economy should have an impact on housing for the poor. It could be argued, in this context, that the recession is relatively recent and there is likely to be a long lead-in to an eventual consequence for the poor. Gilbert (1987) has speculated on the impact of the recession in the context of Latin American cities and has suggested that, despite the wide variation in urban processes, the recession must lead to a growth in the numbers of urban poor and an expansion of those in the more marginal activities of petty-commodity production (since the less marginal and more profitable are usually more dependent on a flourishing manufacturing sector). As far as housing is concerned, Gilbert argues that the current recession has had several effects:

> First, it is reducing employment in the construction sector. Second it had led to increased crowding in the housing stock. Third, it has cut budgets for servicing low-income settlements. Fourth, it has reduced the possibilities for capital accumulation open to local elites. Fifth, it has reduced incomes throughout society and therefore limited the effective demand for housing (ibid.).

Clearly the impact will vary in individual nations or cities, dependent on local circumstances: for example, where strong authoritarian or cultural controls exist, the possibilities of further squatter invasions will be limited and sub-letting or lodging in both slum and squatter areas may expand (Bryant, 1987). Where such controls are not so rigid, the recession may well result in a fresh wave of land invasions.

Yet is the situation so predictable? Investment in land (Angel et al., 1983), construction in general, and housing in particular, often constitutes a haven for capital during an economic recession when investment in commodities or manufacturing offers comparatively poor returns. Evidence for such activity exists across a wide range of contexts, both in space and time. In Colombo, Sri Lanka, for example, tenement housing has long been a popular form of secondary capital investment (Marga Institute, 1976). The recent upsurge of investment in the black townships of Johannesburg, referred to above, also illustrates this point. In similar vein, Johnstone (1978) has evidence that in the Malaysian private sector, multiple interests within companies affect housing investment, flowing into construction only when commodity prices for tin and rubber are low. This means that private-sector housing provision in some countries will operate almost irrespective of market demand, as a secondary source of profits in times of recession.

Nor is this reverse tendency to invest in housing during recession confined to the private sector. In Singapore, for example, it has been a strong commitment to public-sector construction projects which has maintained economic growth during the early years of the recession. Indeed, the growth induced by airport reconstruction and an accelerated public housing programme effectively masked the drastic downturn in the industrial sectors of the economy until the mid-1980s. However, the curtailment of the building programme and a failure to generate sufficient high-tech industrial growth has eventually revealed the real impact of the recession on the republic (Grice and Drakakis-Smith, 1985).

In addition to this broad political-economy rationale for maintaining investment in housing during periods of recession, it is also true that the momentum which has accumulated within social movements related to housing provision has helped to retain some funds that may otherwise have been diverted elsewhere (see Meyerink and Vekemans, 1983; Drakakis-Smith, 1987a). Site and service programmes are, after all, 'sold' to the poor on the basis of community participation, however spurious this may be, and for some governments to forsake such schemes, particularly in times of general difficulties for the poor could create political problems. We must be cautious about overstating this factor, but international capitalism has long been wary of the 'revolutionary' potential of the poor. Moreover, it is ironic that sympathetic reaction to social movements is more likely in democratic states that in the totalitarian ones where the poor are usually worse off.

Obviously there can be no predictable universal reaction to the global recession within the field of low-cost urban housing provision. As with other aspects of housing policy, this will vary according to the blend of local, national and international forces, not only in an economic sense but in social and political terms too. However, as with other basic needs for the poor, the situation needs constant monitoring in order to identify the general trends and common denominators which may be restricting the improvement of the quality of life for the poor. This ought to be an area of increasing research activity, within which the specific foci outlined earlier need particular attention, viz. the nature of the interaction between macro- and micro-policy decisions, particularly the relationship between the central and local state, and the nature of the role of international agencies in shaping the low-cost housing programmes and projects within particular nations and cities. Although major studies have been made over the last decade in the provision of better housing for the urban poor, much remains to be done, particularly if these gains are not to be lost in the reassertion of economic priorities within the current recession.

REFERENCES

Abrams, C. (1964) Man's struggle for shelter in an urbanising world. Faber, London

Abu-Lughod, J. (1981) Strategies for the improvement of different types of lower-income urban settlement in the Arab region. In HABITAT, The residential circumstances of the urban poor in developing countries, Praegar, New York, pp. 116-34

Angel, S. et al. (1983) Land for housing the poor. Select Books, Singapore

Atman, R. (1975) Kampung improvements in Indonesia. Ekistics, 40 (238), 216-20

Baross, P. (1983) The articulation of land supply for popular settlements in Third World cities. In S. Angel et al. (eds), Land for housing the poor, Select Books, Singapore, pp. 180-210

Batley, R. (1977) Expulsion and exclusion in Sao Paulo. Paper presented to a Conference on Access to Housing,

Institute of Development Studies, University of Sussex, Brighton

Bell, G. (1974) Banking on human communities in Brazil. Ekistics, 38 (224), 27-34

Bello, W. et al. (1982) Development debate: the World Bank in the Philippines. Institute for Food and Development Policy, San Francisco

Bryant, J.J. (1987) Renting in Fijian squatter settlements: exploitation or survival strategy. In D.W. Drakakis-Smith (ed.), The urban totem, Croom Helm, London

Burch, D. (1979) Socio-economic variables in the construction industry. In H. Murison and J. Lea (eds), Housing in Third World countries, Macmillan, London, pp. 131-6

Burgess, R. (1979) Informal sector housing? A critique of the Turner school. In R. Bromley (ed.), The urban informal sector, Pergamon, Oxford

------ (1985) The limits of state self-help housing programmes. Development and Change, 16, 271-312

Drakakis-Smith, D.W. (1977) Housing the urban poor in West Malaysia: the role of the private sector. Habitat International, 2 (5-6), 571-84

------ (1978) Housing: an overview. In P.J. Rimmer et al. (eds), Food, shelter and transport in Southeast Asia and the Pacific, Monograph HG12, RSPACS, Australian National University, Canberra, pp. 101-11

------ (1981) Housing, urbanization and the development process. Croom Helm, London

------ (1986) Urbanisation in the developing world. Croom Helm, London

------ (1987a) Urbanization, economic development and social and political change in the semi-periphery: urban housing provision in North Australia. In D.W. Drakakis-Smith (ed.), The urban totem, Croom Helm, London

------ (1987b) Approaches to development. In D.J. Dwyer (ed.), Southeast Asia, Longman, London (in press)

------ (ed.) (1987c) The urban totem: urbanisation and economic development in the periphery. Croom Helm, London (forthcoming)

Gilbert, A. (1986) Self-help housing and state intervention: illustrative reflections on the petty commodity production debate. In D. Drakakis-Smith (ed.), Urbanisation in the developing world, Croom Helm, London, pp. 171-95

------ (1987) Urbanization at the periphery: reflections on

the changing dynamics of housing and employment in Latin American cities. In D. Drakakis-Smith (ed.), The urban totem, Croom Helm, London

------ and Gugler, J. (1982) Cities, poverty and development. Oxford University Press, London

------ and Ward, P. (1982) Low income housing and the state. In A. Gilbert (ed.), Urbanization in contemporary Latin America, Wiley, Chichester, pp. 79-128

------ and Ward, P. (1985) Housing, the state and the poor: policy and practice in Latin American Cities. Cambridge University Press, Cambridge

Grice, K. and Drakakis-Smith, D.W. (1985) World recession and national prosperity: Singapore in the 1980s. Transactions of the Institute of British Geographers, 10 (3), 347-59

Guhr, I. (1983) Cooperatives in State housing programmes: an alternative for low-income groups. In R. Skinner and M. Rodell (eds), People, poverty and shelter, Methuen, London, pp. 80-105

HABITAT (1981) The residential circumstances of the urban poor in developing countries. Praegar, New York

Hardoy, J. and Satterthwaite, D. (1981) Shelter: need and response. Wiley, Chichester

van der Harst, J. (1983) Financing housing in the slums of Karachi. In J.W. Schoorl et al. (eds), Between basti dwellers and bureaucrats, Pergamon, Oxford, pp. 61-8

Harthe-Deneke, A. (1981) Quasi-legal urban land submissions in Latin America. In HABITAT, The residential circumstances of the urban poor in developing countries, Praegar, New York, pp. 82-115

Hendler, P. (1986) 'Capital accumulation, the state and the housing question: the private allocation of residences in African townships on the Witwatersrand, 1980-1985'. Unpublished MA Dissertation, University of the Witwatersrand, Johannesburg

Henry, R.C. (1979) Colonial Karma: factors affecting manpower and training for Third World housing. In H. Murison and J.P. Lea (eds), Housing in Third World countries, Macmillan, London, pp. 111-17

Heppell, M. (1979) A black reality: Aboriginal camps and housing in remote Australia. AIAS, Canberra

Islam, N. (1982) Food consumption expenditure pattern of urban households in Bangladesh. Geojournal, 4, 7-14

Johnstone, M. (1978) Unconventional housing in West Malaysian cities. In P.J. Rimmer et al. (eds), Food, *shelter and transport in Southeast Asia and the Pacific*, Monograph HG12, RSPACS, Australian National University, Canberra, pp. 111-34

------ (1979) 'Problems of access to housing in Peninsular Malaysia'. Unpublished PhD Dissertation, RSPACS, Australian National University, Canberra

Kirke, J. (1984) The provision of infrastructure and utility services. In G. Payne (ed.), *Low income housing in the developing world*, Wiley, Chichester, pp. 233-49

Lea, J.P. (1979) Self-help and autonomy in housing. In H. Murison and J. Lea (eds.), *Housing in Third World countries*, Macmillan, London, pp. 49-54

------ (1983) Customary land tenure and urban housing land. In S. Angel *et al.*, *Land for housing the poor*, Select Books, Singapore, pp. 54-74

Leeds, A. (1981) Lower-income settlement types: processes, structures, policies. In HABITAT, *The residential circumstances of the urban poor in developing countries*, Praegar, New York, pp. 21-61

Linn, J.C. (1983) *Policies for equitable and efficient growth of cities in developing countries*. World Bank/Oxford University Press, Oxford

Lloyd, P. (1980) *The young towns of Lima*. Cambridge University Press, Cambridge

Mabogunje, A. *et al.* (1978) *Shelter provision in developing countries*. Scopell, Wiley, Chichester

Makanas, E.D. (1974-5) Inter-industry analysis of the housing construction industry. *NEDA Journal of Economic Development*, 150-78

Marga Institute (1976) *Housing in Sri Lanka*. Research Study no. 6, Marga Publications, Colombo

Martin, R.J. (1983) Upgrading. In R. Skinner and M. Rodell (eds), *People, poverty and shelter*, Methuen, London, pp. 53-79

Mather, C. and Parnell, S. (1987) Urban renewal in Soweto 1976-1986. In D.W. Drakakis-Smith (ed.), *The urban totem*, Croom Helm

Meyerink, H. and Vekemans, R. (1983) Class structure in a basti. In J.W. Schoorl *et al.* (eds), *Between basti dwellers and bureaucrats*, Pergamon, Oxford, pp. 77-97

Mghweno, J. (1984) Tanzania's surveyed plots programme. In E. Payne (ed.), *Low income housing in the developing world*, Wiley, Chichester, pp. 109-24

Mukherjee, S. and Singh, A.M. (1981) Hierarchical and symbiotic relationships amongst the urban poor:

pavement dwellers in Calcutta. In HABITAT, The residential circumstances of the urban poor in developing countries, Praegar, New York, pp. 126-30

Murison, H. (1979) Indigenous building materials. In H. Murison and J. Lea (eds), Housing in Third World countries, Macmillan, London, pp. 126-30

------ and J. Lea (eds) (1979) Housing in Third World countries. Macmillan, London

Oram, N.D. (1976) Colonial town to Melanesian city: Port Moresby 1884-1974. Australian National University Press, Canberra

Parry, J.P.M. (1984) Building materials and construction systems. In G. Payne (ed.), Low income housing in the developing world, Wiley, Chichester, pp. 249-64

Payne, G. (ed.) (1984) Low income housing in the developing world. Wiley, Chichester

Potter, R.B. (1985) Urbanisation and planning in the Third World. Croom Helm, London

Richards, P. and Thompson, A. (eds) (1984) Basic needs and the urban poor. Croom Helm, London

Rodell, M. (1983) Sites and services and low income housing. In R. Skinner and M. Rodell (eds), People, poverty and shelter, Methuen, London, pp. 21-52

------ and Skinner, R. (1983) Introduction: contemporary self help programmes. In R. Skinner and M. Rodell (eds), People, poverty and shelter, Methuen, London, pp. 1-20

Schoorl, J.W. et al. (eds) (1983) Between basti dwellers and bureaucrats. Pergamon Press, Oxford

Selvarajah, E. (1983) The impact of the ceiling on housing property law on the slum and shanty improvement programme in Sri Lanka. In S. Angel et al. (eds), Land for housing the poor, Select Books, Singapore, pp. 156-78

Sheridan, N. (1979) Energy for the built environment. In H. Murison and J. Lea (eds), Housing in Third World countries, Macmillan, London, pp. 100-10

Siddiqi, I. (1983) The implementation of the Metroville I project in Karachi. In J.W. Schoorl et al. (eds), Between basti dwellers and bureaucrats, Pergamon, Oxford

Skinner, R. and Rodell, M. (eds) (1983) People, poverty and shelter. Methuen, London

Teedon, P. (1987) 'Low cost housing schemes in Harare, Zimbabwe'. PhD Dissertation, Department of Geography, University of Keele (in preparation)

Thiedeke, G. (1979) Manpower, skills and materials. In H.

Murison and J. Lea (eds) Housing in Third World countries, Macmillan, London, pp. 118-25

Tokman, K. (1983) Ankara: procedures for urban upgrading and urban management. In G. Payne (ed.), Low income housing in the developing world, Wiley, Chichester, pp. 89-108

Turner, J.C. (1967) Barriers and channels for housing development in modernizing countries. Journal of the American Institute of Planners, 33, 167-81

------ (1968) Housing priorities, settlement patterns and urban development in modernizing countries. Journal of the American Institute of Planners, 34, 354-63

------ (1969) Uncontrolled urban settlement: problems and policies. In G. Breese (ed.), The city in newly developing countries, Prentice-Hall, Englewood Cliffs, pp. 507-34

Walton, D.S. (1983) The role of international consultants. In G. Payne (ed.), Low income housing in the developing world, Wiley, Chichester, pp. 187-98

Ward, P. (1978) Self-help housing in Mexico City: social and economic determinants of success. Town Planning Review, 49, 38-50

------ (ed.) (1982) Self-help housing: a critique. Mansell, London

------ (1984) Mexico: beyond sites and services. In G. Payne (ed.), Low income housing in the developing world, Wiley, Chichester, pp. 149-58

Wegelin, E.A. and Chantana, C. (1983) Home improvement, housing finance and security of tenure in Bangkok Slums. In S. Angel et al. (eds), Land for housing the poor, Select Books, Singapore, pp. 75-97

Wells, J. (1986) The construction industry in developing countries. Croom Helm, London

Williams, D.G. (1984) The role of international agencies: the World Bank. In G. Payne, Low income housing in the developing world, Wiley, Chichester, pp. 173-86

Young, Y.-M. and Belisle, F. (1986) Third World urban development: agency responses with particular reference to IDRC. In D.W. Drakakis-Smith (ed.), Urbanisation in the developing world, Croom Helm, London, pp. 99-120

Zetter, R. (1984) Land issues in low-income housing. In G. Payne (ed.), Low income housing in the developing world, Wiley, Chichester, pp. 221-32

Chapter Five

WELFARE, CULTURE AND ENVIRONMENT

M. Bell

INTRODUCTION

Contemporary uses of the term welfare are many and varied. Dictionary definitions interpret it as a state of society typically embracing a mental state (happiness), an economic state (good fortune and prosperity) and a social state (health). According to this definition, the social and economic spheres combine to shape the quality of human existence. Today, 'welfare' is frequently extended to include 'not only a state of society but also policy instruments designed to alter that state' (Smith 1975, p. 33). In recent decades, the improvement of human welfare has become an essential responsibility of national governments. In the case of the developing countries, under the influence of international aid donors, the central state plays an active role in the formulation and implementation of policies which directly or indirectly influence human welfare. These cover both the social and economic spheres.

However, defining more precisely appropriate policy measures requires that an abstraction like welfare be given a concrete identity. This is no straightforward matter, since there are no absolutes associated with the concept, only perspectives on it. Nevertheless, it is acknowledged internationally that the quality of human existence depends at least in part on meeting the most basic needs which societies have historically sought to secure - namely, adequate shelter, a clean environment, medical care and nutritional food. Today these are recognised as basic human rights and hence a legal entitlement which is codified internationally.

Moreover, they are tangible, physical needs which governments can be involved in providing through the exercise of bureaucratic administration. Thus, a key responsibility of the contemporary welfare economy is to ensure equality in opportunity of access to both physical and social infrastructure - acceptable minimum standards of housing, purified drinking water, sanitation, immunisation, decentralised primary health care and adequate nutrition. Furthermore, an essential element in their provision is to improve the daily experience of living by reducing the hours spent on back-breaking work like collecting water or food crop production. In consequence, the introduction of appropriate technology is deemed to be a further important welfare intervention to reduce the work burden, notably the labour time and effort expended by women (Ahmed, 1985).

Here, we concentrate on two interventions by national governments and international agencies: improved domestic water supply, and sanitation and medical care. Like all interventions which seek directly to achieve specific welfare goals, each of these is confined to a particular sector within the social sphere - public health and medicine respectively. Each one is the responsibility of professional elites in the public or private domain. In consequence these sectors have become separated from each other and from the sphere of production. Thus, for example, the work of the public health engineer is not linked in practice with the medical profession: the former deals with the physical infrastructure associated with preventive medicine, and the latter with the physical causes of ill-health. Neither one is closely linked with the productive activities which, in rural areas, contribute directly to family nutrition and health. Typically, measures such as land reform or the provision of extension advice fall outside the sphere of welfare. Yet they may be vital to the well-being of rural families by determining their ability to satisfy their own nutritional needs.

Thus, within the structures of the modern nation-state, welfare has become organised on a sectoral basis in development policy and administration. In consequence, welfare as an holistic concept has been obscured. In this chapter three points are emphasised concerning the contemporary welfare economy. Firstly, it is argued that this sectoralism is artificial: both in theory and practice, the apparently separate social spheres are closely interrelated. They are brought together through the overarching political and economic considerations which shape welfare issues as a whole, their place on the policy agenda, and the priorities they command. This interrelationship is also apparent in development practice. Improvements in human health depend upon the proper integration of a range of primary health care measures, including clean water supply, appropriate sanitation, immunisation and better nutrition, as well as curative medicine (Cvjetanovic, 1986). China illustrates the impressive results which can be obtained by adopting an integrated approach. Since 1949, through active health promotion, dramatic improvements have been recorded in the country's health statistics (Hillier and Jewell, 1983).

Secondly, it is argued that the sectoral approach adopted, derived as it is from Western practice, frequently makes little allowance for indigenous welfare systems which persist in many societies. These indigenous systems are shaped by culture and environment. They do not harmonise easily with the imported model, which is typically based on a top-down technocratic approach characterised by a uniformity in style of service and in delivery system. Granted, indigenous systems have undergone important changes as a result of outside influence, but they are not being destroyed. Local conditions affect the response to outside interventions and shape their outcome, with the result that diversities in the provision of, and access to, welfare services persist between societies. Welfare policies and practices have a particular history in each community and culture. There is a need to explore the form of integration, of incompatibility and conflict between external and indigenous welfare systems in different regions and groups. In explaining these variations, studies of welfare need to integrate broad issues of a politico-economic nature with the specific historical, cultural and environmental conditions which shape the welfare economy in specific countries and regions.

Thirdly, by adopting an integrated approach of this kind, some popular development myths are exposed - notably, that complex systems of welfare provision are new; that welfare provision has followed a linear, evolutionary progressive path, and that the Western imported welfare model based on technocratic principles is dominant and unchanged. Indeed, the roots of the contemporary welfare economy lie in non-capitalist societies.

WELFARE IN INDIGENOUS SOCIO-CULTURAL SYSTEMS

In non-capitalist societies, both Western and non-Western, welfare is a social and moral issue (Thomas, 1971). The search for security against a hostile environment, against seasonal underproduction and against disease, have led societies to evolve particular survival strategies. Scott (1976) argues that adaptive flexibility, the capacity to cope in a risky environment, is intrinsic to peasant societies. These survival strategies have involved support mechanisms for individuals, households and communities. They have taken a number of forms, which Scott (ibid.) groups as 'safety first' in agriculture, involving specific technical and agronomic practices to avoid risks, such as crop rotation and the intercropping of plants like sorghums and millets, each with different moisture requirements; 'the norm of reciprocity', characterised by mutual support within households, communal work groups and elite redistribution to the poor; 'the moral economy', based on an expectation of minimal support from the state. Thus, many societies have traditionally had redistributive mechanisms (Watts, 1983) embedded in the socio-cultural system which were sensitive to, and shaped by, the vagaries of environment.

In Botswana, the strongest bonds were within the household and community - the levels to which public health and community medical interventions now seek to return (Schapera, 1953). Supports at this level were inextricably linked with the composition of households, the structure of families and the strength of kin networks. The household continues to be the smallest well-defined social unit in Tswana society. In the past it comprised a man, his wife or wives, their unmarried children and often, in addition, married sons, brothers or daughters, with their respective families. The household would produce most of its own food, clothing and domestic utensils. The family group comprised related households living adjacent to each other in the permanent settlement. 'Its members associate together constantly, cooperate in such major tasks as building and thatching huts, clearing new fields, weeding and reaping, and help one another with gifts or loans of food, livestock and other commodities' (ibid., p. 40). It would be erroneous to assume that welfare supports operated only at local level. State support was also important in times of drought and famine. Thus, the locus of responsibility for individual and household welfare varied between social institutions at different geographical scales at any one time and also over time, shifting to state level in particular in times of dearth. A similar system is described by Watts (1983) in relation to the Hausa in northern Nigeria. Social supports operated similarly both horizontally and vertically within Hausa society and at different spatial scales. Evidence of this kind confounds the welfare myth that only 'modern' as opposed to 'traditional' societies possess complex forms of social organisation capable of administering a system of social welfare beyond the local level (Thomson, 1986, p. 358).

The legitimacy of these survival strategies and the basis on which individuals secured entitlement to them can be contrasted with those in contemporary state welfare. They were based on customary obligations and rights of exchange associated with kinship. Moreover, in the evolution of these strategies economic activities and social obligations were intertwined. Central to the subsistence ethic, and the moral economy of the Hausa, was the 'logic of the gift', the reciprocal and redistributive qualities which bind the peasant fabric. Among the non-Muslim Hausa, kinship and descent groupings provided a collective security which helped to diffuse risks: 'When the grain stores of one household are exhausted, its head may borrow grain from another household (within the same descent group) and repay that grain at harvest without interest' (Watts, 1983, p. 28). Instances also occurred where food redistribution was associated with particular institutions such as sarkin noma and elaborate ceremonies like the harvest festival. Sarkin noma (the king of farming) was elected on the basis of his ability to produce more than 1,000 bundles of grain. This office entailed the redistribution of foodstuffs at the harvest festival, a ritual which had symbolic value in bestowing status and prestige upon the office holder. However, it also had an important practical function as the ultimate defence against famine. Such methods of addressing inequality exist in most societies and regions. In Islamic culture, gift-giving in the form of charity is institutionalised as a religious obligation on the part of the wealthy and office holders.

Welfare extended into the realms of curative and preventive medicine. In many societies there is an indigenous tradition of both, which is regionally specific. Ethno-botanic research in Africa from the early twentieth century has demonstrated the existence of a rich pool of indigenous environmental knowledge which is applied to water purification (Jahn, 1981) and used in the treatment of illnesses. In Africa a wide variety of herbal medicines are derived from local environmental resources (Lester, 1986). For the high veld of Southern Africa, detailed written records have been compiled on the use of tree products in indigenous medicine, including fruits from the Albizia Amara in Zimbabwe to cure coughs and malaria, and the roots of a range of species in the treatment of diarrhoeas (Drummond and Palgrave, 1973; Drummond, 1981).

This medical knowledge was rooted in social organisation often associated with a well-defined social hierarchy and a division of labour by sex. In the realms of health care, MacCormack (1979, 1984) describes two secret societies, or sodalities, in Sierra Leone, the Poro for men and Sande for women. The former have traditionally had responsibility for environmental hygiene, including clean water supplies and the prohibition of indiscriminate defecation, the latter for maternal and child health, nutrition, domestic hygiene and some aspects of herbal medicine. Within the Sande there were five ranked grades, only the few in the highest rank commanded great knowledge, controlled Sande medicine, were effective teachers and authoritative leaders. MacCormack (1979) reports that before hospitals were established, Sande women, as traditional midwives, had a monopoly over child delivery and were aware of the control they exercised over the production of a scarce resource - offspring for their husband's descent groups.

These forms of welfare have been the customary supports on which people have depended for their survival. They have been at the core of the socio-economic system and the source of people's values - their sense of pride in their ability to order their own lives in the face of a hostile environment. Such welfare systems have been modified in response to outside influence, but they persist and continue to shape the local response to external systems. Again, with reference to the Sande, MacCormack (ibid., p. 27) emphasises that where it was a pervasive institution in the seventeenth century, 'it has not dwindled to a wispy relic today but thrives in the very heart and soul of contemporary culture'. After marriage, while virilocal residence tends to separate women from the solidarity of their kin, leaving them fragmented and therefore powerless as a group, by contrast, Sande provides a lifelong organisation wherever women reside.

Similar welfare systems also existed in pre-capitalist Europe (Thomas, 1971). These systems persisted, albeit in modified form, during the industrial revolution, when welfare was provided by private individuals in the form of charity or alms-giving to the poor. Nineteenth-century benefactors felt a moral obligation to distribute essentials to the urban working classes and to those without paid employment (Peabody Trust). This redistribution could, of course, be corrupted, as in the case of Victorian workhouses. Over the last 40 years, the welfare economy in the industrialised countries has undergone further transformations with the advent of the welfare state. The expansion of state welfare services, notably health, education and social security, has taken place within the context of a broad debate over the essential features of modern, post-industrial society (Midgeley, 1984, p. 1). While the effectiveness and efficiency of state welfare provision is not without its critics, much of this critical work has lacked historical depth and, in consequence, has assumed that social welfare has followed an evolutionary progressive model (Thomson, 1986, p. 358). It will be demonstrated below that this same model has been instrumental in shaping international development thought and practice, but that its historical and cultural insensitivity has been particularly harmful in the non-Western world.

EXTERNAL INFLUENCES ON INDIGENOUS WELFARE ECONOMIES

Why do welfare issues become part of the policy agenda of national governments? What priority do they command and what are the effects of welfare interventions on the recipients? In the developing countries, contemporary governmental decisions over the priority attached to different welfare investments and the form they should take are influenced by the national and international political and economic environment within which governments operate. In this the impact of Western thought and presence in these countries has been profound. Traditional welfare systems changed with the creation of a colonial and post-colonial states system, with an increasing involvement by national governments and a professional elite in public health provision, in medicine and in agricultural production and with the formulation and implementation of ideas about 'development'. Changes have therefore been initiated by broader political-economic forces from above and outside. Four points are stressed here.

Firstly, as professionals like public health engineers and physicians trained on Western principles have increasingly established a legal monopoly over welfare provision in the developing countries, so the institutions involved in providing welfare services have changed and responsibility for them has moved away from local level. The public and private sectors have assumed a measure of control over the functions traditionally performed at household and community levels and authority for these institutions has become concentrated at the geographical and political centre of power, the capital city, within a bureaucratic structure of Ministries of Health, Water Affairs etc. Secondly, the integrity of traditional social welfare institutions operating and controlled at local level such as the extended family, indigenous medical practitioners and traditional midwives, has been called into question by professional authority. Indigenous knowledge has been subordinated to the technical expertise derived from Western science and engineering. Hence, power has shifted upwards from communities and so too has the source of community values.

Thirdly, from the perspective of the clients, welfare entitlement has been affected by the import of Western concepts and personnel and by changes in social relationships at community and household levels resulting from the import of an alien economic model. Indigenous welfare supports have been eroded by economic change. With respect to medical care, Feierman (1979, p. 282) points out that as the scale and character of economic production has increased, coupled with population growth and rapid urbanisation, so changes have occurred in the social care of illness and in the content of therapeutic choices. With the introduction of formal medical institutions like clinics and hospitals, so the groups primarily responsible for therapy management - that is, the household and close kin - have witnessed a decline in their influence and control. Indigenous interpretations of illness and disease have been called into question and so too has the health knowledge on which indigenous therapies were based. Parallels can be drawn with the transformation of European medicine at the end of the eighteenth century, when the treatment of the poor became organised in clinics and when both medical teaching and medical care focused on the course of separate diseases (Foucault, 1973). It was these changes which permitted knowledge of disease based on clinical experience to be systematised.

In the developing countries, traditional therapy managing groups have also witnessed a decline in their control over the basic conditions of life and work of their members:

> The much greater spatial dispersal of close relatives in the twentieth century, with some living in large cities and some moving off across the countryside, combined with the increase in occupational differentiation, and in the degree of alienation of social production, mean that many of the most basic conditions of life which affect the health and well being of each individual, are beyond the control of those assisting in social welfare. (Feierman, 1979, p. 282).

Thus, a decline in the intimate control of therapy is closely associated with a decline in local control over the means of subsistence as indigenous modes of production have been transformed.

The resulting strains on particular cultural groups can be illustrated by reference to the Indians of southern Peru. In isolated rural areas, the Western-trained physician is not shunned due to his or her greater success than the local healer in treating certain illnesses (Quintanella, 1976). However, faith in the latter has not been destroyed owing to the great powers s/he is perceived to have in relation to certain maladies over which the physician has apparently no control. This medical pluralism is upset, however, with migration to the cities, where access to a traditional healer is more limited. In consequence, many migrants have a pathological fear of becoming sick in case their illness is related to 'unnatural agents' which physicians do not understand.

Fourthly, the concept of development has brought with it a new imperialism in the realm of ideas and practices into the developing countries through, among other things, the aid business. Development, as a post-1945 phenomenon, can be distinguished from what went before through its close association with the formation of the United Nations and major international aid agencies. Developmental ideas on welfare have been shaped by the Western experience with welfare provision, including attitudes to the welfare state. Thus, trends in welfare provision in the developing countries are frequently portrayed within a modernisation paradigm in a language which implies progressive, evolutionary change, but the discussion tends to be historically ignorant or naive. Thus, what happens now is viewed in the context of what happened during the 1960s and 1970s. In consequence, the role of indigenous systems, their persistence and transformation, is largely ignored.

Welfare as charity or development

Over the last four decades, during which time the concepts of world development and underdevelopment have been explicitly discussed, important changes have taken place in the definition of, and the status attached to, welfare in international debate. These have in turn affected its position on the development agenda of aid donors and national governments. Until the early 1970s, the economic dimension received priority in studies of comparative levels of welfare between countries. This was reflected in the use of GNP *per capita* as the primary or sole indicator. Economic development and welfare were seen as synonymous. The concentration of current and capital expenditure on the productive sectors of the economy and on infrastructure provision was based on the assumption that by initially increasing national income and deferring welfare payments, greater benefits would accrue in the long term from future welfare expenditure. In consequence, the social services provided were urban-orientated and based on a top-down imported model. Total public expenditure on education in the developing countries rose from 2.4 per cent of their collective GNP in 1960 to 4.0 per cent in 1976 (World Bank, 1980). However, enrolments were substantially higher in urban than in rural areas due, in part, to the concentration of facilities in towns. In Columbia, for example, in 1974 school attendance in towns among the age group 6-11 was 62 per cent for the poorest households and 89 per cent for the richest households. Comparable figures in rural areas were 51 and 60 per cent respectively. In the Indian state of Gujarat, in 1972-3 school attendance among boys aged 5-9 in the poorest urban households was 42 and 77 per cent in the richest urban households. In rural areas the equivalent enrolment figures were 22 and 53 per cent respectively (ibid.).

Imported welfare provision directed specifically to the needs of the poor was mainly in the form of charity by private individuals and non-governmental organisations. Clinard (1966) discusses the establishment of social welfare centres in the Indian slums. Operated by private welfare groups, these centres had their origins in the settlement houses established in the nineteenth century for the poor in England and for immigrants in the United States. 'By providing services and exposing slum dwellers to a different set of values and norms', they were designed to incorporate the poor into the dominant culture, that to which the professional social workers and volunteers - most of whom did not live in the area served - subscribed (ibid., p. 101).

However, the priority attached to welfare provision by national governments and the character of welfare services have changed as the concept of development has been reassessed. Indeed, evidence from the social sphere has been instrumental in stimulating this reassessment. During the 1970s evidence from numerous studies, including the much-referenced International Labour Organisation report on Kenya (ILO, 1972), together with the accumulation and publication of comparative social and economic statistics, showed the persistence of grotesque inequalities in health between rich and poor within the developing countries, despite economic growth (World Health Organisation, 1981). Reference was made to the many doleful indicators of poverty, notably high infant mortality rates among the rural and urban poor, despite an overall declining death rate. Sanders (1985, p. 39) reports that in Africa and Asia, infant mortality rates among elite groups are similar to European rates - approximately 20 per thousand live births - while among poor families, the recorded figures reach 300 to 400 for every thousand born alive. In the socially segregated environment of African shanty towns, public health conditions have been described as follows:

> From the sanitary aspect Engel's description of the Manchester slums in 1844 is applicable to these shanty towns. Sanitation is non-existent, and open drains run down what passes for streets. The shanties are built of mud and wattle, old packing cases, or kerosene tins, with tattered blankets as doors. Children crawl among the uncollected rubbish or in the drains. Water has to be fetched from a pump, well, or tap, and may be

contaminated (cited by Sanders, 1985, p. 36).

This kind of observation and statistical evidence, which is now well documented, called into question the limited interpretation of both development and welfare based on narrow economic criteria. With the subsequent inclusion of a range of social measures such as life expectancy and infant mortality into a composite index of development (Morris, 1979), enhancing the quality of life by improving social conditions and meeting basic human needs has become an acceptable development goal in its own right (Streeten, 1981). Welfare interventions have been interpreted, at least by liberal academics and practitioners, as a means by which to reduce inequalities in society and to alleviate squalor and poverty (Midgeley, 1984, pp. 9-10). In the words of the Food and Agriculture Organisation, 'the fundamental purpose of development is individual and social betterment, development of endogenous capabilities and improvement of the living standards of all people' (quoted in Underhill, 1984, p. 1).

Through the influence of aid donors, these changes in attitude and approach have had some effect upon the geographical distribution of services and upon social access. The needs of the poor have been highlighted and renewed attention drawn to hitherto neglected rural and low-income urban environments. In line with this, support for small-scale projects targeted on the poor, long considered to be solely the domain of non-governmental organisations, has become increasingly important among multilateral aid organisations. The European Community, for example, under successive Lomé Conventions, has provided funds for micro-projects. These are grass-roots schemes, normally in rural areas, designed to respond to the needs of local communities and include both social infrastructure and agricultural production. Although they absorb only a meagre proportion of the aid budget, by December 1978 the European Commission had approved the first 28 annual programmes, of which 22 were in sub-Saharan Africa, extending through former British and French West Africa to Sudan, Ethiopia, Kenya, Malawi, Zambia, Madagascar, Lesotho, and Swaziland (EEC, 1980). Early successes with these micro-projects have afforded them a greater priority under Lomé II and Lomé III (O'Neill, 1985).

Welfare in the sense of the provision of services for the benefit of the poor (a traditional view of welfare as charity) has become more than merely a respectable palliative: it has also been seen as developmental. It is difficult to define a 'developmentalist' approach to welfare in precise terms. Loosely, it is an approach in which welfare is seen as a process which is instrumental in bringing about other forms of social, economic or political change (Conyers, 1986, p. 599). In so doing, it challenges the patronising, charitable attitude to poor people which creates dependency and discourages initiative, and replaces it by one in which poor people are seen as active agents in their own development. Thus, the provision of, among other things, appropriate physical and social infrastructure through community participation, has been interpreted as part of a dynamic process which strengthens community institutions and thereby reallocates power from the centre to the periphery. To achieve this, what is interpreted as appropriate infrastructure has also changed. The range of technologies deemed to be acceptable has broadened. Thus, for example, as a way of improving access to safe domestic water supplies, research has been conducted into low-cost solutions, including standposts, handpumps and protected wells (Cairncross et al., 1980).

In summary, a redefinition of welfare has been closely related to, and contributed to, a shift in the interpretation of development in international debate. It has changed from a monolithic concept 'defined by economic criteria, competitive behaviour, external motivation and large-scale redistributive mechanisms to diversified concepts defined by broader social goals, by collaborative behaviour and by endogenous motivation' (Stohr, 1981). Thus, although welfare has continued to involve investments which champion the cause of the poor, they have shifted from being interpreted as merely charity to activities which are inherently developmental. It is also apparent that this change in definition of welfare has been accompanied by an improvement in the status of poor people as active agents in their own development and, with this, a return to the grass-roots, bottom-up approach to welfare provision apparently characteristic of pre-colonial societies. The status of particular kinds of social welfare interventions, notably preventive medicine, has also improved, and with this, a recognition of the contribution which can be made by a range of technologies. With these changes in emphasis, hitherto neglected rural and urban environments have become objects of attention. The notable and important exceptions to this developmentalist trend in welfare provision are hand-outs like supplementary feeding programmes and food aid which remain in the charity mould.

Health care in the development debate - top-down or bottom-up?

A close look at formal medical practice and the public health sector provides evidence of the changes which have taken place in welfare issues at the international level. During the economic growth era of the first development decade in many newly independent African countries, city hospitals, large dams and irrigation schemes based on the Western model were prestige projects and symbols of progress. In the case of Ghana, Twumasi (1981, p. 150) indicates that local doctors who inherited the colonial medical system had an interest in maintaining it. They were the new elite, remaining in the cities and the principal towns where supporting facilities were available, including good schools for their children and an urban life style suited to the 'new colonials'. The only significant form of imported health care directed specifically to the needs of the poor, and which penetrated the countryside and the urban poor, was associated with the Christian medical missions in the tradition of Albert Schweitzer and Mother Teresa.

Today, it has become apparent that clinical, curative medicine - the rational scientific approach to disease - involving costly investments in the form of high-status hospitals in urban areas, is ill-equipped to tackle the diseases which predominate, either in the industrialised or in the developing countries (De Kadt, 1976; Fox, 1986; Lewis, 1986). Sanders (1985, p. xi) describes the position as follows:

> In so far as western professional medicine has had an impact on the mass of the people in underdeveloped countries, it has largely been a negative one, reinforcing a political and economic system that is the root cause of their ill health.

Similar claims have been made for conventional public health practice. The provision of urban infrastructure in the form of piped water supplies and conventional sewerage employing technology modelled on Western standards has similarly proved to be socially and environmentally

Table 5.1: Health and related socio-economic indicators

	Least developed countries	Other developing countries	Developed countries
Number of countries	29	90	37
Total population (millions)	283	3001	1131
Infant mortality rate (per thousand live-born)	160	94	19
Life expectancy (years)	45	60	72
Percentage of new-born with a birth weight of 2500g or more	70	83	93
Coverage by safe water supply (%)	31	41	100
Adult literacy rate (%)	28	55	98
GNP *per capita* (US$)	170	520	6230
Per capita public expenditure on health (US$)	1.7	6.5	244
Public expenditure on health as % of GNP	1.0	1.2	3.9

Note: The figures in the table are weighted averages, based on data for 1980 or for the latest available year.

Source: WHO, 1981

inappropriate to countries in which the majority of people are poor and based in rural areas (Table 5.1). The resulting social and spatial disparities in access to social services have been widely documented (Agarwal *et al.*, 1981).

In response to this evidence the United Nations system has given political respectability to something well known throughout history, the importance of community-based curative and preventive medicine, including clean water and sanitation, in maintaining human health. Evidence for this is found in two major international conferences held during the 1970s. The WHO/UNICEF meeting on Primary Health Care in Alma Ata in 1978 declared that:

> Primary health care is essential health care based on practical, scientifically sound and socially acceptable

> methods and technology, made universally acceptable to individuals and families in the community through their full participation and at a cost that the community and country can afford to maintain at every stage of their development in the spirit of self-reliance and self-determination (WHO/UNICEF, 1978).

Similarly, the United Nations Conference on Human Settlements (HABITAT) held in Vancouver in 1976 declared its abhorrence of a situation in which half of the world's population did not have ready access to clean water. The World Water Conference in Mar del Plata in 1977 endorsed this declaration, with the result that the United Nations created a special ten-year programme devoted to the problems of water supply and sanitation (Table 5.2).

According to Bourne (1984) the creation of the International Drinking Supply and Sanitation Decade with a goal of 'water and sanitation for all by 1990' has forced governments to accord this sector a much higher priority than had previously been the case. More importantly, however, 'it has transformed what was previously a purely technical program within a routine bureaucratic setting into a visible and politically charged initiative where national prestige is at stake' (ibid., p. 7). It has advocated a reallocation of resources from urban to rural areas and from rich to poor. Within many developing countries such priorities have been acceptable to national governments anxious to present a profile which challenges the inequities of colonialism. Appropriate rhetoric has appeared, therefore, in development plans; and with the assistance of international aid, programmes and projects have been initiated. In Botswana, for example, post-independence rural development strategies have concentrated on the provision of social infrastructure (Land, 1987). The Accelerated Rural Development Programme initiated in 1974 sought to provide domestic water supplies, primary health care facilities and primary schools in rural settlements which had been largely neglected prior to independence.

Both of these health interventions encompass a philosophy which extends far beyond the Western technocratic approach. They stress equitable distribution, community involvement, social acceptability, appropriate technology, and disease prevention. The links between the social and technological spheres are deemed to be vital since health behaviour impinges on deeply sensitive cultural

Table 5.2: Numbers of people in developing countries (in millions) to be reached with clean water and sanitation in 1981-90 if the 100 per cent targets of the International Drinking Water Supply and Sanitation Decade are to be achieved (WHO figures based on data supplied by governments)

UN region	Water			Sanitation		
	urban	rural	total	urban	rural	total
Asia and the Pacific (ESCAP: west to Iran)	203	925	1128	355	1136	1491
Latin America (ECLA: incl. Caribbean)	108	110	218	212	120	332
Africa (ECA: incl. N. Africa)	104	310	414	130	342	472
West Asia (ECWA: Arab World excl. Africa)	16	22	38	20	25	45
Europe (ECE: eg. Cyprus, Portugal)	14	21	35	30	30	60
World totals	445	1388	1833	747	1653	2400

Source: Agarwal <u>et al.</u>, 1981

beliefs and practices. Alternative forms of culturally acceptable technology in water supply and sanitation have been devised following the principles of Schumacher (1973), while in the realms of primary health care, similar initiatives have been taken (Valt and Vaughan, 1981, p. 10). This philosophy has received its strongest support from the United Nations system, while non-governmental aid organisations have been most active in its implementation. These institutions have faced two major challenges: firstly, to secure political respectability among donor and recipient governments for forms of health care which are contrary to the symbols and images typically associated with development and progress; secondly, to win support within the engineering and medical professions for approaches which challenge much conventional professional training.

A PROGRESSIVE MODEL OF WELFARE CHANGE?

In the realms of rhetoric at least, welfare provision based on a bottom-up approach has sought to counteract the established top-down model. Due sensitivity has been paid to the needs and cultural preferences of the recipients. It is tempting to interpret the history of the welfare economy in the developing countries according to an evolutionary progressive model as follows: welfare responsibilities have shifted from being an integral part of the socio-cultural system of non-capitalist societies, to a form of benevolent charity for the poor, to the provision of services on the Western model, to, most recently, a form of intervention led by international aid which champions the cause of the poor but in an important developmental way.

Two major criticisms can, however, be made of this progressive interpretation. Firstly, it ignores the power of vested interests in maintaining the status quo. The welfare economy in the developing countries is subject to political-economic influences which extend beyond those of the international aid community. De Kadt (1976, p. 526) claims that:

> many of the world's most tenacious problems would not automatically be solved if individuals suddenly became more rational, or even more moral. Tobacco companies would continue to press their wares upon us, and unhealthy foods would still be produced. The

> pharmaceutical industry would not stop trying to make large profits by marginal improvements in patented drugs, which were then pushed with hard-sell methods on GPs.

The control exercised by multinational companies over governments in the developing countries hinders attempts from either within or outside these countries to achieve rapid social change. The established alliance in Britain between medical and business interests has been compared with that which occurs in the poor countries (Sanders, 1985, p. 130). In the latter, the resultant bias towards curative medicine is particularly severe and unfortunate in view of the economic burden, in the form of expensive technology and drugs, which it places on already overstretched health resources. In Tanzania, there is one drug representative for every four doctors as opposed to one for every 30 in Britain, while three times as much as in Britain is spent by the drug companies on each doctor (ibid., p. 130). The widespread promotion of modern drugs fosters the myth that the solution to illness lies in the purchase and consumption of medications rather than in an improvement in living conditions. The effects on low-income groups are particularly severe. In one town in Brazil it was found that poor families spent some 6 per cent of their monthly income on medicines, while in the Philippines, payments for treatment represented some 10 per cent of a household's meagre income (Melrose, 1982).

For the recipients of welfare provision, there is a second, and equally serious criticism of the progressive model. It eschews the importance of conditions within the developing countries in shaping welfare policies and practices. By assigning only a passive role to internal forces, it ignores the fact that differences exist between national governments in their approaches to the welfare economy and that indigenous systems continue to play a key role in welfare provision.

The political economy of welfare and the role of the state

Significant differences exist between governments in the developing countries in their welfare policies and practices. The model of equality does not apply equally to Chile and South Yemen. National ideology and the power of the state play an important role. Following the Chinese Revolution in 1949, the objective of Chinese health policy was to make health care available to all sectors of the population, to stress preventive medicine and to utilise the knowledge and practices of Chinese traditional medicine (Hillier and Jewell, 1983). During the Cultural Revolution, the large-scale training and deployment of barefoot doctors - rural health workers, part peasant and part paramedic - provided a significant boost to preventive medicine and with it a shift in the balance of health work from urban to rural areas.

Since the political independence of Tanzania and Sri Lanka, progressive social policies have also been followed in line with those now promoted by the aid community. Tanzania's ideology of socialism and self-reliance sought to replace the inherited pattern of unequal development by a policy of decentralisation from Dar es Salaam. Although much criticised in practice as an example of inefficient state capitalism, it has nevertheless succeeded in spreading basic services, notably primary education, health facilities and piped water supply to a large number of rural people. In 1961, 80 per cent of the adult population was illiterate; today, 85 per cent are reported to be able to read and write. In 1961 only 11 per cent of the population had easy access to clean water, while today the figure is reputed to be 50 per cent. The ratio of doctors to population has been reduced from 1:830,000 to 1:26,000. The mortality rate of infants has been halved, while life expectancy for adults has risen from 35 to 51 years (Crowder, 1987).

In the case of Sri Lanka, since independence the national government has provided what for most developing countries is an almost unique package of public welfare programmes with redistributive goals: a rice subsidy, free education, free public health services, and land reform. In consequence, on social criteria such as infant mortality rate (32/1,000), life expectancy at birth (69 years) and nutrition status, the statistics are impressive (Abeyratne and Poleman, 1983). Both Tanzania and Sri Lanka are classified by the World Bank (1986) as low income economies with a <u>per capita</u> gross national product of less than $360. In welfare terms, they would appear to be much 'richer'.

Continuity and transformation in welfare practices

While national governments vary in their priorities attached to welfare provision, by contrast, in the scope and authority that they have assigned to indigenous welfare systems, the approach has been remarkably uniform. The progressive interpretation of welfare promoted by international aid emphasises the need for assistance to the poor through a bottom-up approach. This approach is based on a recognition that biological needs for human survival such as food and health are socially constituted and that indigenous beliefs and customs therefore need to be incorporated into the design of development projects. Control over food production and preparation, for example, is determined by social rules relating to gender which vary between societies (Basse, 1984). Similarly, food preferences, food distribution within the household, and when, how and where it is eaten are all culturally specific (Sen, 1984). A recognition of this diversity is critical to the success of feeding programmes for the nutritionally vulnerable (Wheeler, 1985). The bottom-up approach also emphasises the use of indigenous models of social organisation in project design and implementation. Involving communities in their own development is deemed to be an effective means by which to incorporate cultural diversity. Thus, the Alma Ata Conference on Primary Health Care endorsed the recommendation that governments should 'encourage and ensure full community participation through the effective propagation of relevant information, increased literacy and the development of the necessary institutional arrangements through which communities can assume responsibility for their health and wellbeing' (WHO/UNICEF, 1978). Problems do arise in practice, however, as De Kadt (1982) has argued, by assigning responsibility to individuals and households for their own health and welfare and for those of the community - in essence a return to indigenous systems - it can be a convenient way of letting governments off the hook. Equally, when outsiders, the government official and professional, do become involved, blame for the limited success of what they perceive to be technically sound projects is often placed on the users. Indigenous beliefs and customs are difficult to cope with by technocratic methods, with the result that 'strange and exotic local values and institutions' become a convenient excuse for a lack of community co-operation. In the field of environmental health, sanitation projects seek to improve health by providing a clean physical environment for households. They may involve simple engineering, but what complicates their design and implementation is that they are also projects in social intervention. Human excreta disposal is a sensitive topic in many cultures. Sanitation behaviour frequently varies between communities within the same physical environment, with the result that predetermined rules cannot be applied. In rural Zimbabwe, privacy in defecation is highly valued. Preferred sites are sheltered places away from the homestead. Customary rules regarding relationships between kin affect defecation behaviour. Consequently different kin categories use different defecation sites (Government of Zimbabwe/Interconsult, 1985). In Botswana also, defecation is private although women may be seen to leave in the early morning or late afternoon for a defecation site. The introduction of a sanitation system close to the homestead which required carrying water to flush the latrine challenged this norm and was therefore completely rejected (IDRC, 1980).

Within rigid bureaucratic systems, technical programmes do not have the scope or flexibility to incorporate local variations in cultural preferences and practices. Thus, governments frequently expect bottom-up development to take place within a top-down structure. Werner's (1980, p. 94) work on rural health programmes in Latin America illustrates how national assistance at local level expresses only the dominant values and norms. He concludes that the government-sponsored community-oriented programmes had very little community participation and a great deal of 'handouts, paternalism and superimposed, initiative-destroying norms'. Explaining community non-co-operation in terms of ignorance or apathy reflects a failure on the part of outside agencies to scrutinise in detail what already exists locally. The work of Bakhteati (1987) over a six-year period (1979-85) in a spontaneous settlement of Karachi, Pakistan, indicates how improvements in preventive medicine through a sanitation programme took place from within, by building upon indigenous technologies and models of community organisation. She illustrates how, by an incremental process, other changes followed, including a home-school programme for girls and employment opportunities for women, in spite of the restrictions upon female freedom in a Moslem society. However, this progressive development depended upon an approach by outside agencies which was supportive rather than interventionalist, enabling the community itself to take the initiative by determining the speed of change and the direction in which investments should be channelled.

Coming to terms with indigenous welfare systems represents a critical challenge for both national governments and the international community. Granted, indigenous curative and preventive medicine, and nutritional supports including reciprocity and redistribution, have undergone major transformations, as indeed have people's entitlement to welfare. However, these indigenous systems cannot be ignored: they exist side by side with imported systems and have been shown to sustain a larger population than the latter. Some three-quarters of the population in the developing countries continue to rely on indigenous medicine - precisely that portion denied access to modern medical care (Melrose, 1982). In rural areas, indigenous medicine is the major source of health care, particularly for poor people. The village population of India, for example, spends ten times more on consulting local healers than the government spends on health services (Sutherland, 1978). Even in towns in which primary health care is more accessible, a traditional healer may often be consulted first. A survey in Dacca, Bangladesh, indicated that over 50 per cent of mothers who took their children to a clinic had already sought help from a traditional healer (Melrose, 1982). Traditional healers command status and respect, and continue to shape the local response to health systems introduced from outside. In Sierra Leone, MacCormack (1984) indicates that young maternal and child health aides recruited to teach and supervise traditional midwives - elderly women who had delivered thousands of babies - were effectively ostracised if they were disrespectful to, or disregarded the wishes of, these traditional healers. Thus, indigenous systems are valued not only for their physical availability but also, and more particularly, because of their social acceptability.

In this final section, we focus on indigenous approaches to health. It is argued that popular interpretations of disease transmission - that it is a product of ignorance or apathy - fail to recognise the environmental relations found in each community and culture and which shape approaches to health and disease.

CULTURE, THE ENVIRONMENT AND HUMAN HEALTH

Much human behaviour which impinges on health is based on

an understanding of the environment, not only in a physical sense, but also in its relationship with the social and spiritual realms. The environment may be represented at three analytical levels: physical space - tangible elements on the ground such as the layout of cities and the spatial ordering of family compounds; social space, relating to social and economic organisation, including divisions of labour between the sexes and generations; and metaphysical space, the ordering of the visible and invisible universe to form a cosmology or world system. In many societies, this latter has been shown to prescribe, order and legitimise both social and physical space. An holistic approach to the environment combining the physical world of nature with the social and metaphysical spheres permeates many of the cultural beliefs and customs relating to the causes of illness and disease, water-use patterns, hygiene and sanitation behaviour.

Beliefs about the causes of disease and its treatment

According to Western scientific thought, disease results from a derangement in human function and is defined in physiological or biological terms. Illness, by contrast, is the culturally constructed reaction to disease and discomfort. Studies in widely varying cultures indicate that illness is perceived to be part of the universe of misfortune, including also unemployment, and unprofitable business and even an unfaithful wife (Feierman, 1979). Moreover, illness may be perceived to have multiple causes which lie not only in the material world (one's own physiology), but also in the social and spiritual worlds (for example, sorcery and spirit possession). In the case of diarrhoea, a leading cause of illness and death among young children, it is precisely defined by health workers as 'three or more waterly stools in a twenty-four hour period'. Aetiology is assigned to infectious pathogens and diarrhoea is classified with other infectious diseases. Within indigenous belief systems, however, diarrhoea may not be perceived as a distinct clinical entity with a single set of causes. In north-east Brazil, diarrhoea was thought to be a symptom of any one of five folk illnesses, including 'evil eye', 'spirit intrusion', and teething. Diagnosis was made on the basis of associated symptoms, together with a detailed knowledge of the child's social context (Nations, 1982). In South India, Lozoff et al.

Table 5.3: Alternative medical systems

Cosmopolitan medicine	- derived from Western natural science and the materialist tradition. A concept of medicine 'held today by some individuals in every nation, and not held universally in any one nation'.
Indigenous regional medical systems	- notably, Aryuvedic and Unani associated with the world religions Hinduism and Islam in Asia.
Folk medicine	- a variety of local religious, ritual and herbal therapies found in Africa associated with a range of practitioners including herbalists, bone-setters, mediums, midwives.
Domestic medicine	- remedies used in the home.

Source: After Bichmann, 1979

(1975) found that diarrhoea, linked with vomiting, sunken eyes and a sunken fontanelle, might occur in circumstances associated with ritual pollution (for example, when a mother fed her child after exposure to a woman who had had a miscarriage).

A range of formal medical systems persist in many non-Western countries (and is increasingly penetrating Western medical practice too), which provides alternative therapies to treat the apparently multiple causes of ill-health (Table 5.3). This medical pluralism embodies concepts of illness and health, and theories behind disease and its treatment which are grounded in a socially constructed world view. In the case of Aryuvedic and Unani and a wide variety of folk medicines, their base lies in the spiritual rather than the material realm of Western science and culture.

Pluralism in healing systems is widespread in India and, from the perspective of the sufferer, these alternative systems are closely interrelated (Caldwell et al., 1983). Rural areas contain not only government and private practitioners, but also local specialists on healing and herbs (both unpaid and paid), priests and saints, those who have astrological knowledge and cast horoscopes, and midwives. Moreover, 'in the lives of most villagers, clinics serve as momentary stopping places on the sick man's pilgrimage from one indigenous practitioner to another' (Marriot, 1955). Indigenous beliefs incorporate and extend beyond the realms of Western science. Typically a range of the therapies are sought including the use of herbal medicines, hospital visits, exorcism and ritual purification. Modern medicine succeeds when it assumes the certainty of traditional healers who speak with 'the authority of the supernatural power which is the real agent of cure'.

Indigenous beliefs and customs relating to hygiene and environmental sanitation

Although communities may lack knowledge of Western medical explanations of disease, such as the germ theory, concepts of what is clean and dirty, what is pure and polluting, are appreciated. Both cleanliness and purity are valued in the major world religions, Hinduism and Islam, but their meanings are somewhat different from those used in Western public health practice (Table 5.4). Studies in selected north Indian villages (Khare, 1962) have found that purity and impurity refer to ritual states in life designed to create 'a condition favourable for upward spiral of the soul', while concepts of dirtiness and cleanliness are more directly related to the state of the physical environment. Ritual impurity may be caused by natural happenings such as birth and death, or coming into contact with material objects (human skin) or polluting human beings (untouchables). Cleanliness and dirtiness relate both to hygiene - personal and the physical environment - and to neatness and orderliness. Thus, during cooking, if one's hands become soiled with flour they are 'dirty' and must be washed to become clean. Similarly, scattered articles in a room create disorderly or dirty conditions which must be made orderly by sweeping and returning items to their proper place. At certain times, purity and cleanliness are linked: for example, actions taken to achieve ritual purification, such as the removal of dirt and refuse, result in physical cleanliness. At other times the two concepts are unrelated, as, for example, when mud is used as a cleansing agent for objects or places which have become impure. A Brahmin and

Table 5.4: The ideological categories of purity-impurity, cleanliness-dirtiness in northern India, with examples

(Ritual) purity	(Ritual) impurity	Cleanliness (orderliness)	Dirtiness (disorderliness)
Removal of dirt and filth	Happenings: - Birth - Death - Menstruation	Washing with soap	Hands soiled with ashes, earth, flour
Disposal of refuse	Material objects: - Human skin - Menstrual blood - Spittle - Reproductive fluids - Impure food or water	Articles in their proper place (clothes, baskets, etc.)	Vegetable scraps in courtyard
Use of air, water and sun as cleansing agents (not soap)		Swept courtyard	Scattered articles in a room (clothes, baskets, etc.)
		Picking out flies, excreta, or worms	Flies on food Worms in water
Avoidance of contact with dirty objects	Human beings: - Untouchables		Rodent excreta in flour

a public health worker might agree that in terms of physical quality, the water in a jug is clean. For the former, however, the water may also be greatly polluted and therefore undrinkable because the jug has been touched by a low-caste person (Simpson-Hébert, 1984). Questions of ignorance of environmental health do not arise here. Rather the issue is one of language as it conveys meaning of the environment in its various forms.

Evidence for the value attached to cleanliness and, by implication, environmental sanitation, by communities is found in studies of diarrhoea. In Zimbabwe, it was found that people's perceptions of its causes could be divided into three categories: 'physical', 'social', and 'spiritual' (De Zoysa et al., 1984, p. 733). Of these, a child's physical environment was perceived to be a major cause. Although the germ theory was not explicitly stated, a relationship was identified between the faecal-oral transmission route of diarrhoea and a polluted environment including 'uncovered food', 'dirty water' and 'flies'. Graphic descriptions of pollution quoted were:

> We have to drink the dam water where animals and children bathe and the dirty water makes us ill.
>
> Flies sit on dirt when they eat, then they come on to uncovered foods and spit on to foods which we eat.

The study also showed that social and spiritual causes of diarrhoea were perceived to be important, notably those relating to breast-feeding and the child's fontanelle. Sexual relationships during lactation are taboo in the study area. If a child becomes ill with diarrhoea, it may be interpreted as a sanction for the mother's transgression. Breast-feeding may therefore be terminated at a time when the child is most vulnerable to the loss of fluids and nutrients in the breast-milk. The most entrenched beliefs were those associated with the fontanelle. Both scientific and traditional concepts of a sunken fontanelle describe a life-threatening condition. The scientific interpretation is that it is a sign of dehydration due to severe diarrhoea. In traditional thought it is believed to be a weak area in the young child through which diseases can easily enter when, for example, breast-feeding takes place in the presence of people with a stronger 'influence'. While a child can be protected by a variety of preventive charms worn on the body, an unprotected child may not be strong enough to combat the 'influence'. S/he will be oppressed and the fontanelle will fall. Diarrhoea may or may not be present. Thus, although fontanelle disease was placed in a separate disease category, its association with diarrhoea and the seriousness of the two conditions was recognised.

A threefold classification of the causes of diarrhoea within traditional thought should not, therefore, be interpreted as mutually exclusive. Nor are the elements of the classification representative of divergent approaches to disease:

> Aetiologies deeply grounded in beliefs about social and supernatural control are intermeshed with perceptions about faecal pollution which are consonant with scientific theory about disease transmission. Traditional beliefs about disease causation do not necessarily conflict with scientific theory; they may persist in parallel (ibid., p. 733).

These studies emphasise that the relationship between the individual and the social and spiritual realms is important in explaining the causes of illness and the treatments sought. Pluralism in beliefs is matched by a pluralism in health actions. Furthermore, the environment as it is socially constituted is more than a physical construct. While many societies have a detailed knowledge of the physical environment as a provider of resources for curative and preventive medicine and as a cause of illness, indigenous beliefs and customs relating to illness and health are rooted in an holistic interpretation of the environment which goes beyond the technocratic model.

CONCLUSIONS

Over the last decade, matters of social relevance and public policy have assumed an increasing importance in the geographical literature, and as a consequence, 'welfare', 'quality of life' and 'level of living' have become prominent terms. Within the international arena, the improvement of human welfare through the provision of appropriate services has become an integral part of development and progress. Equality in opportunity of access to what are perceived to be basic human needs (health care, housing, clean water, adequate food) is now an established development goal. Policy priorities designed to achieve this goal have been formulated at the international level by the aid community, and a range of alternative approaches defined. Appropriate strategies have been prescribed to national governments and incorporated into development plans. However, welfare and development are complex and variable concepts. They have been interpreted and applied in many different ways over time between countries and cultures, and equally varied have been the attempts, both formal and informal, to achieve them. This chapter has focused on the form and nature of the relationship between welfare and development. Within the developing countries, important changes have taken place over time in the concept of welfare, in the welfare entitlement of individuals and in the methods by which entitlement systems are maintained. These changes can be related to the political, social and economic transformations which have accompanied Western influence and presence in these countries through colonialism and, more recently, international aid. However, the issues and problems faced by the developing countries are not unique or subordinate to those in the industrialised world. Parallels, both historical and contemporary, can be found. Two concepts have been stressed here, pluralism and holism. Welfare combines the economic and social spheres and incorporates an integrated approach to the environment, combining the physical, social and spiritual realms. The progressive, evolutionary model of welfare change implies a pattern of international convergence. In doing so it ignores the evidence that welfare beliefs and practices vary over space and time and that they are shaped by culture and environment.

ACKNOWLEDGEMENTS

I am grateful to Denis Cosgrove and Neil Roberts for their helpful comments on earlier drafts of this chapter.

REFERENCES

Abeyratne, S. and Poleman, T.T. (1983) Socio-economic determinants of child malnutrition in Sri Lanka: the evidence from Galle and Kalutara Districts.

Department of Agricultural Economics, Cornell University, New York

Agarwal, A., Kimondo, J., Moreno, G. and Tinker, J. (1981) Water, sanitation, health - for all? Earthscan

Ahmed, I. (ed) (1985) Technology and rural women: conceptual and empirical issues. George Allen and Unwin, London

Bakhteari, Q. Ain (1987) 'A strategy for the integrated development of squatter settlements: A Karachi case study'. Unpublished PhD thesis, Loughborough University

Basse, M-T (1984) Women, food and nutrition in Africa: perspectives from Senegal. Food and Nutrition, 10, 65-71

Bichmann, W. (1979) Primary health care and traditional medicine - considering the background of changing health care concepts in Africa. Social Science and Medicine, 13B, 175-82

Bourne, P. (1984) Water and sanitation. Economic and sociological perspectives. Academic Press

Cairncross, S., Carruthers, I., Feachem, R., Bradley, D. and Baldwin, G. (1980) Evaluation for village water supply planning. Wiley, London

Caldwell, J.C., Reddy, P.H. and Caldwell, P. (1983) The social component of mortality decline: an investigation in South India employing alternative methodologies. Population Studies, 37, 185-205

Clinard, M.B. (1966) Slums and community development. Experiments in self-help. Free Press

Conyers, D. (1986) Future directions in development studies. The case of decentralisation. World Development, 14, 593-603

Crowder, M. (1987) Whose dream was it anyway? Twenty-five years of African independence. African Affairs, 86, 7-24

Cvjetanovic, B. (1986) Health effects and impact of water supply and sanitation. World Health Statistical Quarterly, 39, 105-17

De Kadt, E. (1976) Wrong priorities in health. New Society, June, 525-6

------ (1982) Ideology, social policy, health and health services: a field of complex interactions. Social Science and Medicine, 16, 741-52

De Zoysa, I. et al. (1984) Perceptions of childhood diarrhoea and its treatment in rural Zimbabwe. Social Science and

Medicine, 19, 727-34

Drummond, R.B. (1981) Common trees of the central watershed woodlands of Zimbabwe. Natural Resources Board, Zimbabwe

------ and Palgrave, K.C. (1973) Common trees of the Highveld. Longman, London

EEC (1980) Report of the Commission on the Implementation of Micro-projects under the Lomé Convention. Brussels

Feierman, S. (1979) Change in African therapeutic systems. Social Science and Medicine, 13B, 277-84

Foucault, M. (1973) The birth of the clinic. Tavistock Publications, London

Fox, D.M. (1986) Health policies, health politics: the British and American experience, 1911-1965. Princeton University Press

Government of Zimbabwe/Interconsult (1985) Social Studies draft: National Masterplan for rural water supply and sanitation. NORAD

Hillier, S.M. and Jewell, J.A. (1983) Health care and traditional medicine in China 1800-1982. Routledge and Kegan Paul, London

IDRC (1980) Sanitation in developing countries. Proceedings of a workshop on training held in Lobatse, Botswana, 14-20 August 1980

ILO (International Labour Organisation) (1972) Employment, incomes and inequality. A strategy for increasing productive employment in Kenya. ILO, Geneva

Jahn, S. Al A. (1981) Traditional water purification in tropical developing countries. German Agency for Technical Co-operation

Khare, R.S. (1962) Ritual purity and pollution in relation to domestic sanitation. Eastern Anthropologist, 15, 125-49

Land, A. (1987) 'The role of the state in the provision of domestic water supply and sanitation in rural Botswana'. Unpublished PhD thesis, Loughborough University

Lester, C. (1986) 'The social ecology of traditional medicine in the Third World'. Undergraduate dissertation, Dept. of Geography, Loughborough University

Lewis, J. (1986) What price community medicine? The philosophy, practice and politics of public health since 1919. Wheatsheaf, Brighton

Lozoff, B., Kamath, K.R. and Felman, R.A. (1975) Infection and disease in South Indian families: beliefs about

childhood diarrhoea. Human Organisation, 34, 353-8
MacCormack, C. (1979) Sande: the public face of a secret society. In B. Jules-Rosette (ed.), The new religions of Africa, Ablex, pp. 27
------ (1984) Primary health care in Sierra Leone. Social Science and Medicine, 19, 199-208
Marriot, McK., (1955) Western medicine in a village of northern India. In B.J. Paul (ed.), Health, culture and community, Russell Sage Foundation, New York, p. 241
Melrose, D. (1982) Bitter pills. Medicines and the Third World poor. Oxfam
Midgeley, J. (1984) Social security, inequality and the Third World. Wiley, London
Morris, M.D. (1979) Measuring the condition of the world's poor. Pergamon Press, Pergamon Policy Studies, 42, Overseas Development Council
Nations, M.K. (1982) 'Illness of the child: the cultural context of childhood diarrhoea in northeast Brazil'. Unpublished PhD thesis, University of California, Berkeley
O'Neill, B. (1985) Small is beautiful: micro-projects in the new convention. Lomé Briefing no. 21, Brussels
Quintanella, A. (1976) Effect of rural-urban migration on beliefs and attitudes toward disease and medicine in Southern Peru. In F.X. Grollig (ed.), Medical anthropology, Mouton, pp. 393-401
Sanders, D., with Carver, R. (1985) The struggle for health: medicine and the politics of underdevelopment. Macmillan, London
Schapera, I. (1953) The Tswana. International African Institute
Schumacher, E.F. (1973) Small is beautiful: a study of economics as if people mattered. Bond and Briggs
Scott, J. (1976) The moral economy of the peasant. Yale University Press
Sen, A. (1981) Poverty and famines: an essay on entitlement and deprivation. Clarendon Press
------ (1984) Food battles: conflicts in the access to food. Food and Nutrition, 10, 81-9
Simpson-Hébert, M. (1984) Water and sanitation: cultural considerations. In P.G. Bourne (ed.), Water and sanitation: economic and sociological perspectives, Academic Press, pp. 173-98
Smith, D.M. (1975) On the concept of welfare. Area, 7, 33-6
Stohr, W.B. (1981) Development from below: the bottom-up

and periphery - inward development paradigm. In W.B. Stohr, and D.R.F. Taylor (eds), Development from above or below?, Wiley, London pp. 39-72

Streeten (1981) First things first. Meeting basic human needs in developing countries. Oxford University Press

Sutherland, W.D. (1978) 'A systems analysis of a rural primary health centre in India including a study of the integration of indigenous practitioners into the primary health centre'. Dissertation for Master of Community Health, Liverpool School of Tropical Medicine

Thomas, K. (1971) Religion and the decline of magic: studies in popular beliefs in seventeenth and eighteenth century England. Weidenfeld and Nicolson

Thomson, D. (1986) Welfare and the historians. In L. Bonfield et al. (eds), The World we have gained. Histories of population and social structure. Essays presented to Peter Laslett on his seventieth birthday, Blackwell, Oxford pp. 355-78

Twumasi, P.A. (1981) Colonialism and international health: a study in social change in Ghana. Social Science and Medicine, 15B, 147-51

Underhill, H.W. (1984) Small-scale irrigation in Africa in the context of rural development. Food and Agriculture Organisation of the United Nations

Valt, G. and Vaughan, P. (1981) An introduction to the primary health care approach in developing countries. A review with selected annotated references. Ross Institute of Tropical Hygiene, Publication no. 13

Watts, M. (1983) Hazards and crises: a political economy of drought and famine in northern Nigeria. Antipode, 15, 24-34

Werner, D. (1980) Where there is no doctor. Macmillan

Wheeler, E. (1985) To feed or to educate: labelling in targeted nutrition interventions. Development and Change, 16, 475-83

World Bank (1980) World Development Report, 1980. Oxford University Press, Oxford

------ (1986) World Development Report, 1986. Oxford University Press, Oxford

World Health Organisation (1981) Global strategy for health for all by the year 2000. WHO, Geneva

WHO/UNICEF (1978) Alma-Ata: primary health care. Geneva-New York

Chapter Six

REGIONAL DEVELOPMENT

G.P. Hollier

INTRODUCTION

Economic geographers have demonstrated an interest in the issues of regional development within Third World countries only in comparatively recent times. Prior to the late 1960s, regional studies rarely stepped beyond traditional descriptions of regional variations in the distribution of resources to examine the processes and spatial patterns of development. Even in the wider development literature there were few contributions to regional development theory that focused specifically on underdeveloped countries. Myrdal's (1957) model of interregional income inequality, for example, was only implicitly framed in terms of underdeveloped countries, and it was left to others, notably Friedmann (1966), to promote Hirschman's (1958) ideas on the interregional transmission of economic growth. Regional planning had been a reality in many developed countries for several decades before it emerged in the Third World (Slater, 1975; Gilbert, 1976). Today, there are few countries that have not either expressed some regional goals in development or put into practice a strategy designed to facilitate more equitable spatial development of national resources.

The main objectives of this chapter are to identify the critical lines of regional development theory that have emerged within the shifting paradigms of development studies, and to examine the application of the major regional planning strategies modelled on these doctrines. Firstly, however, it is helpful to clarify what is meant by regional development.

REGIONAL DEVELOPMENT AND REGIONAL PLANNING

At its simplest, regional development is the spatial manifestation of development in the economy as a whole. Economic development tends to favour certain geographic areas, creating and sustaining spatial inequalities in income, welfare and material well-being. Whether such disparities are transitory phenomena or more deeply rooted is at the heart of the development debate. If unchecked for long, and governments' temporal horizons are necessarily short, counterbalancing intervention may be called for. Governments may be motivated by the morality of social justice or the need to neutralise political dissent and separatist claims, while presenting to the outside an acceptable face of modernity untarnished by regional problems. Conscious intervention in the process of regional development forms regional planning. It involves the formulation and implementation of regional policies to facilitate a more balanced pattern of regional development within a country. The focus of development planning becomes the region or geographic area rather than the nation as a whole or specific sectors or projects (Conyers, 1985).

Our understanding of what is meant by regional development is at least partly conditioned by what development itself implies. For much of the post-war era, development has been equated with economic growth, measured by annual increases in both total and per capita Gross National Product, and changes in a country's economic structure from primary to secondary and tertiary activities. Attempts to place economic growth within a wider process of social change failed to lift 'development' from the straight-jacket of the narrowly positivist perspective of neoclassical economics. During the 1960s such definitions were increasingly challenged. Seers (1969) argued that there could be no development without a reduction in poverty, unemployment and inequality, a conception of development to which he later added the need for self-reliance (Seers, 1977). Development, it is claimed, should be released from the confines of the universal and unilinear growth paradigm to become culturally relevant, creating the conditions in which human personality can be realised. This call for a rethinking of the goals of development has been reflected in approaches to regional development, contributing to attacks on long-established theories and to the evolution of significant new directions in regional planning philosophy.

THE REGION

How to define and delimit a region is a recurring theme (Grigg, 1967; Conyers, 1985). Regions have been categorised according to their predominant internal characteristics, by the nature of their relationship to other regions in the country, and with respect to their role in the national economy. Friedmann (1966, pp. 41-4), for example, distinguished five kinds of problem region:

(1) core regions, metropolitan areas with high economic growth potential;
(2) upward-transitional areas whose natural endowments and location relative to core regions suggest the possibility of a greatly intensified use of resources;
(3) resource frontier regions, zones of new settlement involving agricultural colonisation or resource exploitation;
(4) downward-transitional areas - old, established regions with stagnant or declining economies; and
(5) regions, such as those along national borders, demanding a specialised development approach.

Subsequent forays into regionalisation have built on this typology (Pedersen, 1975; Stöhr, 1975; Morris, 1981a). Pedersen's classification is based on the regional demographic structure and the economic dimensions of urbanisation and export-base dependence, while Morris argues for a social dimension to account for divisions between modern and traditionally organised societies in developed and less developed regions.

This type of regionalisation has rarely influenced the delimitation of planning regions. They have tended to complement existing administrative boundaries where compatible secondary data bases and existing political structures generally facilitate more expeditious implementation of regional development plans. The desire to assist administrative harmony is well illustrated by the establishment of regional agencies in many Latin American countries (Gilbert, 1974, pp. 247-55). Alternatively, the region may be delimited in response to particular planning requirements to focus on key issues or problems.

Commissions charged with the integrated development of river basins, as in India, Brazil, and especially Mexico

> have offered a way of planning and coordinating public expenditure in a region that was difficult to do through already established ministries and state governments, quite apart from the economic desirability of investments made with a view to making the most efficient use of water (Barkin and King, 1970, p. 93).

A focus on specific problems assumes that the boundaries between problem and non-problem areas can be drawn. Regions based on existing administrative units invariably expose extensive intra-regional inequalities. The use of pre-selected socio-economic variables to identify regions with greater internal homogeneity (Haller, 1982) may help, but a genuinely satisfactory regionalisation is perhaps unattainable beyond the adoption of progressively smaller areas. The question of the desired size of a region has not yet been resolved. Much will depend on the circumstances appropriate to the country concerned, but it is clear that some countries' regions are larger in size and population than many nation states, while the regions in, for example, Costa Rica (Hall, 1984) would be lost in the vastness of the Brazilian North-East.

Intra- and interregional imbalances are not the only forms of spatial inequality to impinge on regional development (Slater, 1975). The bulk of the regional development literature may focus on interregional inequalities between politically appropriate subdivisions of the national territory, but very often the greatest functional differentiation is between urban and rural areas. Spatial distinctions between the city and the countryside may be sharper than between many regions, contributing to the high inequality of overall income distribution in most countries. The physical extent and the productive importance of rural areas in the Third World promotes a case for regarding regional development as essentially rural development, with the rural-urban dichotomy identified as the major regional problem (Rauch, 1982). For Riddell (1985) regional development policy has become the struggle for rural progress. In related vein, Conyers (1985, p. 10) has claimed a role for regional planning in rural development 'as a means of implementing national plans, facilitating popular participation, coordinating sectoral activities within regions, and tackling the problems of spatial inequality'. Regional development, then, concerns both the process of development at the sub-national level, be it inter- or intra-regional, or essentially rural-urban in character, and the adoption of specific planning policies to redress spatial inequalities arising from that regional development.

REGIONAL DEVELOPMENT THEORY

Regional development theory and planning practice are linked through philosophical perspective in establishing doctrines of accepted orthodoxy. For the Third World, these doctrines reflect, for the most part, the transfer of theories, models, and policy guidelines originally formulated in the light of experience in industrialised market economies. Space does not permit more than an outline of the major post-war paradigms of regional development: they have been thoroughly reviewed elsewhere. Keeble (1967) laid bare the salient features of the major economic development models to geographers then largely unfamiliar with such approaches, and Richardson (1973) has surveyed the field of regional economics. Friedmann and Weaver (1979) and Weaver (1981) have charted the evolution of regional planning, stressing paradigm shifts in an emergent regional science discipline, while Stöhr and Tödtling (1977) have shown how limitations in current theory have weakened the conceptual bases on which regional planning rests. Brookfield (1975), Slater (1975), and Gore (1984) have all focused on regional development within a primarily Third World context.

The growth-oriented paradigm of development favoured in the early post-war period stressed the modernisation of the social and economic structures of underdeveloped countries, with little concern for spatial considerations. With continued emphasis on primary exports, coupled to expanded foreign investment, there was general acceptance that inequality, spatial or otherwise, did not in itself represent an inefficient use of resources. Capital concentrated in the hands of an innovative and entrepreneurially dynamic urban-industrial elite provided a plausible strategy for countries striving, in Rostow's (1960) terms, to 'take-off' into self-sustaining economic growth. In the natural order, the excesses of regional and social income inequality would be reduced. Concern over the emergence of regional imbalance seemed misplaced when such colossal problems of national development had still to be overcome. Regional policies were held to be detrimental to economic growth, justifiable only on rather vague grounds of social equity (Alonso, 1968).

The functional response to regional development

This line of reasoning notwithstanding, work on both sides of the Atlantic began to introduce spatial, if not specifically regional dimensions to economic development. In North America, where economic growth theory emphasised the propulsive role of manufacturing industry, it was perhaps inevitable that industrial location studies would influence planning most strongly. Isard's (1956) synthesis of this material (see Friedmann and Weaver, 1979) underlined the polarised, urban-centred nature of the economy, and implied that regional planning should concentrate on city-dominated functional regions. Though clearly a strategy of unequal growth, it established an approach to regional development that urged functional integration - that is, integration between backward and relatively advanced areas by means of the development of urban and infrastructural functions through which growth impulses would ultimately filter to all parts of the space-economy. It stood in marked contrast with the territorial approach of river-basin planning adopted in the United States in the 1930s, which had pioneered regional self-reliance.

Meanwhile, in France, the idea of polarised growth had been advanced with Perroux's (1950, 1955) recognition that growth did not occur everywhere at the same time, but rather manifested itself in points or 'poles' of growth. These growth poles were located not in geographic space but in what he termed abstract economic space, a field of forces in which each pole was a centre of attraction and repulsion. Perroux saw the growth pole as a means of understanding the mechanisms by which development impulses are transmitted across the whole economy. In effect, they are best regarded simply as sectors of an economy within which growth is generated (Darwent, 1969), or as sets of propulsive industries, large in size, dynamic, innovative, and dominant throughout dependent relationships of backward linkages. They become the poles around which the economy clusters.

Separation through language and academic tradition

meant that it was some time before growth-pole notions were taken up at all enthusiastically or explicitly in the Anglo-American literature. By the 1960s, the desire to operationalise growth poles within a planning framework had led to a progressive loosening of the concept to a position of methodological weakness. This did not, however, prevent the growth pole, the extreme manifestation of the functional urban-industrial approach to development, from becoming the major expression of regional planning initiatives in the Third World.

Modelling regional income inequality

Two independently conceived studies examining the mechanisms of polarised development provide a theoretical base and a terminology that have dominated regional development since the 1950s. Myrdal (1957) advanced the belief that once economic development was initiated in a particular favoured locality, it would continue to grow through a process of circular and cumulative causation in such a way as to increase regional inequality. Migration, capital movements, and trade oriented towards the growing region would act as 'backwash effects' on the rest of the country, frustrating their potential for development. On the other hand, demands on the lagging regions for agricultural products and industrial raw materials might stimulate economic growth there too, leading to localised cumulative causation. Such centrifugal forces were termed 'spread effects' by Myrdal, though he doubted their ability to overcome backwash. In effect, Myrdal challenged the conventional equilibrium theory of neoclassical economics that saw time as healing regionally uneven development. Government intervention to redress regional imbalance played no part in Myrdal's basic model, though he recognised that committed regional policy to promote spread effects might well be the only answer to a capitalist system's inherent propensity to live off the creation of inequalities.

In what Brookfield (1975, p. 100) has called a 'spatial aside to a text on growth', Hirschman (1958) presented a structurally similar model of polarised development. Accepting that international and interregional inequalities were inevitable concomitants and conditions of growth, he adopted the terms 'polarisation' and 'trickle-down' where Myrdal had invoked the notions of 'backwash' and 'spread' effects. Whereas Myrdal stressed a course of events that would lead to increased spatial inequality, Hirschman placed more faith in counter-balancing forces, including some element of government assistance, that would eventually restore equilibrium between the growing North and the lagging South.

Both these models have since been categorised as core-periphery conceptions of regional development, but the first clear articulation of core or centre-periphery notions came from Friedmann and Alonso (1964), in explaining the interaction between cities and their surroundings. Rapidly growing centres were said to act 'as suction pumps, pulling in the more dynamic elements from the more static regions', while the rest of the country was 'relegated to a second-class, peripheral position' (p. 3). Influenced by his experience with the Venezuelan government's attempts to integrate the growth pole of Ciudad Guayana into the national space-economy, Friedmann (1966) went on to propose a four-stage model of the unfolding spatial organisation of a previously uninhabited area (Figure 6.1). In a typical pre-industrial phase tending towards economic stagnation, independent settlements function at the centre of their own market areas in a system lacking any urban hierarchical organisation. With advance into a phase of incipient industrialisation, a single strong centre comes to dominate a larger region, exploiting the resources of the periphery. By phase three the primate city impact weakens, in part by the construction of strategically located urban growth centres that reduce the periphery to more manageable pockets of backwardness. This Friedmann saw as a transitional stage on the way to fully fledged spatial organisation with a functionally independent system of cities, hierarchically arranged to cover the whole national territory. With eventual convergence comes national integration, locational efficiency, maximum growth potential and a high degree of interregional balance.

The model gave support to the doctrine of polarised development, yet recognised the need for deliberate locational planning. Growth centres, the spatial embodiment of Perroux's growth poles, were increasingly promoted to counter the forces in a dynamic market economy that seemingly work against any automatic convergence of core and periphery. The model continues to influence regional development despite much critical comment. Gilbert (1982), for example, identified five main weaknesses. Firstly, the

Figure 6.1: Stages in the spatial organisation of a previously unhabited area

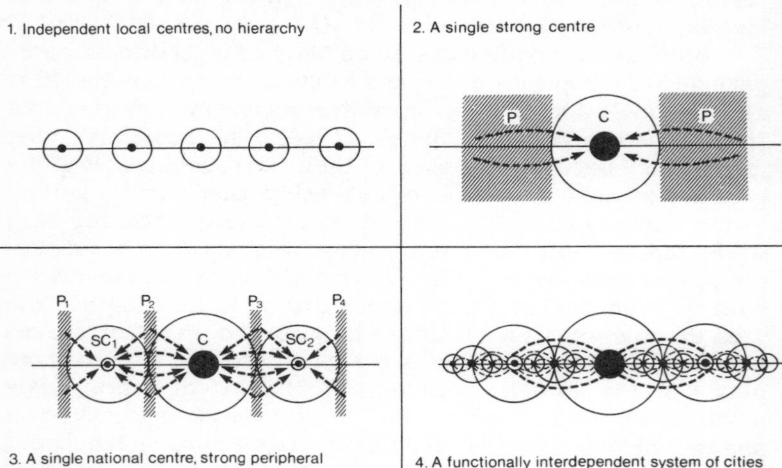

assumption that colonisation will involve minimal contact with indigenous populations is unrealistic. Contact may involve conflict or at least coexistence with alternatively structured systems of spatial organisation. Secondly, the model assumes internal decision-making when in practice foreign influence through investment and patronage have been pervasive in most developing countries. Thirdly, the complexity of power politics at the centre, weak regional representation, and a poorly developed sense of spatial egalitarianism may fail to advance regional policy as either necessary or in the public interest. Fourthly, the model's preoccupation with regional income inequality too easily assumes that extremes of poverty will vanish with regional development. Finally, the model reflects the Rostovian conception of development as a unilineal sequence of stages. Few commentators today, including Friedmann (see Friedmann and Douglass, 1978; Friedmann and Weaver, 1979), would place as much faith in such an ordered progression to well-balanced modernity. On more fundamental grounds, Gore (1984) has criticised Friedmann for failing to present an economic case for why the core-periphery structure of the second phase inhibits growth, beyond its persistence.

GROWTH ORTHODOXY UNDER ATTACK

The period since the late 1960s has seen the growth of an increasingly critical literature in development studies. In part, it stemmed from the poor record of policies devised within the existing paradigm, and a realisation that core regions consolidate their dominance over peripheral regions through self-reinforcing feedback effects (see, for example, Friedmann's (1972) 'general theory' of polarised development). However, the critique extended to the paradigm itself. Several new ideologies sought attention. However disparate their constructs, they weakened the pillars on which the established doctrines of regional policy were founded, though without replacing the crumbling edifice with any unity of direction or purpose.

Convergence or divergence in regional development?

Whether the 'normal' pattern of economic development leads ultimately to a convergence of regional income levels is not easily tested, for most governments have intervened in regional development, if not directly through regional policy, then indirectly through social or sectoral programmes that have some regional impact.

At first, it seemed as though the empirical evidence (Borts and Stein, 1962; Williamson, 1965; El-Shakhs, 1972) favoured Hirschman's (1958) expectation of convergence. Williamson's international cross-section analysis, using a weighted index of regional inequality based on per capita income differentials, identified rising regional income disparity and increasing North-South dualism as typical of early development stages. With maturity though, regional convergence would eradicate severe North-South problems.

Thus, spatial income differentials would trace out graphically an inverted 'U' or bell-shaped curve during the course of economic growth. Both Brazil and Mexico showed signs of this by the late 1950s (Gilbert, 1974) as did Peru between 1946 and 1966 on the basis of certain indicators (Slater, 1975). As Gilbert (1982) later warned, however, convergence in regional incomes may simply represent upper income groups in the disadvantaged regions improving their position relative to similar groups in richer regions, without the mass of low income groups benefiting at all. The extent to which this constitutes 'development' is the subject of critical debate.

Since the early 1970s the pendulum has swung towards Myrdal's (1957) more pessimistic predictions. Gilbert and Goodman (1976) found little support for Williamson's hypothesis in either the original data or more recent sources. Examination of Brazilian statistics at the interregional level and within the North-East region indicate that the spread effects of national and regional policies have failed to secure much improvement in the living conditions of the mass of the population (see also Enders, 1980; Haddad, 1981). Elsewhere in Latin America, the evidence from Venezuela (Jones, 1982), Costa Rica (Brugger, 1982; Hall, 1984), Peru (Hilhorst, 1981), Ecuador (Morris, 1981b) and Chile (Boisier, 1981) all points to persistent and widening regional and interpersonal disparities.

Rondinelli (1980) has sought in vain for signs of 'trickle down' in south and south-east Asia, and Kirk (1981) drew on the Indian literature to argue that economic growth in the core areas is likely to lead to even greater spatial inequality than exists at present. Other Asian case studies in India (Misra and Natraj, 1981), Nepal (Blaikie et al., 1981) and Thailand (Douglass, 1981) have shown increasing regional inequality and widespread poverty, while in Papua New Guinea, Mountjoy (1984) has drawn attention to the complex nature of regional inequalities in an economy much affected by multinational mining corporations. The African literature stresses that development is still at an early stage, and that one might not expect convergence as yet. Certainly, evidence from Filani (1981) highlights the increasing disparity in socio-economic activity between regions in Nigeria, despite some regional shifts in manufacturing industry reported by Oyebanji (1982). The very different development strategies pursued in Ivory Coast (Penouil,

1981) and Tanzania (Lundqvist, 1981) both reveal continuing regional imbalance, but with greater success in countering interpersonal disparity in the latter. Finally, work by Bigsten (1980) in Kenya and Adarkwa (1981) in Ghana confirm patterns of strong and persistent regional inequality.

These examples are not exhaustive, but their evidence does not suggest that sustained trends towards convergence are likely in the near future. Nor is it simply a case of catching up within a similar historical growth process to the point at which maturity and convergence came to characterise the economies of developed countries. The nature of the process of capital accumulation is significantly different (Krebs, 1982). At the very least, it would seem that most underdeveloped countries are starting out with lower per capita incomes and more marked levels of inequality than characterised by advanced countries at supposedly similar stages in their development (Lo and Salih, 1981; Rauch, 1984). Even Williamson (1965) hinted that the experience of the United States and Japan might not be repeated in underdeveloped countries. If regional convergence is to come about, it would seem that government intervention of a more committed kind will be required. Jones's (1982) examination of Venezuela strongly underlines this view. In Malaysia, where authorities have adopted a more determined approach to regional policy, recent assessments indicate some positive effects in the disadvantaged northern states since about 1975 (Kruger, 1982), and some improvement in regional welfare differentials.

Regional development re-appraised

Regional development cannot be treated in separation from the wider tenets of development theory. The main thrust of the attack on the established orthodoxy of polarised growth came from outside the ranks of regional development theorists, amongst the advocates of underdevelopment and dependency theory (see the reviews by Brookfield, 1975; Kitching, 1982; Forbes, 1984). Based on a broader analysis of core-periphery relationships within the international economic system, it was argued by Frank (1967) that the process of capital penetration in the Third World through the expansion of foreign-dominated industry confirmed, not the spread of development, but a reinforced dependence on the metropolitan core. The very act of furthering the integration of the world capitalist system through the kind of benevolent multinational enterprise envisaged by Tinbergen (1977) merely secures control over the means of production at the expense of territorial integrity. The development of the First World is counterposed to the underdevelopment of the Third World.

At the international level, if the condition of backwardness can be attributed to the underdeveloping role of core polarisation and capital penetration rather than inherent locational deficiencies, then the dependent periphery cannot hope to follow the path of development laid down by neoclassical growth theory. Just as dynamic regions in developing countries are linked in an external dependency relationship with the foreign metropolis, so are peripheral regions linked to national cores in satellite relationships (Slater, 1975; Ettema, 1983). The strength of this dependent relationship is such as to limit the likelihood of significant reductions in regional inequality.

Many who had become disenchanted with the established doctrines of regional planning were reluctant to face the political implications of the radical critiques that had raised such fundamental questions about the process of development. They found solace in the call for a redefinition of the meaning of development discussed earlier, a re-appraisal that contributed to the emergence of several approaches to regional development stressing what Friedmann and Weaver (1979) termed territorial integration. Defined as 'the use of an area's resources by its residents to meet their own needs', it challenged the notion of functional integration, 'the narrow exploitation of a region's potentials only because of the role they play in the larger international economy' (Weaver, 1981, p. 93), enshrined in polarised growth.

The concern for territorial integration embodied the call for greater self-reliance, and the promotion of strategies oriented towards meeting basic needs (Ghai et al., 1977; Lee, 1981). The essence of territorial planning is its endogenous character. It thus embraced approaches as diverse as the Chinese model of development, the International Labour Office's World Employment Programme championing the employment generating capacity of new investment, even at the expense of profitability (ILO, 1972; but see also Kitching, 1982), and the agropolitan strategy advanced by Friedmann and Douglass (1978) to secure a more equal and symbiotic relationship between rural and urban areas.

Development from above or below?

The advocacy of territorial approaches to development stems from the desire to contain functional power which is seen to exacerbate social and geographic inequities once freed from the control of territorial authority (Weaver, 1981). To a large extent the functional-territorial dichotomy is mirrored in the opposed planning strategies labelled 'from above', or 'top-down', and 'from below', or 'bottom-up'. The former are the product of neoclassical economic theory, and involve an essentially functional approach to development. They emphasise maximisation of economic growth rates, and strong centralised direction and control of investment. At the heart of such strategies is the notion of polarised growth and the belief that implanted urban-industrial growth centres will most likely restore regional equilibrium. Territorial approaches generally favour strategies of development from below. Where greater self-reliance is urged to break core dominance, satisfaction of basic needs becomes a primary objective. Development programmes are oriented towards problems of poverty and are determined at the lowest feasible territorial scale, motivated and initially controlled from the bottom (Stöhr, 1981).

If the 1960s was the era of the polarised growth paradigm, the period since at least the mid-1970s has seen the abandonment of the universalism of earlier regional development theories for a plurality of approaches (Lo and Salih, 1981). Whether, as Friedmann and Weaver (1979) believe, a new doctrine of regional planning will emerge from some synthesis of these different streams of development thought remains to be seen. They, and a majority of the contributors to Development from above or below? (Stöhr and Taylor, 1981a) nail their flag to the mast of territorial strategies pressing greater rural justice. Rejection of the old order is by no means complete, however. Hansen (1981) continues to see centre-down strategies as more likely to bring about ultimate convergence than untried alternatives, however laudable their intentions. In policy terms, it is clear that growth centres are far from redundant in many Third World countries.

REGIONAL DEVELOPMENT STRATEGIES

In most developing countries where some form of regional development intervention has been attempted, the objective has been to alleviate one or more of three recurring problems: 'regional disparities in development, the excessive size of the national metropoli, and rural-urban inequality' (Gore, 1984, p. 25). Regional policies are fashioned by the perception of these problems, and the mechanisms to solve them as laid down by the prevailing development paradigm. The internal character of the problem regions is less likely to influence the strategy than national commitment and the structure of the regions' administration. Few of the regional typologies considered earlier have influenced official regionalisation for policy purposes, and most attempts to evaluate regional development do not review systematically the programmes employed in each kind of region (e.g. Gilbert, 1974; Morris, 1981a; though see Stöhr (1975) for a notable exception).

It has required a particular combination of circumstances for regional policies to be adopted in most developing countries. The establishment of regional agencies in Latin America, for example, awaited the arrival of regional personalities on the national scene, but owed much to natural disaster forcing government recognition of the regional problem, and then only as unacceptable levels of corruption in relief administration led to scandal and foreign publicity (Stöhr, 1975). Political influence has frequently been the key. Post-independence leaders in many African states fought to secure resources for 'their' regions, creating tensions that contributed, in the case of Nigeria, to civil war. It is no coincidence that the proposed new capital in Ivory Coast is the birthplace of President Houphouet-Boigny. In Mexico, President Miguel Aleman was instrumental in establishing the Papaloapan River Basin Commission on his home ground (Barkin and King, 1970).

Regional goals have increasingly been incorporated into national development plans. Integration of all the regions of a country is seen as an important prerequisite for a general equalisation of incomes (Bigsten, 1980) and as a precaution against political fragmentation. However, for many developing countries, especially in Africa, regional planning continues to be hampered by a lack of resources, effective manpower, and the political will. Bland or disingenuous official statements of tackling regional disparities abound (see, for example, Filani (1981) and Penouil (1981) on development planning in Nigeria and Ivory Coast). In practice, most African nations' planning continues to favour sectoral rather than regional development (Taylor, 1981).

The limitations of space preclude a comprehensive review of regional development policies. It can be argued, as Mabogunje and Faniran's (1977) contributors imply, that any project that improves the income of groups within a region is a form of regional development. Programmes oriented towards agrarian reform and land colonisation in Latin America (see Morris, 1981a, pp. 190-8), and strategies as diverse as the development of new land resources, in-situ rural development and rural urbanisation in Malaysia (Alden and Awang, 1985) are in similar vein. However, they rarely operate within a clearly defined region, still less within a complete regionalisation of the country. They are better classed as agricultural policies, or programmes of infrastructural provision, than regional plans. The colonisation of the Amazon Basin has been motivated less by the desire to develop an underutilised region than by the need to solve population pressures elsewhere, and to avoid land reform, and in any case has done little to reduce regional disparities (Becker, 1985).

The intention here is to concentrate on the three forms of regional development policy that have dominated the spatial planning of Third World countries in the post-war era: river basin programmes; growth-pole strategies; and a range of alternatives reflecting the territorial reorientation of planning initiatives.

River basin programmes

The idea of environmentally sympathetic regional planning (Doornkamp, 1982) - that is, resource development within areas largely defined by physical watersheds - has a certain intuitive appeal. To become a genuine focus of regional development, river basin programmes should adopt a comprehensive approach in which all the resources within a drainage basin are developed together. In practice, financial limitations may impose a phasing of projects which then stall after the initial development of water resources

(Faniran et al., 1977). Where effort is concentrated on particular resources, the outcome may be far removed from integrated regional development. River-basin programmes in the Third World owe much to the example of the Tennessee Valley Authority (TVA). Established in 1933, the TVA started out as a comprehensive programme aimed at power development, flood control, improved navigation, afforestation, agricultural rationalisation, soil conservation, and industrial rehabilitation and diversification. Although subsequently reined in by waning political enthusiasm, notable progress was achieved. Its overriding philosophy, however, shifted from territorially based regional development towards national urban-industrial expansion policy with cheap power the means of promoting economic growth (Friedmann and Weaver, 1979, pp. 73-9).

With technical assistance from TVA engineers it is not surprising that the Damodar Valley Corporation (DVC) in India, launched in 1948, replicated much of the TVA's civil engineering strategy of flood control, irrigation, and power generation. According to Saha (1979, p. 287) 'much of what has been done in the name of planning has actually emerged as by-products of the process of coping with floods'. The physical targets have mostly been met, but they were not conceived within the comprehensive notion of a regional plan. Little is known of the broader effects of the DVC's somewhat disjointed programmes, beyond the clear evidence that cheap electricity has fuelled growth in base industries and heavy engineering to an extent that the region, albeit a few urban locations, will become, if Kirk (1981, p. 196) is correct, 'one of the world's greatest industrial regions of the future'. However, regional development within the valley has focused firmly on urban-industrial advance, and the region was, in any case, one of the two major established core areas of the colonial era.

River-basin programmes as the focus for regional development have been taken further in Mexico than elsewhere. The setting up of River Basin Commissions from 1947 was seen both as a form of integrated water resource investment and as a step in the direction of developing regions away from the Mesa Central. It was conditioned by the belief that they would contribute substantially to the national development effort through improved utilisation of natural resources, as much as to essentially regional development. The political case for co-ordinating government expenditure under semi-independent agencies and for eschewing administrative boundaries and state rivalries in favour of river basins generally made more sense than the underlying economic rationale. The goals varied from little more than flood control and irrigation to a more extensive provision of basic social overhead capital and the development of agricultural, settlement and hydroelectric potential. Even then it proved insufficient for self-sustained regional economic development. The lack of special incentives to encourage the location of new enterprises ensured that there would be no significant shift in industrial investment from existing core areas. It signalled an unwillingness by the Mexican government 'to promote regional development at the expense of national economic growth' (Barkin and King, 1970, p. 246). Furthermore, the emphasis on physical planning gave too little consideration to socio-economic measures. As with developments elsewhere in Latin America (Stöhr, 1975), they lacked the scope for comprehensive planning and failed to bring expected levels of regional development, despite in some cases fulfilling national goals.

In Africa, river basin programmes have been narrowly focused around hydroelectric schemes on, for example, the Volta in Ghana, at Kariba in Zimbabwe, and at Kainji on the Niger. They have involved little more than short-term resettlement measures for displaced populations or the opportunities for irrigated agriculture. Whether the integrated development proposals for the Kafue River Basin in Zambia (Williams, 1979) will amount to more than the regulation of flooding across the Kafue Flats and agricultural investment on the surrounding higher ground remains to be seen. In Nigeria, eleven River Basin Development Authorities (RBDAs) were set up in 1976. On closer inspection, none conformed to the physical parameters of a river basin. Indeed, they were subsequently modified and increased in number to coincide with the administrative boundaries of the 19 states. The designation RBDA has no justification beyond the aim of managing water resources through irrigation in the hope of solving Nigeria's food crisis. In practice, they represent the latest in a growing line of largely unsuccessful attempts to galvanise food production efforts (Okafor, 1985). They cannot be entertained as regional planning mechanisms, however valid agricultural policies might be in the attempt to reduce regional disparities.

A distinction has to be made between those basin

programmes conceived as a means for comprehensive regional development, and those that are merely a regional expression of particular projects. In the latter, the physical fact that a river basin forms a bounded region within which the project is administered does not point to regional planning as such. Disjointed programmes may bring their own rewards but lack the integration necessary to ensure that the regional benefits are more than simply the sum of the project's parts. Further, as Barkin and King (1970, p. 243) argue, basin-wide planning is useful only if 'the interdependence between different investment projects within the basin are substantially greater than those between investments inside the basin and those outside it'. The conclusion must be that frequently they are not.

Growth-pole strategies

The doctrine of polarised growth and the application of growth-pole strategies have dominated Third World regional planning efforts since at least the early 1960s. Although originally framed in non-spatial terms by Perroux, growth poles proved too strong an attraction for those anxious to operationalise the concept within geographic space. Boudeville (1966), for example, identified growth poles with geographical agglomerations of activities, towns possessing a complex of propulsive industries. With his broadening of the definition of growth poles to embrace the idea of a polarised region, growth poles or centres began to look more like the means of explaining urban hierarchies or central place systems (Brookfield, 1975), regardless of how growth is initiated. In a progressive loosening of the Perrouxian concept, therefore, growth poles became urban centres whose economic activities impact on their immediate hinterland, and later, as almost any urban centre witnessing growth (Parr, 1973). This variety in interpretation forced the need to distinguish between growth poles and growth centres, the former tending to be defined in terms of the sectoral activities involved, while the latter identified the form of spatial organisation which embodied the pole (Appalraju and Safier, 1976). For Gilbert (1982, p. 173) the growth centre was, at its simplest, 'an urban complex containing a series of industrial enterprises focused on a dynamic growth industry'.

This spatial transformation has been much discussed

(Darwent, 1969; Hansen, 1972; Moseley, 1974). If one agrees with Lasuen (1969) that there was much that was imprecise, incomplete, partial or quite simply erroneous in Perroux's initial concept, then whatever rationale there was in the original formulation could not help but be weakened in the new growth-centre guise. Nonetheless, the application of growth-pole strategies to the regional problem in many countries came to embody the pre-selection and promotion of development at growth points or centres.

The attraction of the strategy has been its seemingly multifaceted answer to the problems of regional development and core area polarisation. It rested on the belief that certain localities have intrinsic locational advantages for industrial development. Public expenditure would be minimised if population and employment opportunities could be attracted to selected centres of optimal size where the marginal costs of social and economic infrastructure provision are lowest. Spatially concentrated investment in dynamic new industries in existing or newly founded growth centres large enough to realise agglomeration economies was expected to lead, though improved linkage prospects, to wider regional economic growth. Furthermore, these centres were seen not only as more likely to stimulate agricultural and rural activities than distant metropolitan centres, thus establishing the conditions under which accelerated regional economic growth might occur, but also as counter-attractions in their own right to the congested metropolitan core. As a mechanism for reducing both regional imbalance and the problems of the centre, the overall strategy of spatial deconcentration fulfilled national as much as regional interests. This concentrated decentralisation came to represent the extension of the polarised development process into formerly disadvantaged regions.

The progress of growth-pole applications in the Third World has been charted by Appalraju and Safier (1976) and Darkoh (1977), while Conroy (1973) and Richardson and Richardson (1975) have examined the experience of several Latin American countries amongst the first to embrace growth-pole policies. The more recent adoption in Asia (Salih et al., 1978; Kruger, 1982; Misra and Prantilla, 1983) highlights the continued intuitive appeal of a policy that promises to secure a seemingly unlimited set of regional objectives. Although these aims have frequently been vague, unstructured and divorced from any notion of the actual mechanisms of polarised growth, most growth-pole policies have focused on one or more of three main concerns.

The stimulation of economic growth in lagging regions. Following the Perrouxian logic of identifying a dominant sector to form the pole around which the economy clusters, a solution to regional underdevelopment has been sought in locations where the growth mechanism can be triggered either by the exploitation of natural resources for industrial production, or direct investment in key propulsive industries. The experience of Ciudad Guayana in Venezuela, based on electric power generation and heavy metal manufacturing, is perhaps the classic case of its kind. In itself and for the national economy as a whole it has been quite successful, but as with the short-lived Chilean growth-centre strategy of the late 1960s in which the car assembly industry was located in Arica and petrochemicals in Punta Arenas, it has developed as an enclave, draining resources from its hinterland (Stöhr, 1975).

The relief of pressure in congested metropolitan areas. The extreme polarisation of urban-industrial functions is a characteristic feature of the geography of Latin America. Concern for economic decentralisation became an overriding consideration in Brazilian regional development plans during the 1970s (Cunningham, 1976). Growth centres based on the exploitation of agricultural or mineral wealth were expected to be sufficiently attractive to divert migrants who otherwise would have headed for the congested South-East. With programmes too weakly focused to divert investment from the core, they have failed in their primary role of counter-balancing urbanisation. In Peru, following the military takeover in 1968 plans were drawn up that aimed both to utilise the country's natural resources through the establishment of integrated propulsive industries and to create points of counter-attraction to the overconcentration of industrial development in Lima-Callao. Although the policies embodied in the zones of concentrated action set up in 1971 contained no specific reference to growth centres, it was clear that industrial decentralisation to designated centres remained a key regional objective. The policy may have lasted longer than Colombia's similar measures to counteract the growth of Bogota, but the drive and political will had evaporated by the mid-1970s as Lima's role in decision making was powerfully asserted (Hilhorst, 1981).

In East Africa both Kenya and Tanzania adopted growth-centre policies to counter their problems of primate city concentration. Tanzania's flirtation in the 1970s with nine up-country growth centres promoted to aid industrial and urban decentralisation from Dar-es-Salaam was all too brief. Kenya, however, remains committed to spreading urbanisation though without seriously restricting growth in Nairobi. Concern over congestion in the capital city has led several countries (e.g. Malawi, Nigeria, Tanzania, Ivory Coast and Pakistan) to follow Brazil's example in planning, if not fully implementing, a relocation of the capital city. A lack of funds, fluctuating political support, and the reluctance of administrative staff and industrial enterprises to locate in the new growth centres does not bode well.

The development of the urban hierarchy. The urban hierarchy in most developing countries reveals significant gaps, mostly at intermediate levels. The creation of growth centres to occupy this middle ground might help to draw investment from the primate city while establishing functional poles at the regional level. For Johnson (1970, p. 219) the solution to the parasitism of the city-size distribution was to be found 'in selecting rural sites that have better than average prospects of becoming future agro-urban communities'. These rural growth centres would be the means 'to bring country-dwelling people into a market economy in a more complete and truly functional way'. His promotion of the commercialisation of small-scale peasant agriculture through the creation of a network of small market towns to stimulate demand and provide geographically accessible marketing incentives, is designed to achieve the 'fuller participation by all groups of people' and to 'elicit the vital forces and latent creativity that reside in an area's human resources' (p. 247).

This approach represented a shift from the urban-industrial growth-pole orientation of the 1960s to a more general concern in the 1970s for national systems of growth centres that might include all levels in the hierarchy down to rural communities. In Kenya, the 1970-4 Development Plan not only designated seven major urban growth centres to counter Nairobi and Mombasa, but also a range of urban, rural, market and local centres in descending order of size and importance. A more general urban development strategy has since emerged (Richardson, 1980) in which the principal aim has been to fill the gaps in the service provision deemed appropriate for each level in the hierarchy (McClintock, 1985). Since the mid-1970s, growth centres have featured among Malaysia's varied strategies to promote regional development (Kruger, 1982). The stated objectives include familiar themes such as the development of depressed areas and the relief of metropolitan congestion. Their primary function has been to encourage rural-urban migration to intermediate centres for economic, social and political reasons. It is perhaps too soon to expect much progress, though according to Alden and Awang (1985), the centres are already failing to meet their targets.

Clearly, some of these growth-centre policies bear only a passing resemblance to the original growth-pole notion. Few have met with much success, though as Gore (1984, p. 118) has stressed, 'the empirical evaluation of growth pole strategies is far from convincing', there being no satisfactory method for measuring their impact. Although criticism can often be countered by claiming inappropriate application or too hasty abandonment, several factors account for the demise in the 1970s of a regional development strategy embraced with such enthusiasm a decade earlier.

Criticisms include, firstly, an ideological rejection on the grounds that growth-pole strategies imply an alignment of the national economy with a world pattern of dominant poles of growth. This, it is claimed, is untenable, for integration in the world economy through the actions of multinational corporations is founded on an unequal and essentially exploitative base (Lo and Salih, 1981). The desire to exploit resources gives some credence to the argument, yet it has been the reluctance of foreign companies to risk locations beyond the existing core regions that is both the reality of metropolitan congestion and the poor performance of growth centres.

Secondly, theoretical objections have been raised over the uncritical transfer of a theory originally expressed within the context of Western Europe and now applied to situations in the Third World where the urban and industrial structures and the conditions of entrepreneurial activity are wholly different. A range of cultural, economic, political and administrative barriers prevent any automatic transfer of spatial strategies (Gilbert, 1976). Richardsón and Richardson (1975), for example, cite 16 factors which make an unmodified transference of growth-pole theory inappropriate to Latin America. Thirdly, political objections rest on the very nature of spatially selective policy. A narrow selection of growth centres that may be economically sound leaves politicians in unfavoured regions demanding similar preferential treatment. A proliferation of centres to avoid both dissent and the risk of committing too much to too few locations that might fail and thus represent a misallocation of resources may weaken the likelihood of achieving propulsive forces within the region. In short, through under-investment it becomes a self-defeating strategy.

Fourthly, a number of practical or technical considerations have limited the effectiveness of growth centres. There has been no concensus on the desired or optimal size of a growth centre, nor on how many should be designated. The literature is replete with policies or proposals of growth centres from villages to the scale of a major metropolis. If the chosen locations are too small or in the case of groups of towns too disparate in nature to become a coherent integrated economic unit, they will, as in Argentina (Morris, 1981a), fail to generate the propulsive momentum required. Too large, they will continue to reflect the existing highly concentrated spatial patterns of growth. There have also been problems with the propulsive industries established to create the development impulses to drive the regional economy. Being too capital-intensive, they generated insufficient employment opportunities. Furthermore, they failed to transmit growth throughout their region. Ciudad Guayana and Arica became islands of relative prosperity in seas of underdevelopment. The poorly developed regional urban structure clearly hindered spread effects, as it did in Malaysia (Robinson and Salih, 1971) and Colombia (Gilbert, 1975), while any multiplier effects were externally derived. The very idea that modern technological industry can be decentralised to the benefit of rural areas is somewhat fanciful (Lo and Salih, 1981), unless accompanied by other, more fundamental changes in the agricultural economy (Gilbert, 1982). Practical problems of application are linked to political considerations. Growth centres have been conceived within and constrained by a set of national planning objectives that ultimately subordinate regional plans to the vested interests of the centre and ensure no more than passive implementation of policies that threaten to divert resources. In Hilhorst's (1981, p. 431) words, 'Lima became the centre, and almost nothing could be done elsewhere unless Lima had agreed'.

The weakness of policy measures is also the first line of defence for the still optimistic advocates of growth poles in regional development. How can they be written off when there have been so few governments willing to implement and finance policy with conviction? If the centres have been too small or too many; if growth centre strategies have been isolated from other development policies; and if there has been no attempt to curb central area growth, is success to be expected? Hansen (1981) argues strongly that growth-pole policies have simply not been tried properly. For this reason, Gilbert (1982) believes they should not be totally discarded. They have continued relevance in Mexico and Venezuela, while Saudi Arabia (Felemban, 1980), Ecuador (Morris, 1981b) and several South-East Asian countries (Misra and Prantilla, 1983) appear to have embarked on growth-pole strategies just as the bulk of planning opinion has moved on. A second line of defence expounded by Richardson (1976) is that all too often, growth-centre strategies were abandoned within a decade when they needed a 15-25 year time horizon. Net positive spillovers should not be expected around a growth pole in the early years of implementation, though Appalraju and Safier (1976) believe that some spatial modifications should have occurred. One suspects, with Stöhr and Taylor (1981b), that development policies which see little or no reduction in disparities after a decade or two are ineffective in political and socio-economic terms.

In the event, detailed analysis of the varied criticisms was sidestepped as new and alternative theories emerged, implying as Gore (1984, p. 121) has put it, 'that the strategy will not work because the assumptions about the regional development process on which it is based are wrong'. For Riddell (1985) the fact that such eminent former proponents as Friedmann have now rejected growth centres is sufficient proof of their conceptual and practical inadequacy.

Alternative approaches to regional planning

The range of alternative strategies that grew from the disenchantment with established regional planning doctrines has been termed neo-populist by Kitching (1982), and categorised as a re-affirmation of territorial integrity by Friedmann and Weaver (1979). They are well chronicled by Gore (1984, pp. 146-71), who links them in a shared 're-evaluation of what planning "development" means and what the general goals of a national development strategy should be' (p. 145). These new concerns include 'preserving equality and community, fostering the evolution of small-scale enterprise, promoting peasant agriculture, engaging "the people" in the development process, and removing the bias towards big cities, large-scale industries, and centralised forms of organization' (p. 149).

The problem of rural development within regional planning had been explored by Johnson (1970). His examination of the spatial organisation of developing countries based on his Indian planning experience was as much a beginning to the territorial alternative as a final twist to the growth-pole concept. Rondinelli and others working for the United States Agency for International Development in the mid-1970s developed Johnson's ideas as a basis for regional policies that would promote greater social and geographic equity (Rondinelli and Ruddle, 1978). They argue that market towns, rural service centres and intermediate cities, and the services, facilities, infrastructure and productive activities associated with them, can and do accelerate rural development, but that in most Third World countries the settlement system is too poorly articulated and insufficiently integrated to alleviate rural poverty. To achieve integrated regional development, Rondinelli (1985), p. 435) considers it as crucial 'to place investments in a pattern of "decentralized concentration", that is, in strategically located settlements that are dispersed geographically'.

What is proposed is a method of regional analysis based on the promotion of urban functions in rural development (UFRD). Data collected on regional resources, the settlement system, existing functional distributions, and the linkages among settlements can reveal areas with poor access to town-based functions, and gaps in the distribution of specific services, facilities and infrastructure when mapped and analysed. The ultimate objective of UFRD is not to produce comprehensive regional plans, but, through the provision of regional profiles of the distribution of functions and resources, to inject into sectoral and economic planning a spatial dimension that could improve access to needed services, facilities and resources for the rural poor. The methodology has been tested in the Philippines, Burkino Faso, Malawi, Ecuador, and Bolivia (see, especially, Rondinelli and Evans, 1983). It is not yet clear whether the proposed transformation of communities through a well-articulated spatial structure of settlements interacting through physical, economic, technological, social, administrative and organisational linkages can be properly financed and implemented, let alone lead to more equitable growth.

There is little here, however, to accommodate the dependency theorist who would argue that greater rural-urban integration through the extension of dendritic market systems and better articulated spatial structures merely serves to further satellite status. Disengagement from the world economy as advocated by Slater (1975) has not been much explored by regional planners, but the strategy of selective regional closure advanced by Stμhr and Tμdtling (1977) might be more feasible. Selective regional closure through supply- and demand-side policies as well as political and administrative decentralisation might be managed by controlling the raw material or commodity transfers which drain investible surplus from the countryside. Devolution of some of the decision-making powers on commodity or factor transfers (capital, technology) currently 'vested in functionally organised (vertical) units back to territorially organised (horizontal) units at different spatial scales' (Stμhr and Tμdtling, 1977, p. 35) may avoid the underemployment of regional resources, natural or human, and major external dependencies. The aim is to reduce leakages - the transfer of resources for urban reinvestment - and create self-sufficient, self-reliant territorial units that minimise backwash effects from core areas while locking in whatever development impulses spread from outside (Stμhr, 1981; Lo and Salih, 1981).

Selective spatial closure is one of several conditions for the agropolitan development strategy proposed by Friedmann and Douglass (1978), extended by Friedmann and Weaver (1979) (Table 6.1), and given a political-territorial framework for rural development in Asia by Friedmann (1981). The objectives of development are stated in terms of self-reliance, expanding employment opportunities, appropriate environmental management, satisfaction of basic needs, and balanced rural-urban development. For Friedmann (1981, p. 248), 'rural development must be

Table 6.1: The conditions for agropolitan development

Basic conditions	1. Selective territorial closure 2. Communalisation of productive wealth 3. Equalisation of access to the bases for the accumulation of social power.
Territorial framework	1. Agropolitan districts (15,000-60,000 people) 2. Agropolitan neighbourhoods in cities
Economic expansion	1. Diversifying the territorial economy 2. Maximum physical development constrained by the need for conservation 3. Expanding regional and interregional (domestic) markets 4. Following principles of self-finance 5. Promoting social learning
Role of the state	1. Agropolitan district is self-governing 2. The central state is strong

Source: Forbes, 1984, p. 133; based on Friedmann and Weaver, 1979, pp. 194-201

centrally guided but locally based'. Thus, there should be a substantial devolution of power to achieve self-government over a territory, the agropolitan district,

> large enough to meet most of the basic needs of the population out of its own resources ... (yet) small enough so that the entire population of the area might have reasonable physical access to the center for political decision-making, planning and administration.

Of overwhelming importance is the need to empower the people, not, according to Friedmann, by smashing the power of existing elites, as in China, but by enlarging the local political assembly to allow for the expression and reconciliation of a variety of class and functional interests.

Agropolitan development is perhaps the most

coherently presented alternative strategy for regional economic planning to have emerged. Accordingly, it has received more attention than other less focused strategies of development 'from below', in which rural questions feature more prominently. Not even agropolitan strategies have, as yet, been much subjected to planning policy. Conyers (1985) has commented briefly on district approaches to rural development in countries as diverse as India, Sri Lanka, Bangladesh, South Korea and Malaysia, and in four African countries: Kenya, Tanzania, Botswana and Lesotho. In India, the idea of decentralised planning dates back to the Community Development programmes of the 1950s, and the promotion of multi-level planning in the early 1960s. District and block planning re-emerged in the 1970s as a means of reviving the concept of planning 'from below' (Misra and Natraj, 1981), but the methodology of multi-level planning is still at a formative stage (Sundaram, 1983). In South-East Asia, several countries have pursued policies which seem related to agropolitan development but without devolution of political power being seriously affected, or with more than lip-service given to land reform, regarded by Friedmann and Douglass (1978) as a quintessential step in a practical programme of implementation.

It is too soon to know whether development 'from below' by means of strategies adopted under even the most appropriate conditions (see, for example, Stöhr, 1981) will be more successful in reducing regional disparities than increased urban-ward migration from the underdeveloped rural periphery. So long as such programmes are underfunded and compete for resources and political support with large-scale, urban-industrial projects they are unlikely to prosper. If they remain subjected to remote decision making and top-down planning the prospects of successful implementation are poor. Most developing countries do not possess the political structures at top or bottom with the will or vision to press such policies. Centralised political elites are hostile to a devolution of power (Hansen, 1981), while 'grass-roots' planning has been frustrated by the drain of local talent and political leadership to the core areas (Kirk, 1981). The idea of self-governing units, and indeed the whole organisational dimension, is but weakly developed. Above all, the agropolitan approach to regional development planning has been criticised as presenting too idealistic a conception of society (Conyers, 1985), and in danger of becoming as nebulous as the growth centres it sought to replace (Gore, 1984).

CONCLUSION

This chapter began with a presentation of the major theoretical and philosophical shifts in regional development thinking in the post-war era, and progressed to an account of the most significant types of regional development policy employed in the Third World. There has been no shortage of regional planning: what has been lacking is a genuine determination to go beyond well-meaning declarations of intent, and to place regional goals above national ones. Regional plans aimed at reducing spatial inequalities have too frequently been undermined by national policies that have further concentrated investment in core areas. If the nature of the development model or the style of administration is at fault, then regional policy can do little to resolve regional problems. One can agree with Gilbert (1982, p. 194) that 'to bring about more balanced regional development and a fairer share of wealth ... requires a national commitment to a modified development programme in which equality occupies a much more dominant position'. To date, there have been few spatial policies which have been at all effective in helping the poor in the poorest regions.

It is beyond the scope of this chapter to make recommendations for the kinds of policy that should be employed to tackle the regional problems of developing countries. The degree of generalisation across the varied circumstances of Africa, Asia, and Latin America that this would entail urges caution, for no one model is universally applicable. The lesson from growth poles is that no single strategy can solve all manner of regional problems. A more particular attempt to suggest directions has been Morris's (1981a) outline of policies relevant to each major type of region found in Latin America. There is much to be said for his general call for replacing the predominantly single-purpose, single-region bodies that have characterised much of Latin American regional development with comprehensive multi-regional planning, involving national agencies prepared to promote policies for the weaker regions that 'support truly regional (that is, intra-regional) development rather than the development of regions to fulfil national aims' (p. 202), whatever the short-term disadvantages.

It would be wrong to imply that the paradigm shift towards territorial integration and development 'from below' has been complete. For Webster (1980) the new models merely represent a reshuffling of old paradigms. A significant section of intellectual opinion may have been converted, and although these ideas have filtered through to hard-pressed planning departments in developing countries, their realisation in planning practice is as yet only partial. It is not unreasonable to suggest that regional development in the Third World is in a state of some confusion. Much has been tried, albeit not fully or properly implemented, and the track record is poor. Regional planning has tended to fall into disrepute (Ekistics, 1980), but with the call for a return to basic principles such as the rights to equal opportunities and a voice in decision making, one can discern a shift in the focus of regional development over the past decade. There has been a move away from macro-level regional planning towards a regionally based, comprehensively planned project policy (Rambousek, 1982) in which planning and evaluation is based on more soundly researched data. Like Rondinelli's UFRD methodology, regional development planning would become an analytical tool to identify the goals of the people who live in the region.

It may be that what is created in the closing years of the twentieth century is less of a regional development policy as normally defined, as a plurality of approaches that include development projects that have regional effects. Through a firmer commitment to rural development and small-scale industrialisation, they might promote the kind of autonomy, self-reliance and decentralisation that selective regional closure and agropolitan development advocate. Many of the alternatives are philosophically appealing but remain vague and unattractive to those planners and their political masters who cherish the publicity of high-profile and lavishly funded projects without concern or responsibility for their almost inevitable failure to bring 'development' to more than a minority. The reality is that few Third World governments seem genuinely prepared to weaken their influence or control by decentralising power to semi-autonomous bodies in unseen quarters of the periphery.

REFERENCES

Adarkwa, K. (1981) A spatio-temporal study of regional inequalities in Ghana. African Urban Notes, 11, 39-64

Alden, J.D. and Awang, A.H. (1985) Regional development planning in Malaysia. Regional Studies, 19, 495-508

Alonso, W. (1968) Urban and regional imbalances in economic development. Economic Development and Cultural Change, 17, 1-14

Appalraju, J. and Safier, M. (1976) Growth-centre strategies in less-developed countries. In A. Gilbert (ed.), Development planning and spatial structure, Wiley, London, pp. 143-67

Barkin, D. and King, T. (1970) Regional economic development: the river basin approach in Mexico. Cambridge University Press, Cambridge

Becker, B.K. (1985) The state and the land question of the frontier - a geopolitical perspective. Geojournal, 11, 7-15

Bigsten, A. (1980) Regional inequality and development: a case study of Kenya. Gower, Farnborough

Blaikie, P. et al. (1981) Nepal: the crisis of regional planning in a double dependent periphery. In W.B. Stöhr and D.R.F. Taylor (eds), Development from above or below?, Wiley, Chichester, pp. 231-58

Boisier, S. (1981) Chile: continuity and change - variations of centre-down strategies under different political regimes. In W.B. Stöhr and D.R.F. Taylor (eds), Development from above or below?, Wiley, Chichester, pp. 401-26

Borts, G.H. and Stein, J.L. (1962) Economic growth in a free economy. Columbia University Press, New York

Boudeville, J.R´ (1966) Problems of regional economic planning. Edinburgh University Press, Edinburgh

Brookfield, H.C. (1975) Interdependent development. Methuen, London

Brugger, E.A. (1982) Regional policy in Costa Rica: the problems of implementation. Geoforum, 13, 177-92

Conroy, M.E. (1973) Rejection of growth center strategy in Latin American regional development planning. Land Economics, 49, 371-80

Conyers, D. (1985) Rural regional planning: towards an operational theory. Progress in Planning, 23, 1-66

Cunningham, S.M. (1976) Planning Brazilian regional development during the 1970s. Geography, 61, 163-7

Darkoh, M.B.H. (1977) Growth poles and growth centres with special reference to developing countries - a critique. Journal of Tropical Geography, 44, 12-22

Darwent, D.F. (1969) Growth poles and growth centres in regional planning: a review. Environment and Planning, 1, 5-31

Doornkamp, J.C. (1982) The physical basis for planning in the Third World. IV: regional planning. Third World Planning Review, 4, 111-18

Douglass, M. (1981) Thailand: territorial dissolution and alternative regional development for the Central Plains. In W.B. Stöhr and D.R.F. Taylor (eds), Development from above or below?, Wiley, Chichester, pp. 183-208

Ekistics (1980) The editor's page. Ekistics, 47 (284), 311

El-Shakhs, S. (1972) City systems, primacy and development. Journal of Developing Areas, 7, 11-36

Enders, W.T. (1980) Regional disparities in industrial growth in Brazil. Economic Geography, 56, 300-10

Ettema, W. (1983) The centre-periphery perspective in development geography. Tijdschrift voor Economische en Sociale Geografie, 74, 107-19

Faniran, A. et al., (1977) Water resources development process and design: case study of the Oshun River catchment. In A.L. Mabogunje and A. Faniran (eds), Regional planning and national development in tropical Africa, Ibadan University Press, Ibadan, pp. 202-12

Felemban, A.H.S. (1980) Regional development planning as an essential part of regional development: a case study in Saudi Arabia. Ekistics, 47 (284), 360-8

Filani, M.O. (1981) Nigeria: the need to modify centre-down development planning. In W.B. Stöhr and D.R.F. Taylor (eds), Development from above or below?, Wiley, Chichester, 283-304

Forbes, D.K. (1984) The geography of underdevelopment: a critical survey. Croom Helm, London

Frank, A.G. (1967) Capitalism and underdevelopment in Latin America. Monthly Review Press, New York

Friedmann, J. (1966) Regional development policy: a case study of Venezuela. MIT Press, Cambridge, MA

------ (1972) A general theory of polarized development. In N.M. Hansen (ed), Growth centres in regional economic development, Free Press, New York, pp. 82-107

------ (1981) The active community: toward a political-

territorial framework for rural development in Asia. Economic Development and Cultural Change, 29, 235-61
------ and Alonso, W. (eds) (1964) Regional development and planning: a reader. MIT Press, Cambridge, MA
------ and Douglass, M. (1978) Agropolitan development: towards a new strategy for regional planning in Asia. In F. Lo and K. Salih (eds), Growth pole strategy and regional development policy: Asian experiences and alternative strategies, Pergamon, Oxford, pp. 163-92
------ and Weaver, C. (1979) Territory and function: the evolution of regional planning. Edward Arnold, London
Ghai, D.P. et al. (eds) (1977) The basic needs approach to development. International Labour Office, Geneva
Gilbert, A. (1974) Latin America development: a geographical perspective. Penguin, Harmondsworth
------ (1975) A note on the incidence of development in the vicinity of a growth centre. Regional Studies, 9, 325-33
------ (1976) Introduction. In A. Gilbert (ed.), Development planning and spatial structure, Wiley, London, pp. 1-19
------ (1982) Urban and regional systems: a suitable case for treatment? In A. Gilbert and J. Gugler, Cities, poverty and development: urbanization in the Third World, Oxford University Press, Oxford, pp. 162-97
------ and Goodman, D. (1976) Regional income disparities and economic development: a critique. In A. Gilbert (ed.), Development planning and spatial structure, Wiley, London, pp. 113-41
Gore, C. (1984) Regions in question: space, development theory and regional policy. Methuen, London
Grigg, D. (1967) Regions, models and classes. In R.J. Chorley and P. Haggett (eds), Models in geography, Methuen, London, pp. 461-509
Haddad, P.R. (1981) Brazil: economic efficiency and the disintegration of peripheral regions. In W.B. Stöhr and D.R.F. Taylor (eds), Development from above or below?, Wiley, Chichester, pp. 379-400
Hall, C. (1984) Regional inequalities in well-being in Costa Rica. Geographical Review, 74, 48-62
Haller, A.O. (1982) A socio-economic regionalization of Brazil. Geographical Review, 72, 450-64
Hansen, N.M. (ed.) (1972) Growth centers in regional economic development. The Free Press, New York
------ (1981) Development from above: the centre-down development paradigm. In W.B. Stöhr and D.R.F.

Taylor (eds), Development from above or below?, Wiley, Chichester, pp. 15-38

Hilhorst, J.G.M. (1981) Peru: regional planning 1968-78: frustrated bottom-up aspirations in a technocratic military setting. In W.B. Stöhr and D.R.F. Taylor (eds), Development from above or below?, Wiley, Chichester, pp. 427-50

Hirschman, A.O. (1958) The strategy of economic development, Yale University Press, New Haven

International Labour Office (1972) Employment, incomes, and equality: a strategy for increasing productive employment in Kenya. ILO, Geneva

Isard, W. (1956) Location and space economy. Wiley, New York

Johnson, E.A.J. (1970) The organization of space in developing countries. Harvard University Press, Cambridge

Jones, R.C. (1982) Regional income inequalities and government investment in Venezuela. Journal of Developing Areas, 16, 373-90

Keeble, D.E. (1967) Models of economic development. In R.J. Chorley and P. Haggett (eds), Models in geography, Methuen, London, pp. 243-302

Kirk, W. (1981) Cores and peripheries: the problems of regional inequality in the development of southern Asia. Geography, 66, 188-201

Kitching, G. (1982) Development and underdevelopment in historical perspective. Methuen, London

Krebs, G. (1982) Regional inequalities during the process of national economic development: a critical approach. Geoforum, 13, 71-81

Kruger, K. (1982) Regional policy in Malaysia. Geoforum, 13, 133-49

Lasuen, J.R´ (1969) On growth poles. Urban Studies, 6, 137-61

Lee, E. (1981) Basic needs strategies: a frustrated response to development from below? In W.B. Stöhr and D.R.F. Taylor (eds), Development from above or below?, Wiley, Chichester, pp. 107-22

Lo, F. and Salih, K. (1981) Growth poles, agropolitan development, and polarization reversal: the debate and search for alternatives. In W.B. Stöhr and D.R.F. Taylor (eds), Development from above or below?, Wiley, Chichester, pp. 123-52

Lundqvist, J. (1981) Tanzania: socialist ideology,

bureaucratic reality and development from below. In W.B. Stöhr and D.R.F. Taylor (eds), Development from above or below?, Wiley, Chichester, pp. 329-49

Mabogunje, A.L. and Faniran, A. (eds) (1977) Regional planning and national development in tropical Africa. Ibadan University Press, Ibadan

McClintock, H. (1985) Some African experience with regional planning implementation with particular reference to Malawi (and Kenya). Public Administration and Development, 5, 289-308

Misra, R.P. and Natraj, V.K. (1981) India: blending central and grass-roots planning. In W.B. Stöhr and D.R.F. Taylor (eds), Development from above or below?, Wiley, Chichester, pp. 259-79

------ and Prantilla, E.B. (1983) Emerging trends in regional development planning: the Southeast Asian experiences. In L. Chatterjee and P. Nijkamp (eds), Urban and regional policy analysis in developing countries, Gower, Aldershot, pp. 19-41

Morris, A.S. (1981a) Latin America: economic development and regional differentiation. Hutchinson, London

------ (1981b) Spatial and sectoral bias in regional development: Ecuador. Tijdschrift voor Economische en Sociale Geografie, 72, 279-87

Moseley, M. (1974) Growth centres in spatial planning. Pergamon, Oxford

Mountjoy, A.B´ (1984) Core-periphery, government and multinational: a Papua New Guinea example. Geography, 69, 234-43

Myrdal, G. (1957) Economic theory and underdeveloped regions. Duckworth, London

Okafor, F.C. (1985) River basin management and food crisis in Nigeria. Geoforum, 16, 413-21

Oyebanji, J.O. (1982) Regional shifts in Nigerian manufacturing. Urban Studies, 19, 261-75

Parr, J.B. (1973) Growth poles, regional development and central place theory. Papers and Proceedings of the Regional Science Association, 31, 172-212

Pedersen, P.O. (1975) Urban-regional development in South America: a process of diffusion and integration. Mouton, The Hague

Penouil, M. (1981) Ivory Coast: an adaptive centre-down approach in transition. In W.B. Stöhr and D.R.F. Taylor (eds), Development from above or below?, Wiley, Chichester, 305-28

Perroux, F. (1950) Economic space, theory and applications. Quarterly Journal of Economics, 64, 89-104
------ (1955) Note sur la notion des pôles de croissance. Cahiers de l'Institute de Science Economique Appliqué, Série D, no. 8. Reprinted in I. Livingstone (ed.) (1971), Economic policy for development, Harmondsworth, Penguin, pp. 278-89
Rambousek, W.H. (1982) Regional policy in Cameroon: the case of planning without facts. Geoforum, 13, 163-75
Rauch, T. (1982) Regional policy in Nigeria. Geoforum, 13, 151-61
------ (1984) An accumulation theory approach to the explanation of regional disparities in underdeveloped countries. Geoforum, 15, 209-29
Richardson, H.W. (1973) Regional growth theory. Macmillan, London
------ (1976) Growth pole spillovers: the dynamics of backwash and spread. Regional Studies, 10, 1-9
------ (1980) An urban development strategy for Kenya. Journal of Developing Areas, 15, 97-118
------ and Richardson, M. (1975) The relevance of growth center strategies to Latin America. Economic Geography, 51, 163-78
Riddell, R. (1985) Regional development policy: the struggle for rural progress in low-income nations. Gower, Aldershot
Robinson, G. and Salih, K.B. (1971) The spread of development around Kuala Lumpur: a methodology for an exploratory test of some assumptions of the growth pole model. Regional Studies, 5, 303-14
Rondinelli, D.A. (1980) Balanced urbanization, regional integration and development planning in Asia. Ekistics, 47 (284), 331-9
------ (1985) Equity, growth, and development: regional analysis in developing countries. Journal of the American Planning Association, 51, 434-48
------ and Evans, H. (1983) Integrated regional development planning: linking urban centres and rural areas in Bolivia. World Development, 11, 31-53
------ and Ruddle, K. (1978) Urbanization and rural development: a spatial policy for equitable growth. Praeger, New York
Rostow, W.W. (1960) The stages of economic growth: a non-Communist manifesto. Cambridge University Press, Cambridge

Saha, S.K. (1979) River-basin planning in the Damodar valley of India. Geographical Review, 69, 273-87

Salih, K. et al. (1978) Decentralization policy, growth pole approach, and resource frontier development: a synthesis of the response in four South-east Asian countries. In F. Lo and K. Salih (eds), Growth pole strategy and regional development policy: Asian experience and alternative approaches, Pergamon, Oxford, pp. 79-119

Seers, D. (1969) The meaning of development. International Development Review, 11, 2-6

------ (1977) The new meaning of development. International Development Review, 19, 2-7

Slater, D. (1975) Underdevelopment and spatial inequality. Approaches to the problems of regional planning in the Third World. Progress in Planning, 4, 97-167

Stöhr, W.B. (1975) Regional development experiences and prospects in Latin America. Mouton, The Hague

------ (1981) Development from below: the bottom-up and periphery-inward development paradigm. In W.B. Stöhr and D.R.F. Taylor (eds), Development from above or below?, Wiley, Chichester, pp. 39-72

------ and Taylor, D.R.F. (eds) (1981a) Development from above or below? The dialectics of regional planning in developing countries. Wiley, Chichester

------ (1981b) Introduction. In W.B. Stöhr and D.R.F. Taylor (eds), Development from above or below?, Wiley, Chichester, pp. 1-12

Stöhr, W.B. and Tödtling, F. (1977) Spatial equity, some antitheses to current regional development strategy. Papers and Proceedings of the Regional Science Association, 38, 33-53

Sundaram, K.V. (1983) Multilevel planning: the Indian experience. In L. Chatterjee and P. Nijkamp (eds), Urban and regional policy analysis in developing countries, Gower, Aldershot, pp. 43-53

Taylor, D.R.F. (1981) Some observations on theory and practice in regional development in Africa. In A. Kinklinski (ed.), Polarized development and regional policies: tribute to Jacques Boudeville, Mouton, The Hague, pp. 327-39

Tinbergen, J. (co-ordinator) (1977) RIO: reshaping the international order. A report to the Club of Rome. Longman, London

Weaver, C. (1981) Development theory and the regional

question: a critique of spatial planning and its detractors. In W.B. Stöhr and D.R.F. Taylor (eds), Development from above or below?, Wiley, Chichester, pp. 73-105

Webster, D.R. (1980) Regional development planning: persistent paradigm or new consensus. Ekistics, 47 (284), 343-5

Williams, G.J. (1979) A future for the Flats. Standard Chartered Review, April, 2-6

Williamson, J.G. (1965) Regional inequality and the process of national development: a description of the patterns. Economic Development and Cultural Change, 13, 3-45

Chapter Seven

POWER, POLITICS AND SOCIETY

R.B. Potter and J.A. Binns

INTRODUCTION

It is a reflection of just how much ruling paradigms have changed in geography as a whole, and in the geographical study of developing areas in particular, that it hardly seems necessary to begin a chapter such as this with a spirited justification of the need to study the roles played by power, politics and social organisation in the development process. Even in a matter of ten or fifteen years ago, this would not have been the case. In the early 1970s, whilst few would have been surprised to see included chapters on agriculture, population and demography, economics, housing and regional development in a book on Third World geography, eyebrows would undoubtedly have been raised in some quarters at one dealing specifically with the general topic of power, politics and society. Quite simply, at that juncture, many would have taken the view that such matters are the province of the political scientist, sociologist or the social and economic historian, and most certainly not that of the geographer.

During the 1960s and 1970s, the natural affinities of geography were substantially with the natural sciences, statistics and increasingly, psychology. The quantitative revolution, which had emphasised empiricism and positivism as the principal modes of academic enquiry, had resulted in the positing of endless models and theories purporting to explain the generalities of the way in which the world was structured, along with a plethora of detailed spatial descriptions. A further characteristic of this work was its tendency to focus on micro-spatial-scale patterns and, less frequently, processes operating at this scale. Work at higher levels of spatial resolution was generally noticeable by virtue of its absence. The implication was perhaps that the scope and intentions of such work were akin to the overthrown paradigm of regional geography.

It seems reasonable to suppose that the disenchantment of a growing band of practitioners at the level of explanation and understanding achieved by such spatial myopia led to the quite rapid reorientation of geography as a whole during the decade of the 1970s. Initially, this movement took the subject towards behavioural, and then humanistic perspectives, and secondly - and more saliently in the context of the present discussion - towards radical and structuralist approaches (Gregory, 1978; Johnston, 1979, 1983; see also Forbes, 1984, Chapter 4). Whilst it is distractingly difficult and perhaps not too productive to attribute paradigm shifts to particular sub-specialisms within any discipline it is undoubtedly the case that historical geography, urban social geography, political geography and the geographical study of development were in the vanguard of these changes. In the ensuing period, radical, and in particular Marxist and neo-Marxist, approaches have become increasingly popular and dominant. The appearance in 1969 of Antipode: A Radical Journal of Geography was a watershed in this respect (see also Peet, 1977a, 1977b). The publication of Harvey's Social justice and the city, and Castells's The urban question, in 1973 and 1977 respectively, was highly pertinent in this regard. At the present time, therefore, the cognate disciplines of geography have changed and are very much sociology, politics and political economy, and there have been heavy infusions of social theory, including the work of Giddens (1979, 1981, 1982a, 1982b). These moves, which may be counterpointed with the development of a 'critical' approach in the social sciences taken as a whole, have brought considerations of social structure, power relations and political science very much into the centre of the geographer's remit.

However, it is possible to suggest that hand in hand with these paradigm changes, which have served to stimulate a concern with such fundamental socio-political issues, a change in the scale at which environmental research is carried out has taken place. The structuralist approach stresses the need to consider what generally are unseen processes. As such, recent work has largely led to the rejection of space as a causal variable in social change and structure, a point that will be developed more fully in the next section. It has also been associated with a questioning of the academic worth of descriptions of spatial structure and behavioural decision-making, seeing these as the outward signs of a spatial fetish which has led to the intellectual desert of the 'spatial fallacy' (see, for instance, Reiser, 1973; Cox, 1981).

There has certainly been a tendency for macro-spatial, global analyses, and even aspatial analyses, to dominate much of human geography, including development studies, since the late 1970s. Approaches owe much to dependency theories of underdevelopment along the lines suggested by Frank (1967), and world-system views such as that of Wallerstein (1974, 1979, 1980). Likewise, much work has focused on the extraction and articulation of social surplus product, including, for example, the consideration of the relation between major societal epochs and stages of urban-industrial development (Johnston, 1980). At the national scale, the formerly neglected topic of the nature of the state and its role in promoting and perpetuating underdevelopment and inequalities in income, housing, and other socio-economic arenas, has become an important focus of investigation. Johnston (1982), for example, has argued that despite the growth of interest in political geography through the 1960s and 1970s, during this time the subject lacked a sensible treatment of the very element it sought to understand - namely, the state. Whilst this lack of attention has increasingly been put to rights with respect to topics such as urbanisation in cities of the developed world (Castells, 1978; Saunders, 1979, 1982; Dear and Scott, 1981; Short, 1984), as Gilbert and Healey (1985, p. 4) have recently observed, few studies on the Third World have 'explicitly examined the nature of the state in their analyses of urban planning and decision making'. What certainly must be accepted is that early analyses of governmental intervention have generally implied a model of the liberal, pluralist state acting on all occasions in the best interests of society as a whole, and not prescribed sections of it. A central concern here, and one which will be elaborated subsequently in this chapter, is that it is manifestly the case that the state may operate in such a way as to benefit particular groups or classes within society. The concept of the 'class state' is one that arises from considerations such as these and leads to the question of how particular societies have become divided into distinct class groups, the power that elites and other groups making up society possess, and the manner in which wealth is distributed among them.

However, there is another aspect of these disciplinary changes that has brought a further set of disappointments for the geographer concerned with development issues. Specifically, this is what appears to have become the increasingly parochial character of geography during the last two decades (Brookfield, 1975; Farmer, 1983). This point has been argued strongly in relation to the consideration of Third World cities within the overall field of urban geography by Potter (1985, Chapter 1). Saliently, several authors have recently taken up this theme, with, for example, Johnston (1984) exorting geographers to remember that 'the world is their oyster' as a result of the fact that:

> ... British geographers are becoming increasingly parochial in their outlook. The proportion of the University and Polytechnic human geographers who have detailed field experience of non-European and non-North American areas is small, and is probably smaller now than three decades ago. We are geographically - and certainly linguistically - myopic ... (ibid., p. 444).

In an essentially similar vein, Porteous (1986) has emphasised what he sees as the need for 'intimate' as opposed to remote sensing of cultures other than our own. In his spirited plea, Porteous reminds geographers that they 'have a duty to report on the state of the world at the microscopic, as well as the telescopic, level. We cannot leave it all to journalists and the National Geographic' (p. 251).

Bearing in mind these wider contextual changes in the epistemology of the subject, in the present chapter we seek to comment broadly both on the ways in which geographers have in the past studied politics, power and social organisation with respect to Third World development issues, and equally on the ways in which they might contribute to their study in the future. In so doing, we shall present and elaborate the argument that although the stress placed on matters such as global power relations, dependency theory, the internal dynamics of social surplus production and distribution have been critical in promoting the realistic study of the geography of Third World development, it has tended to result in a surprising neglect of local, micro-spatial aspects of power relations, political issues and social structure. In this connection we generally concur with Porteous that for too long the emphasis has been on the telescope and not on the microscope. We shall argue that it is now high time for work which translates the macro-societal perspective into a micro-level concern which emphasises environmental and ecological factors in the evolution of distinctive landscapes and modes of production, and likewise the study of the ways in which resources are distributed and employed within society. Within such study, a strong spatial interest would emerge, so that, as suggested by Johnston (1984, p. 444), once more 'Geography is about local variability within a general context'.

POWER, POLITICS, SOCIETY AND SPACE

In this section, we endeavour to develop the initial argument presented previously, principally by taking a closer look at the significance of power, politics and society as variables of concern to geographers studying Third World regions. Before doing so, however, there are two definitional matters that require brief clarification.

The first concerns use of the label 'Third World', a description that has recently troubled a number of authors (O'Connor, 1976; Auty, 1979; Wolf-Phillips, 1979; Mountjoy, 1980; Ward, 1980; Crush and Riddell, 1980). The origin of the term was as a category to describe those newly independent and non-aligned nations, in contrast to the free market First World and the centrally planned Second World, in the immediate post-war period. However, the fact that countries within this group are varied in all respects is increasingly obvious, so that even a division of them into several sub-groups fails to cater adequately for their heterogeneity. What can be said in general terms is just as true of the political constitutions of such nations, varying from parliamentary democracies to military dictatorships and revolutionary states such as China, Vietnam, Cuba and Kampuchea (Clapham, 1985).

Secondly, change in consensus views of the meaning of development is highly germane to the present account, and can be related to the brief review of changing academic paradigms provided above. In the immediate post-war period, development was primarily seen as synonymous with economic growth and income generation. The prevailing ideology was one of modernisation, basically arguing that the transfer of technology, capital and expertise from developed to less developed countries would slowly but inevitably set them on the royal road to 'development'. Rostow's (1960) stage model of economic growth, sub-titled 'a capitalist manifesto', was typical of the approach adopted at that time. Since then, however, other yardsticks and philosophies of development have emerged. In an important statement, Brookfield (1975) defined development as gradual change brought about by the creation and expansion of an interdependent world system, whether or not such change was ultimately for the better. Other, more socially and politically oriented criteria of development have subsequently emerged, and it has been observed by Seers (1979) and Bromley and Bromley (1982, Chapter 1) that development can just as easily be regarded as improvement, modernisation, increasing welfare and the enhancement of the quality of life. In this respect, the existence of democratic constitutions, the maintenance of civil rights, and freedom from political, religious and cultural oppression can be seen as important ingredients of the development process. The degree to which power is centralised in the state or concentrated with certain groups become crucial factors, stressing the relevance to geography of studies of politics, power and society.

The concepts of power, politics and society are obviously closely interrelated, making it difficult to define them separately. Starting with power, Haralambos (1980) cites the Weberian definition of it as the chance that people have of realising their own will in a communal action, even against the resistance of others who are also participating in that action. Effectively, therefore, power may be defined as the extent to which an individual or group is able to get its own way in social and other activities, and the degree to which they can control their own destiny. Thus, poverty represents a lack of power in the market place. Customarily, sociologists distinguish between two forms of power: firstly, that which is legitimate, referred to as authority; and secondly, power which is not regarded as legitimate by those subject to it, and which is commonly denoted as coercion (ibid.). Even so basic a definition of power provides some indications of its geographical connotations: for example, groups with power will have the chance to further their own sectoral interests. The poor may well need to enlist the services of powerful patrons to assist them in the quest for even quite basic facilities and environmental resources, an example of this being provided by the lobbying for services within squatter settlements. Corruption, violence and graft can all be seen as important variables in promoting decisions about various phenomena. Thus, many decisions which fundamentally influence the environment are taken on the basis of political expediency and involve the activities of various power brokers.

Power is of course exercised in different arenas. Politics is not just about the ideological struggle between capitalism and communism, but about power relationships at all levels. As Chinoy (1969) notes, although power and authority are found in all areas of social life, they are most clearly focused in the state. Dowse and Hughes (1972) maintain that politics occurs when there are differences in power, and thereby imply that to all intents and purposes, power and politics are one and the same thing. This viewpoint is also reflected in Clapham's (1985, p. 1) recent introduction to Third World politics, in which it is argued that 'Politics everywhere, in its essentials, is much the same. People do not greatly differ. They want security, wealth and the power through which to get them' (emphasis added). All forms of social organisation essential for the pursual of both group and individual societal goals and the management of competing interests produce inequalities of power.

Finally, social organisation reflects the disposition of power between different groups, distinguished by factors such as occupation, prestige, age, race, ethnicity, family status and the like. Social stratification, whereby groups are effectively ranked one above another, is the expression of differences in wealth, prestige and power. Marx differentiated between two main social classes, the capitalists or bourgeoisie, who own the factors of production, and the workers or proletariat, who have only their labour to sell. Family structure, gender roles, social norms, and marriage are all further aspects of social organisation that witness the existence of power and authority and which ultimately are therefore likely to be intimately related to the use of resources and the structure of the environment.

We can relate the geographical importance of these factors to the argument that has been presented earlier in this chapter. In much of the geography carried out prior to

Figure 7.1: Socio-political organisation and spatial organisation (reproduced from Potter, 1985)

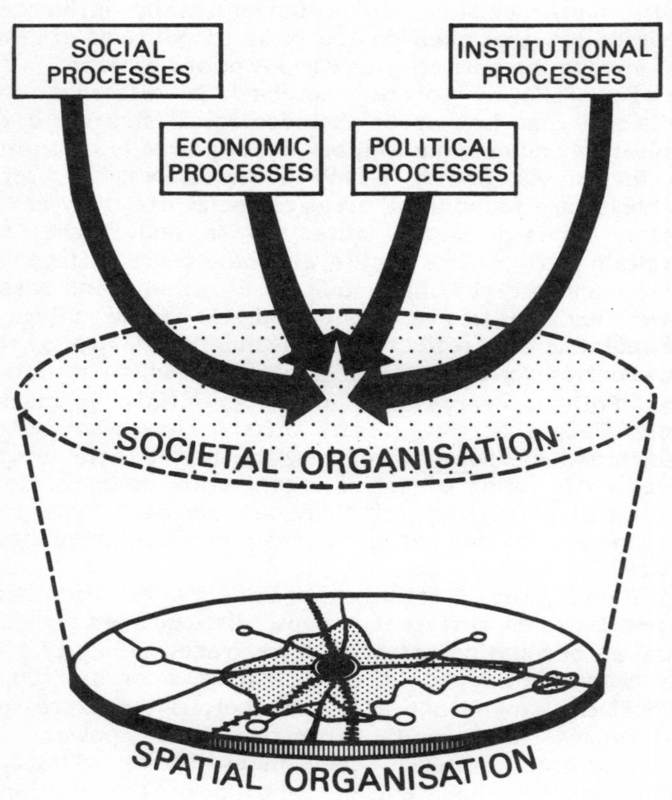

1960, attention was placed firmly on the environment, and little was directed towards the processes influencing space. A major benefit of the extensive adoption of a political economy approach in the last decade has been that attention has been directed to the formative political, economic, social and institutional processes which influence societal organisation. This is illustrated by Figure 7.1. These structural processes operate in space but are not spatial processes *per se*. It also has to be recognised, however, that even aspatial decisions are eventually enacted in space, so that the study of space and place in relation to key political, social and economic agents is an essential task of geography. Thus, the focus on the spatial milieux of the individual, the family or household unit, the community or region is highly germane to geographical enquiry. In so far as political economy perspectives have been taken by some as a 'party line', suggesting that micro-spatial empirical research is irrelevant, *status quo* oriented and positively reductionist, it can be posited that they have caused an imbalance in the geographical study of developing areas.

Some interesting corroborations of this line of reasoning are provided by recent development writings from a variety of perspectives. Examining the study of politics in Third World countries, Clapham (1985) argues that neither what he terms 'internationalist' nor 'nationalist' approaches are adequate by themselves in fostering an understanding of Third World politics. The former follows a strict Marxist-Leninist approach and tends to explain everything in terms of the incorporation of Third World countries into the global order. Clapham suggests that while such an approach has been invaluable in a number of respects, it has had the unfortunate consequence that all too often the salience of internal, or essentially national, factors has been largely ignored. In many Third World countries the state has become very powerful, serving to centralise decision making and to bolster class interests. To argue that all forms of change are the direct outcome of global dependency relations is to ignore the importance of the power of the state, both legitimate and otherwise. This is an argument which has recently found some popularity in the study of the underdevelopment of certain African and Caribbean countries, and which will be considered in detail in the next section (Sandbrook, 1985; Vendovato, 1986).

Similar arguments have been presented by Forbes (1984), who makes a strong plea for the geographical study of underdevelopment in which the focus of attention is moved towards the analysis of social relations, in order to:

> ... shift the focus of research on underdevelopment away from its close links with political economy. It would be absurd to argue that economic processes are still not central to 'economic development', but the humanist critique of political economy stresses the way in which economic problems are embedded in social relations which are far from well understood. To be more specific, a narrowly economistic viewpoint, with its characteristic focus on production, needs to be incorporated into and supplanted by an approach which emphasises social reproduction (ibid., pp. viii-ix).

The argument is elaborated by Forbes, suggesting that development geographers should turn their attention toward the analysis of social relations and social reproduction and the significance of space as an active agent of social processes.

Finally, writing from a somewhat different perspective, Bell and Roberts (1986, p. 3) have issued a strong plea for work embracing and combining the traditional perspectives of both physical and human geographers within the field of development studies. They also present the argument that although the domination of the political economy approach in development geography has brought many benefits, 'it has been accompanied by a declining awareness of the environmental dimension which was traditionally central to geographical studies in the tropics (e.g. Gourou, 1953)'. The reawakening of an interest in the 'environmental' aspects of development is thus seen by Bell and Roberts as an important theme for future work.

These recent arguments can be related to, and serve to substantiate, the central theme of the present account - namely, that there is a role for the study of the particularities of place and space in the geographical study of development. We are not rejecting political economy or structuralist approaches; rather, we are arguing that their very success has shown the need for detailed micro-studies that may well have strong empirical, and even positivistic, bases. In rounded terms, the focus of such work will be on the environment, the ways in which it is managed and regulated, the ways in which it is being changed and transformed, or alternatively, kept as it has always been.

Political, social and power relations will be central to such studies which are concerned with environmental and resource impacts of societal organisation. Just like classical economic theories that ignore space and regard the world as the outcome of events of incalculable complexity which ostensibly occur on a pinhead, so social theories have tended to assume a spatial void. Whilst there may be relatively few social processes per se, all processes operate in a three-dimensional world.

Thus, geographers do have a distinctive role to play in furthering our understanding of power, politics and society in the Third World. Whilst it may be the case that geographers are interested in much the same issues as sociologists, political scientists and economic historians, their focus should be distinctive, at least in part. How do power, politics and society influence the distribution and use of resources and the structure and transformation of local and whole environments? An excellent example of this type of approach and one that certainly dovetails well with the comments of Bell and Roberts (1986) and Forbes (1984) is provided by Blaikie's recent book on the political economy of soil erosion (Blaikie, 1985). The geographer is also concerned with the influence of these variables on entire landscapes. Here, the dependency theory approach may be of vital importance, showing, for instance, the power of multinational companies in tourism or manufacturing industry. Other examples include the power of the state, both national and local, in distributing resources. The active agents of power and political support in client-patron relations are also important, as are factors such as corruption at all levels, again from the state down to the individual. Factors such as gender and age roles in employment and household tasks are further interesting examples of the important issues that require detailed study in geographical work conducted along these lines.

It follows from the argument presented so far in this chapter that the variables of power, politics and society are virtually all-embracing, and further, that they operate at all scales from that of the nation and the region, through to the settlement, the household and the individual. In the remainder of the chapter, therefore, we seek to examine issues and provide illustrative examples of this approach. This is done at two basic levels - firstly, in relation to socio-political and power relations at the national and regional scales, and secondly, with regard to settlements, households and individuals. With such a wide canvas, although an effort is made to draw on examples and illustrations from a variety of world regions, the emphasis is placed on the geographical regions with which authors have most familiarity - namely, Africa and the Caribbean.

NATIONAL-REGIONAL CONSIDERATIONS OF POWER, POLITICS AND SOCIETY

At the global scale, it can, of course, be appreciated that the very process of economic development has much to do with the political battle between the contrasting claims of capitalism and communism as the two major competing modes of production and social organisation. This is well-exemplified within the Caribbean which, as Uncle Sam's Backyard, has traditionally been seen under the hegemony of the United States (Payne, 1984). The Cuban Revolution of 1959, the coming to power in Jamaica of Michael Manley's Peoples' National Party in 1972, but more particularly, the coup in Grenada in 1979 which brought to power the Peoples' Revolutionary Government under the leadership of Maurice Bishop, all serve to illustrate the vital role that is played by political ideologies and their implied social systems in the global geography of development and change.

There are two conflations of concepts that necessitate the geographical study of politics, power and society at the macro spatial scale. These are, firstly, the essential correspondence which exists between political, economic and social elites; and secondly, the close interrelations between social and spatial inequalities. At this scale, questions of global dependency and its expression in regional structuring are of particular salience, and these are topics in which geographers have generally shown considerable interest, as witnessed by Chapters 1 and 6 of this book, as well as in the two previous sections to this chapter. In the present section, we follow the arguments presented in the previous review, and urge strongly that geographers should keep to the fore the spatial connotations of power, politics and societal organisation, however aspatial the processes themselves may ostensibly appear.

The inequitable distribution of wealth and power within developing countries has been a major topic for discussion in geographical studies of Third World countries. In the majority of developing countries, the top 10 per cent of households account for over 35 per cent of incomes, whilst the lowest 20 per cent earn less than 4.2 per cent, and even the lowest 40 per cent, a meagre 12.7 per cent (Gilbert and Gugler, 1982, quoting the World Bank, 1980). Similar statistics for a range of industrialised countries, show that the top 10 per cent of households took 26.3 per cent of income and the lowest 20 and 40 per cent, 6.0 and 17.8 per cent respectively. It is generally recognised that disparities in income, welfare and quality of life are far greater in Third World countries and are showing every sign of increasing rather than decreasing, despite years of development planning. Generally, it appears that social polarisation is becoming further accentuated, for in the majority of Third World countries, there is an elite group which is separated by a diminutive and under-represented middle class from a massive and all too often increasingly impoverished and dispossessed proletariat, both urban and rural.

Under such circumstances, political elites frequently exhibit a close congruence with both commercial and social elites, and indeed, members of these groups are often drawn from the same families. Nepotism may be rife and forms of patron-clientism endemic in all walks of life, as will be exemplified in the next section. Recent examples are not hard to find, whether they be Papa Doc and Baby Doc Duvalier, a family empire which ruled the impoverished Caribbean island of Haiti for over 28 years, or the repressive rein of Ferdinand Marcos in the Philippines. Frequently, under such dictatorial rule, corruption, repression and violence such as that inflicted by the Tontons Macoutes in Haiti, are the norms setting limits on the possibilities of development and change for the majority of the population.

In conditions such as these, the state may merely act in ways which are designed to ensure the maintenance of the interests of the ruling group, and not those of the general populace. Miliband (1977), for example, refers to the 'state for itself', to explain the all too frequent situation where those in powerful positions effectively use their influence to further their own economic interests and those of their families, friends and close associates. Gilbert and Healey (1985) have used the examples of Nigeria and Venezuela to suggest that many oil-rich Third World states, a form of 'bureaucratic-authoritarianism' pertains, in which the state uses the revenues obtained from oil to provide itself with

Figure 7.2: Race and social class in Caribbean slave societies

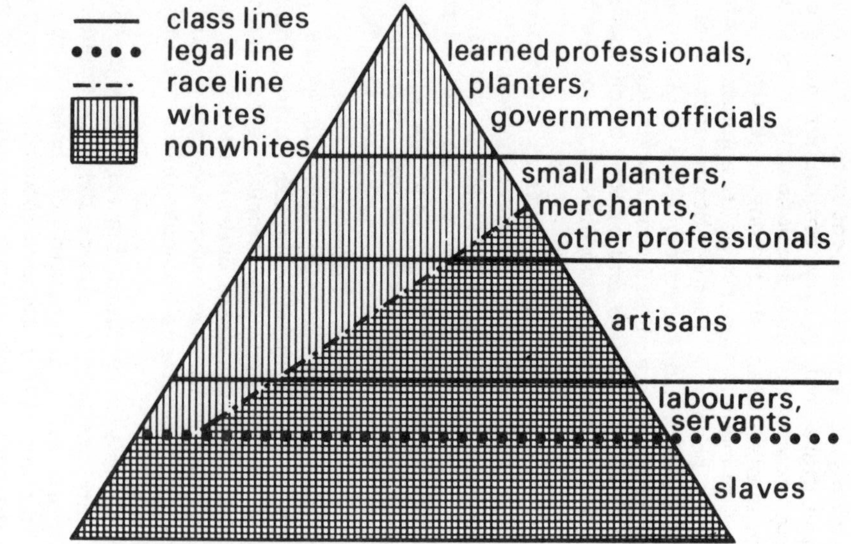

Source: redrawn from Augelli and West, 1976

inordinate power. This offers the state a large measure of patronage and control to maintain its position and that of its ruling groups, and helps to explain many of the policies adopted, such as turning a blind eye to squatting, whilst such activities are officially illegal. A more humane and democratic system should undoubtedly recognise squatting as the outcome of the highly uneven distribution of land and income and would seek to implement structural reforms.

Interpretations of development issues are frequently drawn along these lines. Sandbrook (1985), for example, in viewing the politics of Africa's economic stagnation, argues forcefully that personal rule or neopatrimonialism has produced a variety of irrationalities, such as political instability, systematic corruption and maladministration, which have impeded growth. More recently, with regard to the persistent underdevelopment of the Dominican Republic, Vedovato (1986) seeks to chronicle how successive regimes have been more committed to increasing their own personal wealth than fostering the genuine development of their people. In so doing, he stresses the concept of the 'predatory state' or the 'kleptocracy' in which public office is used for the illegal acquisition of wealth. The examination of internal national obstacles to growth, as opposed to the barriers created by the webs of dependency, highlights the salience of the assertion, made in the present account, that geographers need to focus upon local as well as global aspects of the socio-politics of underdevelopment.

Some less extreme examples of the influences of power, politics and social structure on patterns of development and geographical change are provided by the dependent economies of the Caribbean. Colonial plantations created grossly inegalitarian social hierarchies, such that during the seventeenth century, two major strata existed, the free whites and their black slaves (Clarke, 1985, 1986). By the mid-eighteenth century, miscegenation had become more common, especially in the French territories, and a coloured element was formed which also occupied an intermediate class position. Figure 7.2 demonstrates vividly the essential correspondence between class, colour and legal lines before emancipation. Although today, colour and class are not matched exactly in all Caribbean societies (Lowenthal, 1972), in many there is a persisting 'white bias', with status and power being inversely related to group size. As Clarke (1986) notes, the persistence of economic and political elitism, bolstered by colonialism, shored up the framework

Figure 7.3: Urban and rural class systems in the Caribbean

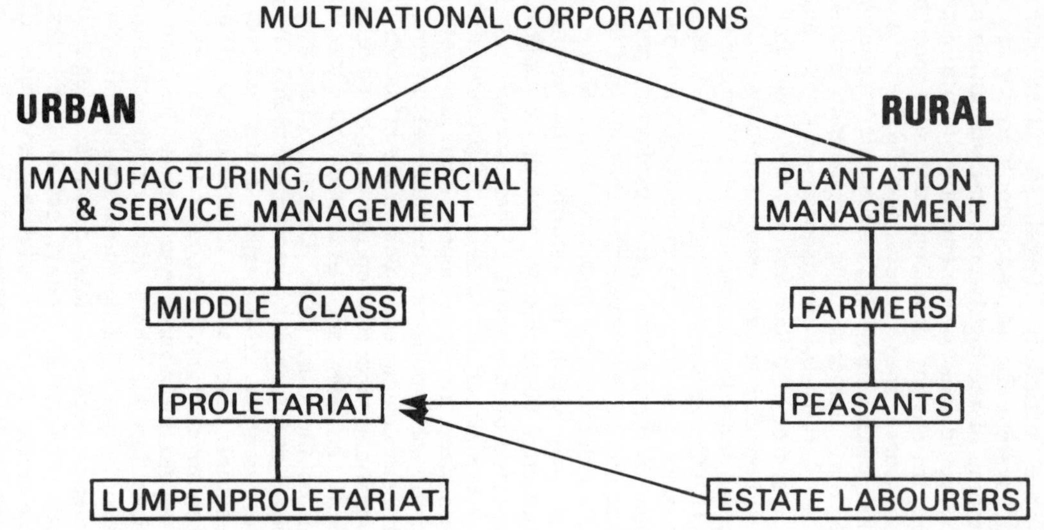

Source: redrawn from Cross, 1979

of the culturally pluralistic contemporary Caribbean. Even in the early twentieth century, property restrictions in the Commonwealth Caribbean meant that less than 10 per cent of the total population had the right to vote, and 'politics remained an elite pastime, and the reins of government were held in expatriate hands' (ibid., p. 23).

In Barbados, although the white population has been displaced from political power, this group still owns virtually all the sugar estates, rum distilleries, major business firms, import and export agencies and large retail stores (see Potter and Dann, 1986, in connection with the latter). This has occurred through the processes of legislation, corporatisation and intermarriage, interlocking directorships and a virtual colour bar (Barrow, 1983). Thus, contemporary social commentators on development issues in Barbados still stress the importance of elitist cleavages between whites and blacks (Layne, 1979; Karch, 1985). Lowenthal (1972, p. 82-3) stresses that 'coastal tourist resorts divide Barbados physically between the white elite (and tourists) and black folk', and socio-political factors are of great importance in relation to these pressing geographical development issues (see also Potter, 1983, 1986). However, the family ownership of economic organisation is the exception in most other Caribbean countries, rather than the norm as in Barbados. In fact, in examining contemporary Caribbean societies, Cross (1979) argues that the structure of Caribbean class systems, both urban and rural, has come to be dominated by multinational corporations in the manner summarised in Figure 7.3. Thus, the rural upper stratum of plantation management is added to its urban counterpart of manufacturing, commercial and service management, followed by a middle-class urban group. The two systems are linked principally through the displaced peasantry and estate workers who are forced to migrate to the city due to a lack of rural opportunity. This thesis has recently been elaborated in the Caribbean context by Kowalewski (1982), who emphasises that the increasing involvement of multinationals in Third World countries means that intra-national inequalities are inseparable from international ones. Thus, for example, multinational expatriates form an essential elite element in the Caribbean class system, whilst millions of First World tourists, although coming and going individually, remain as a permanent elite group on the islands.

From a geographical-environmental perspective, the

salient point is that such inequalities are not manifest merely between different social groups, but also spatially - that is to say, between regions, particularly rural and urban areas, and the primate city and lower levels of the settlement hierarchy. Indeed, in commenting on Frank's theory of dependency, Roxborough (1979) notes that one of its distinctive characteristics is the conflation of spatial entities and social classes. The same is also true of Lipton's thesis on 'urban bias' (Lipton, 1977, 1982) in which urban and rural are viewed in almost exclusively social class terms. Although a number of authors have criticised such a gross association, it does emphasise the need for studies of the spatial connotations of such social duality, in the manner stressed previously in this account. Indeed, dependency theory stresses that surplus value is appropriated from agricultural workers on plantations and from peasant farmers to be concentrated in regional metropolitan centres. By stages, such expropriated capital is moved from the regional metropoli to the national primate capital and from thence to the world metropoles. The mercantile model of Vance (1970) is an excellent corrective to central place theory and sets out to map the outcome of these developments. However, this did not prevent Harvey (1973) arguing that the Christallerian steps found in the settlement fabrics of many nations mirror the concentration of social surplus product. Similarly, Armstrong and McGee (1985) have recently written of Asian and Latin American cities as 'theatres of accumulation'. Indeed, political responses to capital accumulation, uneven development and regional problems are another area of prime concern to geographers.

Policies of continued concentration are generally associated with capitalist philosophies of development, even in situations in which concentration is effectively merely reduced and moved elsewhere, as in growth-pole theory and practice. Not surprisingly, therefore, Friedmann and Weaver (1979, p. 186) have described growth poles as the 'handmaiden of transnational capital', and as being entirely attuned to the needs of transnational corporations. Recently, such right-of-centre views of spatial development planning have come to be incorporated under the general heading 'development from above'. Socialist-inspired approaches, on the other hand, tend to emphasise 'development from below' (Stöhr and Taylor, 1981). This philosophy is associated with 'agropolitan' principles of development which involve the meeting of basic needs at the local scale and the nurturing of industrial enterprises in rural, non-agglomerated settings. Examples of the implementation of this type of approach are afforded by China, Vietnam, Korea, Sri Lanka, Bangladesh, Pakistan and Tanzania. The introduction of any form of bottom-up planning is effectively premissed on the establishment of greater public participation in the overall processes of planning and change, and this is an important issue that also connotes power, politics and social structure in a geographical context. Specifically, the critical point is the degree to which ordinary citizens have the power to influence decision making and environmental change between their periodic visits to the ballot box (see Conyers, 1982; Potter, 1985, pp. 148-65). It is in this sense that geographical development work is increasingly recognising that all planning is inherently political.

LOCAL-HOUSEHOLD CONSIDERATIONS OF POWER, POLITICS AND SOCIETY

The case has already been made that too little recent geographical work has focused on detailed empirical circumstances. Accordingly, the remainder of this chapter stresses the vital importance of power, politics and society at the level of the locality and the household. Overall, the philosophy of the approach emphasised concurs with Porteous's (1986) call for the replacement of 'remote' sensing by its more 'intimate' counterparts.

Thirty or more years ago, other disciplines had much to learn from a well-established tradition of detailed empirical studies undertaken by geographers in the former colonies (see, for example, Forde, 1934; Gourou, 1953; Buchanan and Pugh, 1955). However, the more recent preoccupation with general theories and parochial locations has meant that in some respects, geography has been overtaken by methodological advances in cognate disciplines such as economics, sociology and political science (see also Dickenson, 1986). These disciplines are increasingly following the lead of anthropology in focusing upon detailed investigation of particular communities and their constituent households, in an attempt to elucidate important decision-making processes and relationships governing the manipulation of resources, which ultimately determine the well-being of individuals.

This trend has received considerable impetus in Third World countries from the failure of many of the 'top-down' development programmes mentioned in the last section, with their obsession for aggregate, national-level data and plan formulation and an inadequate, and frequently inaccurate, understanding of indigenous production and social systems. The aspirations of individuals and communities have often remained unfulfilled and little progress has been achieved in either increasing national wealth or distributing it more fairly (Potter, 1985). The magnitude of poverty and inequality in the Third World has scarcely diminished and some would argue that little real development has occurred (Seers, 1979).

Considerable progress has, however, been made recently in interpreting rural food production systems and in appreciating the intricacies of peasants' knowledge of the environmental and economic settings in which they operate. Through 'farming systems research', valuable information is now being channelled into rural development programmes which, as a consequence, have much better chances of improving rural living standards. The starting point for such programmes is increasingly the diagnostic survey of existing farming systems, which identifies constraints and farmers' aspirations prior to formulating appropriate biological, mechanical or economic innovations (Richards, 1985, 1986; Norman and Baker, 1986). In Third World towns and cities too, detailed local-level investigation is illuminating such processes as housing improvement and consolidation, and the nature of informal-sector employment (Bromley and Gerry, 1979; Ward, 1982). In fact, the household is becoming the focus of an increasing volume of social-scientific investigation, since an appreciation of relationships within and between households reveals the importance of power and politics in controlling the access to important productive resources. In many Third World societies the household is the basic unit of production and consumption and the focus of decision making.

Defining Third World households can be difficult, since they may vary in size and composition and can bear little resemblance to the typical Western family unit. It is important at the outset to be aware of some of the problems associated with household analysis. Firstly, it would be quite inappropriate to assume that households, whether urban or rural, are independent and isolated. Important links between households and within communities must be appreciated. It is also necessary to understand the historical and political context within which particular households operate, including the effects of national and international policies. The dynamics of households must be evaluated, since all households pass through a developmental cycle during which their size and composition may change, and the important ratio between workers and dependents also varies. Finally, within particular households there may be persons whose resources are not always common to the household as a whole. Husbands and wives may have very different responsibilities and separate incomes. Amongst the matrilineal Ashanti of Ghana, for example, a man and his wife may live separately, remaining in their respective compounds of birth after marriage and earning individual incomes in farming and trade. Wives send food to their husbands each evening and receive contributions, in return for the care and education of the children (Guyer, 1981).

It may be necessary to make a clear distinction between work units, consumption units and investment units, which may or may not be confined to one household. Household analysis should also take note of relationships between households (inter-household), as well as the composition of, and relationships within, individual households (intra-household). It should also be remembered that individual households are encompassed within larger 'supra-household' networks (Peters, 1986). Provided one is aware of these problems, then the household is a vital focus of geographical study, seeking to enhance our understanding of the interaction of power, politics and society at the local level in Third World countries.

As discussed in the previous section, one of the most striking features of both Third World rural and urban communities is the inequalities which exist between households. Detailed analysis can identify rich and poor households and can interpret the underlying causes of such inequality. At one end of the social spectrum are the elites - strong, powerful, well-off in assets and income, often having close links with government officials, police, large landowners and traders. The power wielded by Third World elites is clearly illustrated by the Creoles of Sierra Leone and the Americo-Liberians of Liberia, two urban-based minorities who settled as freed slaves on the West African coast in the eighteenth and nineteenth centuries. Both groups became involved in politics, commerce and the professions, the Christian missions providing education and a basis for community life which still persists today. Sierra Leone's Creoles identified closely with European traditions and values, adopting European-style dress, houses and eating habits. Whilst the Creoles gradually lost their dominant position in Sierra Leone after the Second World War, in Liberia a relatively small group of Americo-Liberian households retained a tight grip on politics, society and economy until their True Whig Party was ousted from power in a bloody military coup in April 1980. It was not uncommon before 1980 for the appointment of a leading Americo-Liberian family member to an important post to be quickly followed by the promotion of his relatives to subordinate positions (Liebenow, 1962). The patronage network was reinforced by a multiplicity of voluntary associations, notably the Episcopalian Church and the Freemasons, membership of which enhanced an individual's status and influential contacts. Although during the rule of the Americo-Liberian elite, Liberia had one of the most rapid rates of economic growth in the world, the benefits were concentrated in relatively few hands, whilst the rural hinterland remained impoverished and undeveloped (Clower, 1966).

In sharp contrast, poor households in the Third World usually have little power, are frequently uneducated and ignorant of the law, having few assets apart from their labour. Such vulnerable households are frequently exploited, as they are locked into cycles of debt and are dependent on the assistance of one or more richer patrons. Some 10-20 per cent of people in low-income countries fall into a more desperate group of the 'ultra-poor', whose households eat below 80 per cent of the FAO/WHO weight-adjusted dietary energy requirements, despite spending at least 80 per cent of their income on food. According to Lipton (1968, p. 4), the ultra-poor '... are often landless, have high child/adult ratios, ... are often incapacitated by hunger and illness and show higher unemployment, being often dependent on casual labour'.

Pryer's (1986) study of a slum in Khulna, Bangladesh describes some ultra-poor urban households whose only productive asset is their labour, yet the prevalence of malnutrition and chronic illness resulted in many days' earnings being lost through physical incapacity. Highly seasonal, poorly paid employment is common, with men entering petty trade, hawking and rickshaw-pulling, whilst women undertake domestic work or home-based piece-rate work, or are sometimes forced into illegal black marketing. With such low, unreliable and seasonal incomes, households attempted to achieve a diverse employment profile with as many family members working as possible. However, all families in the slum were deeply indebted to landlords, employers, shopkeepers and neighbours, with debts averaging 173 per cent of monthly income. One woman's household was totally dependent on her patron for its survival, and in return, she accepted wages at under 50 per cent of the market rate for her domestic service. With a precarious nutritional status below the recommended daily intake, the ultimate sacrifice was to cut back further on food intake or to go without food entirely (ibid.).

Stratification within some communities is long-established and handed down from one generation to the next. Amongst the Tuareg of the inland Niger delta of Mali, the rich and poor households of today broadly correspond with the noble and slave groups of the past. Tuareg recognise five status groups, relating to the days, less than half a century ago, when domestic slavery was commonplace. The highest status group, making up only 1 per cent of the population, comprises descendants of the traditional warrior aristocracy, with three other non-slave groups consisting of Islamic scholars, blacksmiths and artisans. By far the largest group of households comprises the ex-slaves or Eklan, descended from captives seized by Tuareg warriors or purchased by free Tuareg in the pre-colonial era. Whereas the Eklan are negroid, their former masters are fair-skinned and almost Arab in appearance (Swift et al., 1984). Historically, Eklan were the poorest members of Tuareg society and today they remain poor and are often forced into dependence on their former masters. Intermarriage between Eklan and Tuareg is forbidden so that nobles, by marrying other nobles, do not allow their livestock to pass into the hands of the Eklan. The former slaves merely have their labour, and traditionally grew food crops for their pastoral masters, though since the abolition of slavery some have left the delta to farm elsewhere, whilst others work as porters or casual labourers in Mopti and other delta towns. Though traditionally excluded from owning livestock, some Eklan have gradually managed to build up herds with their cash earnings (ibid.).

Long-established systems of stratification are also apparent in many Indian villages where the caste system persists. Access to key factors such as land has enabled higher castes to become richer and more dominant, whilst those with little land are obliged to sell to the large landowners, eventually becoming landless labourers. In his study of five villages in Karnataka, India, Mishra (1982) found that 82 per cent of the large farm households came from the dominant caste, the Vokkaligas, whereas at the other end of the social spectrum, some 70 per cent of the landless households were from scheduled castes. Dominant castes, through their powerful position, have better access to rural credit and have benefited from government development programmes promoting dairying and sericulture (production of raw silk). These groups are also able to hire labour, invest in well irrigation and farm machinery, and market substantial crop surpluses. In short, what has happened is the emergence of a modern class-based village society within the framework of its traditional caste-based social structure.

The importance of access to land as a determinant of social stratification is also apparent in rural Bangladesh, where Hartmann and Boyce (1981) recognise five strata. At one end of the spectrum are the landlords and rich peasants who own land and hire labour, whilst at the other extreme are labourers who have no land, draft animals or agricultural implements. Hartmann and Boyce conclude that the small minority of rural farmers who own over half the country's farmland are, '... at the apex of the structure of power in rural Bangladesh; the political economy of the countryside is controlled by them' (ibid., p. 182). As in Karnataka, India, so in Bangladesh, the wealthy landowners are in a better position to obtain government credit to invest in irrigation, fertiliser and machinery. The landless poor must resort to local moneylenders (who may be large landowners) for credit, and interest rates of more than 100 per cent a year are common. Social stratification is thus reinforced, increasing the inequalities between rich and poor (ibid.).

Land ownership is a major factor in household status and wealth in Hausaland, northern Nigeria. In the village of Batagarawa, Hill (1982) classified the 171 farming units into four economic groups such that the 17 farming units in Group 1 owned 28 per cent of the mapped acreage, whilst the 41 farming units in Group 4 only owned 11 per cent of the land. To enable consolidation, richer farmers with insufficient inherited farmland often bought manured farmland from poorer farmers adjacent to their plots.

Differential access to land is also a major factor

determining the pronounced residential segregation which is a feature of so many Third World cities. In Bogota, Colombia, for example, the elites have deserted the city centre since the early 1900s, when the success of Colombian coffee brought wealth to a small number of families. They have moved into high-class peripheral suburbs to the north of the city, and in so doing have 'leapfrogged' middle-income residential groups developing around previous elite locations (Dwyer, 1975).

In Venezuelan cities, high-income groups can bid for the most desirable residential land, and because of their powerful links with the ruling authorities, they are able to ensure the costly installation of services such as water, sewerage and electricity. The poor, however, with little or no power, must often obtain land for housing illegally through organised invasions of public land. Their land is invariably badly located and unserviced and dwelling consolidation can be a slow process with limited financial resources. Access to land and services often depends on complex links with the original invaders, politicians and the state. Gilbert and Healey (1985) show how support for land invasions in Venezuela sometimes came from the highest level of government, such as between 1945 and 1948, and again after 1957. The closer an election, then the more likely it is that an invasion will be permitted. Individual lots of land acquire exchange value and commodity status as families living in invasion settlements are forced to pay the original invaders, a process which may involve the buying of an 'improvement' made to the invaded land (ibid.).

The importance of politics and patronage is also clearly revealed in Moser's study of the Indio Guayas barrio in Guayaquil, Ecuador, where low-income households can take a long time to consolidate their homes. An elder son first establishes a house for himself and his nuclear family, and then, after a few months or so, he will acquire a second plot in the same street or close by, to which he will move, bringing his mother and/or siblings to the older plot, which he transfers to their name. In such a way, extended families congregate in the barrio (Moser, 1982). Resources for the provision of infrastructure are sometimes allocated by political leaders to those settlements which can promise electoral support. Frequently, inhabitants form themselves into self-help committees to petition for infrastructure, and the committees are then co-opted by political parties. The settlements are therefore dependent on the capitalist sector for the provision of services and the land market is '... controlled by the interests of large scale capitalism, articulated through the conflicting interests of different political parties' (ibid., p. 185).

Transactions and relationships between different households, political parties and the state are infused with what Hyden (1986, p. 26) calls 'the politics of affection'. Patronage relations can be seen in land investment and the way business enterprises hire people from their own household or home area and invest money back in their village base. Even civil servants get involved in this patronage, 'particularly as they must serve politicians whose legitimacy is almost wholly dependent on being able to mobilize ... inter-personal networks' (ibid.). Where poor people do participate in elections, their vote is frequently controlled by local elites through patronage or even coercion, in return for some limited material rewards. The existence and persistence of the extended family network alongside colonially implanted public institutions where officials control and allocate considerable resources and power seems to encourage the 'politics of affection', by which large sums of public money may be invested in 'political maintenance'. The patron-client relationship is usually an informal and unequal reciprocal relationship, since the patron is more powerful and may have numerous clients, such that the leverage that any of them can exert on the patron is limited. In turn, many patrons are themselves clients to higher-placed patrons.

The sort of patronage which exists today in Third World countries could well be descended from the redistributive mechanisms which existed in many rural societies in the pre-colonial period. The Hausa of northern Nigeria, for example, elected the '*sarkin noma*' (king of farmers) because of his ability to produce large amounts of food, some of which was distributed at ceremonies and at times of shortage. Many pastoral communities, living precarious existences in marginal environments, also display complex redistributive mechanisms which serve to ameliorate the problems of households in difficulty. Amongst the Tuareg, despite their marked social stratification, it is possible for poor households with few animals to join others for production and consumption purposes and to remain economically viable (Swift *et al.*, 1984). Similarly, in Botswana, Norman and Baker (1986) have shown how households without draft animals can gain access through hiring, borrowing, co-operative resource-sharing agreements, or family help. The main disadvantage of this is that households dependent on such assistance may have less control over the timing of ploughing than other households which own their traction and equipment (ibid.). In many Third World societies, particularly in rural areas, the exchange of gifts is still common. In Hausaland, Nigeria, exchanges in the form of money, food, cloth or other items is common, particularly at marriages, naming-ceremonies and funerals (Matlon, 1979). In north-eastern Sierra Leone, amongst the Kono people, it is common to present gifts, usually of food, to secret-society elders at the time of their children's initiation (Binns, 1981).

Transactions involving labour are also widespread, particularly among farming communities. Labour shortages in parts of rural Africa can cause bottlenecks at certain points in the cultivation cycle, placing severe limitations on farm size. In eastern Sierra Leone, for example, 67 out of 100 farmers interviewed by Binns (1981) identified labour as such a constraint. Households with large family labour reserves could avoid such bottlenecks and expand their holdings, given the plentiful supply of land. Others with insufficient family labour were forced to hire workers, which could be costly. Alternatively, they might make a smaller farm, resulting in seasonal food shortages, particularly in the period immediately before the harvest, the so-called 'hungry season'. Influential chiefs and elders could count upon large voluntary work-groups for certain difficult farm jobs such as clearing and hoeing, the labourers being rewarded with plentiful supplies of food and palm wine (ibid.).

In their study of dryland agriculture in India, Harriss et al. (1984) found that during any given year, certain families were net 'hirers-in', whilst other families depended mainly on labouring for others and were net 'hirers-out'. 'Hirers-out' relied heavily on their female members' earnings, whilst 'hirers-in' had considerable influence over the food market through sales of surplus grain, and were therefore less dependent than most on purchasing food for consumption.

Links between town and farm are strong in many Third World countries, a feature which is reflected in the importance of off-farm employment in determining the wealth of rural households. Migration on a daily, seasonal or longer-term basis to towns, mines, industrial centres or plantations is a widespread phenomenon, particularly amongst young males.

Wealth generated by the development of diamond mining in Sierra Leone since the 1930s has played an important role in the expansion and increased market-orientation of agriculture and has resulted in some farming households being much better-off than others. Farmers with adequate labour have created larger farms and sell more produce in the markets of the rapidly growing mining towns. Wealth generated from crop sales has been invested in hired labour, coffee, cocoa and citrus plantations, as well as contributing to better living standards, notably in the form of home improvements and payment of school fees and medical bills (Binns, 1982).

The importance of off-farm employment with strong rural-urban links is also apparent in regions such as northern Nigeria, particularly in the vicinity of large cities like Kano. As Williams (1981, p. 33) comments, 'The richest people in Hausa villages derive their wealth from their position within an urban-based political hierarchy or their commercial links to urban traders'. The flow of people, money, goods and ideas between rural and urban areas in the Third World has recently generated much interest among social scientists (see for example, Moore, 1984; Swindell and Sutherland, 1985).

Geographical studies of households must also take particular account of the salience of intra-household characteristics. Within individual households, there is often a marked division of labour; and furthermore, decision-making responsibilities between men and women, and the allocation of productive resources, together with the status of the household, may change through time. Two particular aspects of intra-household relationships will be considered here - namely, the question of gender, and the idea that all households go through a life-cycle of development.

Since 1970, increasing interest has been shown in the question of gender in Third World societies. This is most appropriate since in rural Africa, for example, women are the primary food producers, contributing 70 per cent of the time spent on food production, 100 per cent of the time of food processing, 50 per cent on food storage and animal husbandry, 60 per cent on marketing, 90 per cent of all beer brewing, 90 per cent of the time spent fetching water and 80 per cent of the time obtaining fuel (Lewis, 1984). Household labour in rural Africa is not an undifferentiated commodity, with responsibilities for specific crops, operations, off-farm production and unpaid household maintenance tasks usually being allocated among different household members according to sex, age, abilities, experience and status (McKee, 1986).

Amongst the Mandinka of The Gambia, for example, women traditionally grow rice in tidal swamps, whilst men cultivate sorghum, millet and groundnuts on free-draining upland areas. Since the expansion of groundnut production in the late nineteenth century, men have increasingly concentrated on this crop, often at the expense of sorghum and millet, while women continue to grow the subsistence rice crop. Increasing revenues from cash crop production have brought the men more power in Mandinka families, which, together with the spread of Islam, has reinforced male supremacy in both the community and household (Dey, 1981).

Cash-crop cultivation in Third World countries has frequently transformed family labour profiles, whilst the neglect of food crops in favour of cash crops has led to nutritional problems within some households. Women and children are often most affected by such changes. A recent study found that in an area of cocoa and coffee production in south-eastern Sierra Leone, there was a very poor nutritional situation, with infant mortality rates as high as 294 per thousand (Longhurst, 1983). In many poor households, the rationale seems to be that of distributing food according to the earning potential of the members, so that, as Dey (1981), p. 117) comments, 'the (Mandinka) women serving the food feel obliged to give more of the nutritious sauces, meat and fish to the men, while they satisfy their hunger on the bulky but lower calorie staple'. As a result, calorie intakes amongst Mandinka children and pregnant and lactating women were sometimes only 75-80 per cent of WHO/FAO standard requirements, whilst men usually had broadly adequate intakes.

In these parts of rural Africa where male out-migration is well established, the burden of tending the farm as well as the family invariably falls upon the women. In Botswana, within easy reach of work opportunities in South Africa, even by the early 1930s, it was exceedingly rare for married couples to live together for more than two months at a time (Schapera, 1971). Skinner's work amongst the Mossi of Upper Volta (now Burkina Faso), carried out in the 1950s, revealed that about 20 per cent of male migrants did not return home to join their families for the rainy season planting (Skinner,

1965). Male migration has placed great strains on intra-household relationships as the women have been forced to spend more time on cultivation, in addition to performing their traditional tasks of child care and cooking (White, 1984).

In rural Botswana, Peters (1986) suggests that the woman is the central lynchpin in the diversified activities of Tswana households. In the absence of their menfolk, women reinforce important inter-household links by participating in short-term work parties for weeding and harvesting. Female-headed households suffer from lower crop outputs and welfare levels and have sometimes withdrawn from cultivation completely, with drastic effects on the household. Recent work on gender in relation to labour inputs, cash cropping and nutrition has done much to identify intra-household relationships and pinpoint the vulnerable elements within households.

Another important intra-household characteristic is the cycle of development through which all households pass, in particular the changing ratio between producers and consumers. This cycle and its important implications were investigated in some detail by Chayanov (1966). Each household may be analysed in relation to its stage of development in a similar manner to Hedges's (1963) analysis of farm firms in developed economies. Hedges identified three stages: learning, maturity and optimum performance, and post-maturity, during which the manager's effectiveness declines. Each stage of the household life-cycle has different associated characteristics, problems and pressures. Households with a preponderance of young children or elderly members will have smaller pools of family labour to draw upon, with possible implications for the size of farm, income and the household's status within the community. Recently-established, young households may be particularly disadvantaged since potentially important patron-client relationships may still be in an embryonic form. It can take some time for households to consolidate their position within communities. Matlon's study of farming households in northern Nigeria revealed that amongst the poorest 30 per cent of households, there was greater representation of households with heads of 60 years or older and with heads of less than 25 years (Matlon, 1979).

There is pressing need for geographical investigations into intra-household social characteristics and relationships such as those which surround gender and household stage of development. The life-cycle model itself implies a degree of dynamism within households, which must be fully appreciated if effective measures to improve the well-being of Third World households are to be identified, targeted and implemented.

Finally in this section, we turn briefly to consider some of the effects of Third World development schemes on intra- and inter-household relationships. Third World countries are littered with failed schemes, their lack of success frequently being attributable to inadequate prior understanding of community and household structures, relationships, constraints and aspirations. Schemes which have aimed to reduce inequalities have often ended up reinforcing them, creating tensions within communities and households. As Heyer et al. (1981, p. 10) comment in relation to Africa, '.... there appears to be little foundation for the assumption that the activities of rural development programmes lead to the improvement of the welfare of the rural population, let alone the rural poor'.

Some schemes actually exclude the poorest and most vulnerable households and target more progressive and wealthy ones. The Funtua Integrated Rural Development Project, for example, established in northern Nigeria in 1974, selected 'progressive farmers' with the expectation that improving farming methods would ultimately 'trickle down' to the remainder of the rural population. However, Mabogunje and Gana (1981, p. 57) conclude that 'no doubt incipient class differentiation within the rural community was reinforced if not initiated' by the scheme. In Sierra Leone, where integrated projects form the major thrust of rural development efforts, whole areas, communities and households have been excluded. The Eastern Area Project, for example, required participating farmers to have at least three acres of swamp for rice cultivation, and in so doing excluded many of the poorest households (Binns, 1977; Binns and Funnell, 1983).

In the Cauca valley, Colombia, where a once productive and ecologically well-adjusted traditional food production system existed, large-scale farming and sugar plantations have been encouraged with World Bank and US finance. By 1970, some 80 per cent of the cultivable land was owned by four sugar plantations, whilst peasants were encouraged to uproot perennial crops and develop machine-based agriculture in open fields. These developments have had wide-ranging effects as peasant farmers have been induced to work for low wages on the plantations. Income per unit of land has decreased on peasant farms and the uprooting of trees has rendered the new crops of soya, corn and beans liable to flood damage, resulting in substantial losses of investment. Many women and children earn money by gleaning the fields of the large capitalist farms after harvest. Small farmers have been forced to sell their land due to indebtedness, and class differences have widened as a larger proportion of people enter the plantation labour force and work on large capitalist farms. Relationships between households have become more commercialised and the sexual division of labour has changed, diminishing women's power (Taussig, 1982).

There is no shortage of case studies illustrating the effects of development schemes on intra-household characteristics, most notably the position of women. In The Gambia, for example, women were excluded from swamp rice development projects, although they had a wealth of accumulated knowledge from their traditional cultivation of swamp rice. As so often happens in rural development schemes, only the male household heads were contacted by technical teams and invited to participate, and The Gambian women were excluded from owning irrigated land and receiving credit. However, female labour, particularly for transplanting and weeding, was crucial for the success of the projects (Dey, 1981).

Another irrigation project, the Mwea scheme in Kenya, also excluded women from being legal tenants and contractors, with all cash returns going to the men, in spite of significant female labour inputs. No land was set aside for growing household or women's crops, it being assumed that families would eat rice and purchase other foods. Consequently, household nutrition deteriorated and over one-third of children aged 1-5 years are less than 80 per cent of the standard weight for their ages. High rates of desertion by wives are reported since they have neither the cash nor the time to complete what they essentially regard as their tasks within the family group (Lewis, 1984).

CONCLUDING COMMENTS

In this review we have sought to stress the benefits which we feel might accrue to geographical studies of developing countries if, after a period during which a strongly macro-spatial, even global, orientation has been in vogue, greater emphasis is placed on the study of particular circumstances and events. Once again, in order that this call should not be misinterpreted, we stress that such an approach should dovetail with macro-spatial and aspatial theories of development. The reorientation that we should like to see would stress the importance of the particularities of space, place and environment in the study of social relations, politics and the nexus of power relations.

In the latter half of this chapter, for example, the focus has been on the significance of power, politics and society within and between households. During the post-war period, many development projects in Third World countries have signally failed to take into account the importance of inter- and intra-household structures and relationships. Whether in an urban or a rural setting, planners and development administrators must appreciate the aspirations of, and constraints faced by, households and communities before devising measures which can be correctly targeted. Most importantly, planners, developers and politicians should be fully aware of the ramifications of power, politics and society from the level of the nation, to that of the individual within a particular household.

Geographers, with their broad training, are in a good position to carry out such diagnostic surveys, appreciating the environmental, social, political and economic contexts within which household decisions concerning the allocation of productive resources are made. The framework of people-environment relationships, utilised by geographers in the past, is well suited to achieving an understanding of the dynamics of livelihood systems in Third World nations.

REFERENCES

Armstrong, W. and McGee, T. (1985) Theatres of accumulation: studies in Asian and Latin American urbanisation. Methuen, London

Augelli, J.P. and West, R.C. (1976) Middle America: its lands and peoples. Prentice-Hall, Englewood Cliffs

Auty, R.M. (1979) Worlds within the Third World. Area, 11, 232-5

Barrow, C. (1983) Ownership and control of resources in Barbados, 1834 to the present. Social and Economic Studies, 32, 83-120

Bell, M. and Roberts, N. (1986) Development theory and practice in human and physical geography. Area, 18, 3-8
Binns, J.A. (1977) Integrated agricultural development: a case study from Sierra Leone. Oxford Polytechnic, Discussion Papers in Geography, 6
------ (1981) 'The dynamics of Third World food production systems: an evaluation of change and development in the rural economy of Sierra Leone'. Unpublished PhD thesis, University of Birmingham
------ (1982) The changing impact of diamond mining in Sierra Leone. University of Sussex, Research Papers in Geography, 9
------ and Funnell, D.C. (1983) Geography and integrated rural development. Geografiska Annaler, 65B, 57-63
Blaikie, P. (1985) The political economy of soil erosion in developing countries. Longman, London
Bromley, R.D.F. and Bromley, R. (1982) South American development: a geographical introduction. Cambridge University Press, Cambridge
Bromley, R. and Gerry, C. (eds) (1979) Casual work and poverty in Third World. Wiley, New York
Brookfield, H. (1975) Interdependent development. Methuen, London
Buchanan, K.M. and Pugh, J.C. (1955) Land and people of Nigeria. London
Castells, M. (1977) The urban question: a Marxist approach. Arnold, London
------ (1978) City, class and power. Macmillan, London
Chayanov, A.V. (1966) The theory of peasant economics. Irwin, London
Chinoy, E. (1969) Society: an introduction to sociology. 2nd edn. Random House, New York
Clapham, C. (1985) Third World politics: an introduction. Croom Helm, London
Clarke, C. (1985) Pluralism and plural societies: Caribbean perspectives. In C. Clarke, D. Ley and C. Peach (eds), Geography and ethnic pluralism, George Allen and Unwin, London, pp. 51-86
------ (1986) Sovereignty, dependency and social change in the Caribbean. In South America, Central America and the Caribbean, 1986, Europa Publications, London
Clower, R.W. (1966) Growth without development: an economic survey of Liberia. Northwestern University Press
Conyers, D. (1982) An introduction to social planning in the

Third World. Wiley, Chichester
Cox, R. (1981) Bourgeois thought and the behavioural geography debate. In R. Cox and R.G. Golledge (eds), Behavioural geography revisited. Methuen, London
Cross, M. (1979) Urbanization and urban growth in the Caribbean. Cambridge University Press, Cambridge
Crush, J.S. and Riddell, J.B. (1980) Third World misunderstanding? Area, 12, 204-6
Dear, M. and Scott, A.J. (eds) (1981) Urbanization and urban planning in capitalist society. Methuen, London
Dey, J. (1981) Gambian women: unequal partners in rice development projects? Journal of Development Studies, 17, 109-22
Dickenson, J.P. (1986) Irreverence and another new geography. Area, 18, 52-4
Dowse, R.E. and Hughes, J.A. (1972) Political sociology. John Wiley, London
Dwyer, D.J. (1975) People and housing in Third World cities. Longman, London
Farmer, B.H. (1983) British geographers overseas, 1933-1983. Transactions of the Institute of British Geographers, New Series, 8, 70-9
Forbes, D.K. (1984) The geography of underdevelopment: a critical survey. Croom Helm, London
Forde, C.D. (1934) Habitat, economy and society. Methuen, London
Frank, A.G. (1967) Capitalism and underdevelopment in Latin America. Monthly Review Press, New York
Friedmann, J. and Weaver, C. (1979) Territory and Function: the evolution of regional planning. Arnold, London
Giddens, A. (1979) Central problems in social theory: action, structure and contradiction in social analysis. Macmillan, London
------ (1981) A contemporary critique of historical materialism, vol. I: Power, property and the state. Macmillan, London
------ (1982a) Profiles and critiques in social theory. Macmillan, London
------ (1982b) Sociology: a brief but critical introduction. Macmillan, London
Gilbert, A. and Gugler, J. (1982) Cities, poverty and development: urbanization in the Third World. Oxford University Press, London
------ and Healey, P. (1985) The political economy of land: urban development in an oil economy. Gower, Aldershot

Gourou, P. (1953) The tropical world. Longman, London

Gregory, D. (1978) Ideology, science and human geography. Hutchinson, London

Guyer, J. (1981) Household and community in African studies. Final Report presented to the Social Science Research Council, London

Haralambos, M. (1980) Sociology: themes and perspectives. University Tutorial Press, Slough

Harriss, B., Chapman, G., McLean, W., Shears, E. and Watson, J.E. (1984) Exchange relations and poverty in dryland agriculture. Concept, Delhi

Hartmann, E. and Boyce, J.K. (1981) Needless hunger: poverty and power in rural Bangladesh. In R.E. Galli (ed), The political economy of rural development: peasants, international capital and the state, State University of New York Press, Albany, pp. 175-210

Harvey, D. (1973) Social justice and the city. Arnold, London

Hedges, T.R. (1963) Farm management decisions. Prentice-Hall, Englewood Cliffs

Heyer, J., Roberts, P. and Williams, G. (1981) Rural development in tropical Africa. Macmillan, London

Hill, P. (1982) Rural Hausa: a village and a setting. Cambridge University Press, Cambridge

Hyden, G. (1986) The invisible economy of smallholder agriculture in Africa. In J.L. Moock (ed.), Understanding Africa's rural households and farming systems, Westview Press, Boulder Colorado, pp. 11-35

Johnston, R.J. (1979) Geography and geographers: Anglo-American human geography since 1945. Arnold, London

------ (1980) City and society: an outline for urban geography. Penguin, Harmondsworth

------ (1982) Geography and the state: an essay in political geography. Macmillan, London

------ (1983) Philosophy and human geography: an introduction to contemporary approaches. Arnold, London

------ (1984) The world is our oyster. Transactions of the Institute of British Geographers, New Series, 9, 443-59

Karch, C. (1985) Class formation and class and race relations in the West Indies. In Dale L. Johnson (ed.), Middle class in dependent countries, Beverley Hills, Sage, pp. 107-36

Kowalewski, D. (1982) Transnational corporations and Caribbean inequalities. Praeger, New York

Layne, A. (1979) Race, class and development in Barbados.

Caribbean Quarterly, 25, 40-51
Lewis, B. (1984) The impact of development policies. In M.J. Hay and S. Stichter (eds), African women south of the Sahara, Longman, London, pp. 170-87
Liebenow, J.G. (1962) Liberia. In G.M. Carter (ed.), African one-party states, Cornell University Press
Lipton, M. (1968) Seasonality and ultra-poverty. Institute of Development Studies Bulletin, 17, 4-8
------ (1977) Why poor people stay poor: urban bias in world development. Temple Smith, London
------ (1982) Why poor people stay poor. In J. Harriss (ed.), Rural development: theories of peasant economy and agrarian change, Hutchinson, London, pp. 66-81
Longhurst, R. (1983) Integrating nutrition into agricultural and rural development projects: an application of the FAO methodology. In R. Longhurst (ed.), Nutritional impact of agricultural projects, London
Lowenthal, D. (1972) West Indian societies. Oxford University Press, London
Mabogunje, A.L. and Gana, J. (1981) Rural development in Nigeria: a case study of the Funtua Integrated Rural Development Project, Kaduna State, Nigeria. University of Ibadan, Nigeria
McKee, K. (1986) Household analysis as an aid to farming systems research: methodological issues. In J.L. Moock (ed.), Understanding Africa's rural households and farming systems, Westview Press, Boulder, Colorado, pp. 188-98
Matlon, P.J. (1979) Income distribution among farmers in northern Nigeria: empirical results and policy implications. Michigan State University, African Rural Economy Paper, 18
Miliband, R. (1977) Marxism and politics. Oxford University Press, London
Mishra, G.P. (1982) Dynamics of rural development in village India. Ashish Publishing House, New Delhi
Moore, M.P. (1984) Political economy and the rural-urban divide, 1767-1981. Journal of Development Studies, 20, 5-27
Moser, C.O.N. (1982) A home of one's own: squatter housing strategies in Guayaquil, Ecuador. In A. Gilbert, J.E. Hardoy and R. Ramirez (eds), Urbanization in contemporary Latin America, Wiley, Chichester, pp. 159-90
Mountjoy, A.B. (1980) Worlds without end. Third World

Quarterly, 2, 753-7
Norman, D.W. and Baker, D.C. (1986) Components of farming systems research, FSR credibility and experiences in Botswana. In J.L. Moock, (ed.), Understanding Africa's rural households and farming systems, Westview Press, Boulder Colorado, pp. 36-57
O'Connor, A.M. (1976) Third World or one world? Area, 8, 269-71
Payne, A. (1984) The international crisis in the Caribbean. Croom Helm, London
Peet, R. (1977a) Radical geography. Methuen, London
------ (1977b) The development of radical geography in the United States. Progress in Human Geography, 1, 64-87
Peters, PE. (1986) Household management in Botswana: cattle crops v wage labour. In J.L. Moock (ed.), Understanding Africa's rural households and farming systems, Westview Press, Boulder, Colorado, pp. 133-55
Porteous, J.D. (1986) Intimate sensing. Area, 18, 250-1
Potter, R.B. (1983) Tourism and development: the case of Barbados, West Indies. Geography, 68, 46-50
------ (1985) Urbanisation and planning in the Third World: spatial perceptions and public participation. Croom Helm, London and St Martin's Press, New York
------ (1986) Spatial inequalities in Barbados. Transactions of the Institute of British Geographers, New Series, 11, 183-98
------ and Dann, G. (1986) 'Core-periphery relations and retail change in a developing country: the case of Barbados, West Indies'. Paper presented at the Symposium on Commercial change organised by the International Geographical Union Study Group on the Geography of Commercial Activities, Barcelona
Pryer, J. (1986) 'Production and reproduction of malnutrition in an urban slum in Khulna, Bangladesh. Paper presented at the Annual Conference of the Institute of British Geographers, Reading
Ramphal, S.S. (1985) A world turned upside down. Geography, 70, 193-205
Reiser, R.L. (1973) The territorial illusion and behavioural sink: critical notes on behavioural geography. Antipode, 5, 52-7
Richards, P. (1985) Indigenous agricultural revolution. Hutchinson, London
------ (1986) Coping with hunger. Allen and Unwin, London
Rostow, W.W. (1960) The stages of economic growth: a non-

communist manifesto. Cambridge University Press, Cambridge
Roxborough, I. (1979) Theories of underdevelopment. Macmillan, London
Sandbrook, R. (1985) The politics of Africa's economic stagnation. Cambridge University Press, Cambridge
Saunders, P. (1979) Urban politics: a sociological interpretation. Weidenfield and Nicholson, London
------ (1981) Social theory and the urban question. Hutchinson, London
Schapera, I. (1971) Married life in an African tribe. Penguin, Harmondsworth
Seers, D. (1979) The meaning of development. In D. Lehmann (ed.), Development theory: four critical studies, Cass Reprint
Short, J.R. (1984) The urban arena: capital, state and community in contemporary Britain. Macmillan, London
Skinner, E.P. (1965) Labour migration among the Mossi of the Upper Volta. In H. Kuper (ed.), Urbanization and migration in West Africa, Berkeley, California
Stöhr, W. and Taylor, D. (1981) Development from above or below? The dialectics of regional planning in developing countries. Wiley, Chichester
Swift, J., Winter, M. and Fowler, C. (1984) Production systems in central Mali: the pastoral Tuareg of the Inner Niger Delta. ILCA Arid and Semi-Arid Zone Programme Report
Swindell, K. and Sutherland, A.M.D. (1985) 'Farming on the fringe: agrarian change at the rural-urban interface'. Paper presented at the Commonwealth Geographical Bureau Workshop on Spatial Inequalities in Developing Countries, Bayero University Kano, Nigeria
Taussig, M. (1982) Peasant economics and the development of capitalist agriculture in the Cavca Valley, Colombia. In J. Harriss (ed.), Rural development: theories of peasant economy and agrarian change, Hutchinson, London
Vance, J.E. (1970) The merchant's world: the geography of wholesaling. Prentice-Hall, Englewood Cliffs
Vedovato, C. (1986) Politics, foreign trade and economic development: a study of the Dominican Republic. Croom Helm, London
Wallerstein, I. (1974) The modern world-system: capitalist agricultural and the origins of the European World-economy in the sixteenth century. Academic Press,

New York
------ (1979) The capitalist world-economy. Cambridge University Press, Cambridge
------- (1980) The modern world-system II. Mercantilism and the consolidation of the European world-economy, 1600-1750. Academic Press, New York
Ward, B. (1980) First, second, third and fourth worlds. The Economist, 18th May, 65-73
Ward, P.M. (ed.) (1982) Self-help housing: a critique. Mansell, London
White, L. (1984) Women in the changing African family. In M.J. Hay and S. Stichter (eds), African women south of the Sahara, Longman, London, pp. 53-68
Williams, G. (1981) Inequalities in rural Nigeria. University of East Anglia, Development Studies Occasional Paper, 16
Wolf-Phillips, L. (1979) Why Third World? Third World Quarterly, 1, 105-13
World Bank (1980) World Development Report, 1980. World Bank, New York

Part II

REGIONAL STUDIES

Chapter Eight

CONTEMPORARY ISSUES IN LATIN AMERICA

R.N. Gwynne

In being asked to discuss the contemporary issues upon which geographers should be focusing in their present and future research on the highly complex and variable continent of Latin America, innumerable questions of what to include face the writer. Two basic problems spring up at the outset. Firstly, for a contributor who believes in the interdisciplinary concept of 'Area Studies', to what extent should reference be made to continent-wide issues being discussed in economics, sociology or political science? Secondly, to what extent should the chapter refer to research methodologies and studies made by geographers from outside the United Kingdom? Should reference be made, for example, to the contributions of French geographers, with their impressive synthetic studies of individual regions or sub-regions of Latin America, such as Grenier's recent (1984) extensive regional monograph on Chiloé, the large island to the south of mainland Chile. As with Grenier's study, the regional approach is often highly appropriate for the study of an impoverished area in which the utilisation of local resources for local needs is still the major characteristic of the regional economy. Although many such regions exist in Latin America, the approach becomes distinctly less applicable to regions in which considerable interaction occurs with other regions - such as metropolitan, highly urbanised or export-economy regions.

Many North American geographers also carry on the regional tradition (James, 1969), although the perceived need for quantification and a greater openness to the contributions of other disciplines have wrought certain changes. Should reference be made to the growing army of

Latin American geographers, many of whom have followed the French and Spanish regional paradigm of geography (Cunhill, 1978), but where increasingly a more interdisciplinarian and systematic approach is taking hold (Ortiz, 1983; Bodini, 1985; Romero et al., 1983)? Indeed, in Becker's recent review of Brazilian geography, scarce mention is made of the regional tradition, despite the admission that 'regional monographs predominated in early studies, from the mid-1960s to the mid-1970s' (Becker, 1986, p. 171). Apart from reference to issues in physical geography and epistemology, Becker sees four crucial themes to be focused upon by Brazilian geographers in the late 1980s: urban issues, the capitalisation of rural areas, labour mobility, and the frontier.

The potential scope of the theme is therefore vast, and any single essay can only tackle but a small proportion of the issues that could be defined as geographical. The issues to which I will refer have some similarity with those mentioned by Becker in that they mainly lie in the field of economic and social geography. However, while Becker remarkably ignores the international dimension of Brazil's development at a time when that country's room for manouevre is sorely constrained by problems of international debt, this chapter will first concentrate on reviewing these international processes that so crucially affect the economic patterns of development within Latin American countries. Secondly, attention will be drawn to the mode of economic growth that Latin America is pursuing, before some important research issues in urban development for the last decade of the twentieth century are examined.

LATIN AMERICA IN THE WORLD ECONOMY

The flood of theoretical writing (with even some empirical work) in the 1970s on dependency in Latin America has at least verified one fundamental point about that continent's development - that the operation of the world capitalist system significantly affects both the pace and rhythm of economic growth in each Latin American country. As a result, the mode of insertion of each Latin American country into the world economy constitutes a fundamental aspect of study. Furthermore, this is by no means an issue of historical research. The last half of the twentieth century has seen a rapidly increasing integration and interdependence of world economies, partly due to the revolution in world communications. As a result, fundamental change in one part of the world economic system rapidly reverberates throughout the rest of that system - one good recent example being that of the high American dollar between 1982 and 1985, and its sudden and substantial impact on world capital flows and trade. Indeed, any examination of Latin America's present position within the world capitalist system must start by referring to these two major contemporary issues of the world economy - trade and capital flows.

Since the early nineteenth century and the independence of most Latin American countries from Spanish rule, trade between Latin America and the industrialised countries has maintained a very distinctive pattern. On the one hand, there has been the export of fuels, minerals, metals, agricultural and other primary products from Latin America, while on the other, there has been the import of manufactured products into Latin America. Tables 8.1 and 8.2 demonstrate that such a pattern was still strongly in evidence in 1965. In that year, primary commodities accounted for more than 80 per cent of exports, and manufactures and machinery were responsible for more than 50 per cent of imports in all Latin American countries.

By 1983, the situation had not changed substantially (see Tables 8.1 and 8.2). In the Andean countries, for example, primary commodities still account for over 80 per cent of total exports and manufactures for over 70 per cent of imports. The small size of markets of most Andean countries and the failure of the Andean Group to foster manufacturing trade between individual countries has meant that the restructuring of export trade has been limited to the primary sector - Ecuador shifting from exporting agricultural goods to petroleum, and Chile changing from an extreme dependence on copper exports to a more diversified structure of primary exports - agricultural, forestry, fish and other mineral products. Similarly, and apart from Uruguay, little restructuring of exports from primary to manufactured goods has occurred in the River Plate countries. Evidence of a modest restructuring of trade can be perceived in Central America, where the success of the Central American Market in promoting industrial growth and manufacturing trade between member countries has

Table 8.1: Changes in the structure of Latin American exports, 1965-83 (Percentage share of merchandise exports)

	Fuels, minerals and metals			Other primary commodities			Manufactures		
	1965	1983		1965	1983		1965	1983	
Brazil	9	15		83	44		9	40	
Oil-exporting countries									
Venezuela	97	97	(1981)	1	0	(1981)	2	3	(1981)
Mexico	22	64		62	9		17	27	
Ecuador	2	64		96	33		3	3	
Andean countries									
Bolivia	93	86	(1979)	3	11	(1979)	4	3	(1979)
Peru	45	69		54	17		1	14	
Colombia	18	15		75	66		6	19	
Chile	89	65	(1981)	7	27	(1981)	5	8	(1981)
Central American countries									
Honduras	6	7		90	84		4	7	
El Salvador	2	5		81	55		17	40	
Nicaragua	4	1		90	91		5	7	
Guatemala	0	2	(1981)	86	69	(1981)	14	29	(1981)
Costa Rica	0	1		84	71		16	28	
Panama	-	23		-	64		-	13	

Table 8.1: continued

	Fuels, minerals and metals			Other primary commodities			Manufactures		
	1965	1983		1965	1983		1965	1983	
River Plate countries									
Argentina	1	6		93	78		5	15	
Paraguay	0	0	(1979)	92	88	(1979)	8	12	(1979)
Uruguay	0	0		95	70		5	29	

Note: Figures which are underlined correspond to 1982, not 1983

Source: World Bank, Development Reports, 1983-6. Oxford University Press, New York

Table 8.2: Changes in the structure of Latin American imports, 1965-83 (Percentage share of merchandise exports)

	Fuels			Food and other primary commodities			Manufactures and machinery		
	1965	1983		1965	1983		1965	1983	
Brazil	21	56		29	12		50	32	
Oil-exporting countries									
Venezuela	1	1	(1981)	17	21	(1981)	82	78	(1981)
Mexico	2	3		15	23		83	74	
Ecuador	9	2		14	11		77	88	
Andean countries									
Bolivia	1	2		22	13		76	85	
Peru	3	2		22	21		75	77	
Colombia	1	13		18	16		80	71	
Chile	6	15	(1981)	29	15	(1981)	65	70	(1981)
Central American countries									
Honduras	6	22		12	12		82	65	
El Salvador	5	25		19	21		76	54	
Nicaragua	5	23		14	13		81	63	
Guatemala	7	38	(1981)	13	9	(1981)	79	53	(1981)
Costa Rica	5	20		11	12		83	68	
Panama	-	27		-	10		-	63	

Table 8.2: continued

	Fuels		Food and other primary commodities		Manufactures and machinery	
	1965	1983	1965	1983	1965	1983
River Plate countries						
Argentina	10	10	27	14	63	75
Paraguay	14	24	16	13	70	63
Uruguay	17	36	23	13	60	51

Note: Figures which are underlined correspond to 1982, not 1983

Source: World Bank, World Development Reports, 1985-6. Oxford University Press, New York

Figure 8.1: Composite commodity price index, 1948-82

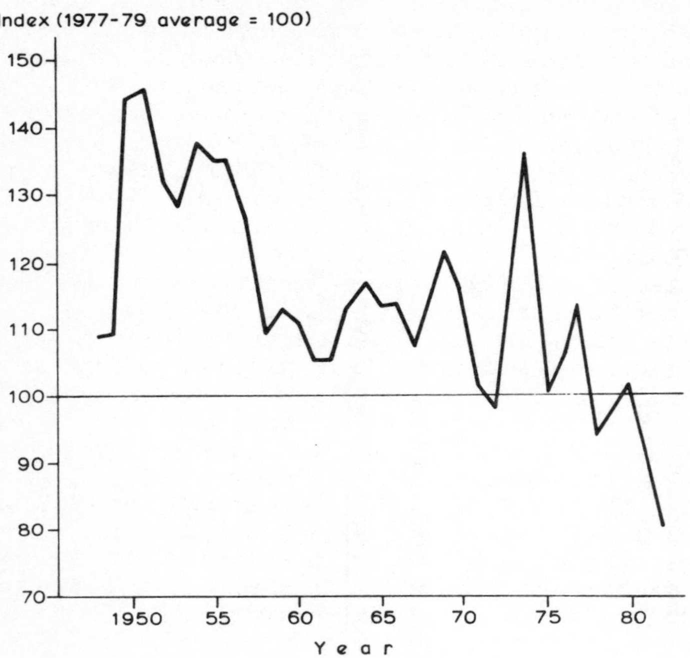

The graph shows non-oil commodity prices as measured by the price of manufactures imported by developing countries. The commodities are coffee, cocoa, tea, maize, rice, wheat, sorghum, soybeans, groundnuts, palm oil, coconut oil, copra, groundnut oil, soybean meal, sugar, beef, bananas, oranges, cotton, jute, rubber, tobacco, logs, copper, tin, nickel, bauxite, aluminium, iron ore, manganese ore, lead, zinc, and phosphate rock.

Source: World Bank (1983) World Development Report, 1983, Oxford University Press, New York

been noted (Gwynne, 1985, p. 75). As can be seen from Table 8.1, only three countries truly benefited from the scheme in being able to expand regional exports of manufactures - El Salvador, Guatemala and Costa Rica. Furthermore, Table 8.2 demonstrates that the need to boost manufactured exports in these three countries was closely linked to the need to finance increased imports of fuel, and particularly petroleum, in these energy-deficient countries. However, by far the most significant reorientation of trade has occurred in Brazil, where by the mid-1980s, about 50 per cent of exports came to be derived from the manufacturing sector, and only 30 per cent of imports corresponded to machinery and manufactures.

Related to this aggregate empirical evidence, one fundamental question must be asked. Why is such dependence on primary exports a disadvantage for the late 1980s and 1990s? One important consideration is that there has been a fall in the share of less developed countries in world exports of primary products other than oil. Less developed countries accounted for less than 36 per cent of exports of raw materials in 1978, compared with a share of almost 44 per cent in 1960. A second consideration is that prices for commodities have been both declining and unstable in the last 30 years. Figure 8.1 shows that non-fuel commodities have declined in value by over a third from the early 1950s to the early 1980s: between 1980 and 1986, non-oil commodity prices have fallen by 26 per cent in dollar terms, or by an astonishing 23 per cent relative to the price of manufactures. The shift in the terms of trade between 1980 and 1984 for each Latin American country is shown in Table 8.3. Considerable price instability has existed alongside this general decline, as price behaviour in the early 1970s demonstrated. In certain commodities, the decline (and instability) of prices has been greater than the composite average. In 1985, the real value of copper was less than 20 per cent of its 1970 value, a fact that has caused considerable stress for Latin America's two copper exporters, Chile (now the world's largest copper exporter) and Peru (see Table 8.3).

Instability and declining prices have even hit the commodity that leading commentators thought disobeyed the rules of the international market in the late 1970s. Oil prices increased from $11 a barrel in 1974 to $40 a barrel in 1980, only to plummet back to the 1974 level in 1986 and then to go even lower. It could be argued that the impact of declining and volatile prices on Latin America's oil-exporting countries (Mexico, Venezuela and Ecuador) has been even more dramatic than the comparative case of metal-exporting countries - simply because of the accepted wisdom of the late 1970s that oil prices would carry on

Table 8.3: Shifts in the terms of trade for Latin American countries during the 1980s recession (1980-4) (1980 = 100)

	1984 Index
Brazil	103
Venezuela	99
Mexico	100
Ecuador	98
Bolivia	91
Peru	84
Colombia	97
Chile	80
Honduras	93
El Salvador	72
Nicaragua	70
Guatemala	80
Costa Rica	103
Panama	84
Argentina	97
Paraguay	95
Uruguay	85

Source: World Bank, World Development Report, 1986. Oxford University Press, New York

increasing for the rest of the twentieth century. Based on this premiss, both Venezuela and Mexico borrowed heavily on international capital markets in order to finance a crash programme of capital-intensive industrialisation. The fall in oil prices has therefore not only caused severe trading problems in 1986, due to the halving in the value of exports, but also exacerbated an already severe debt problem.

While countries dependent on primary exports of minerals and metals have been faced with exacerbated terms of trade in the 1980s due to low world commodity prices, countries dependent on agricultural exports have not only encountered low world prices but also restrictive trade barriers from the industrial countries. As a result, and according to the World Bank (1986), 'most of the world's food exports are grown in industrial countries, where the costs of food production are high, and consumed in developing ones, where the costs are lower'. In Latin America, such restrictions on world food trade have been particularly harmful for Argentina and Uruguay, where over 70 per cent of exports are composed of food products that directly compete in world markets with the subsidised surpluses of the industrialised countries.

Such severe problems in world commodity trade have not allowed Latin American countries to reduce the impact of their other international constraint - indebtedness. The only partial exception is that of Brazil, due to that country's high proportion of manufactured exports, which have been better able to maintain their international prices. The Latin American debt crisis had its origins in the unusually large inflow of capital during the late 1970s, much of it through commercial banks lending short-term maturities at floating rates. This left them very vulnerable to the events of the 1980s. The monetary and fiscal policies pursued by industrial countries after 1979 drove world interest rates up. By 1982, oil-importing Latin American countries were paying a nominal rate of interest of around 13 per cent for commercial loans (while their export prices declined by 5 per cent). Figure 8.2 demonstrates the relationship between _per capita_ debt and _per capita_ income for most Latin American countries. It constitutes a dramatic indication of the burden of debt on some Latin American economies. For two of the poorer Latin American countries (Nicaragua and Bolivia), _per capita_ debt is actually greater than _per capita_ income. For three middle-income countries (Chile, Costa Rica and Peru), _per capita_ incomes are at about the same level as _per capita_ debt. On this criterion, the major Latin American debtor in absolute terms, Brazil, seems better off, with _per capita_ income over two times greater than _per capita_ debt. Therefore, Figure 8.2 demonstrates that the effect of the debt burden is not restricted to the large absolute debtors; often the smaller, less publicised debtors have economies that are constrained considerably more.

The economic development of Latin American countries is presently being severely affected by low export prices, trade restrictions in industrial countries, high debts and interest rates and the associated features of capital outflow, constraints on imports and a depreciating exchange rate. Prebisch's (1962) assertion that the true nature of the dependent relationship between Latin America (the

Figure 8.2: Latin American per capita debt, 1983

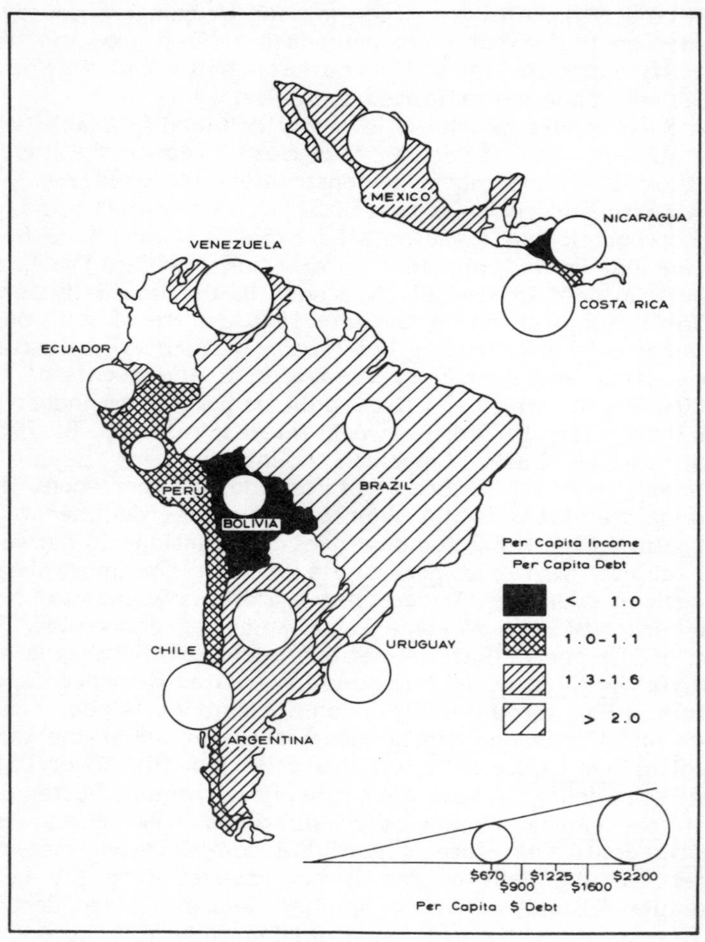

periphery) and the industrialised countries (the centre) is most obvious and acute during times of recession is most definitely being proved during the 1980s. Indeed, the differential impact of the 1980s recession on industrialised and less developed countries brings into focus again the work of researchers in the Economic Commission for Latin America (ECLA) and their spatial analysis of centre-periphery relationships. Two ECLA researchers, Pinto and Knakal, reviewed the centre-periphery model in 1973 and argued that it had been affected by 'two parallel and contradictory processes which could be called relative marginalisation and dependent insertion'. Relative marginalisation of the periphery referred to the latter's declining share of world trade and especially its declining share of primary-product trade. The major feature of post-Second World War trade has been the growth in trade between the industrialised countries of the centre itself. In this way, Pinto and Knakal saw the Latin American periphery as decreasingly 'necessary' for the centre in terms of providing primary products and as a market for goods and services. Alongside marginalisation, however, the contradictory process of dependent insertion has occurred. However, the concept of dependency has come to have a much wider context than that of trade. It referred, rather, to a broader set of structural relations between Latin America and other peripheral countries, on the one hand, and the countries of the industrialised centre on the other. In terms of economic dependency, the processes of technology transfer, capital flows and direct investment by multinational corporations were seen as further examples of dependency relationships. Dependency also became associated with social and cultural processes, however (Palma, 1978).

The relationships of marginalisation and dependency between the Latin American periphery and the industrialised centre are still crucial areas for both empirical and theoretical research, particularly after the recession of the 1980s has demonstrated again the fundamental nature of the centre-periphery model - the centre active, the periphery passive and reflexive. The development of each Latin American country is now heavily constrained by international forces over which it has little influence. Within this pessimistic external context, what form of economic growth can be achieved?

ISSUES OF PRODUCTION AND EMPLOYMENT

One conclusion to be drawn from the differential economic performance of Latin American countries during the recession is that the one country able to have shifted the majority of its exports from primary to manufactured products, Brazil, has been better able to reduce the impact of external constraints on its economic development. Such a conclusion ties in with Prebisch's policy recommendation that only through industrialisation would Latin American countries effectively reduce the impact of external constraints on their development. What are the prospects for growth in the manufacturing sector in Latin America?

Firstly, it is necessary to examine some of the characteristics of Latin American industry. Industrialisation in Latin America has traditionally been promoted in order to supply the national market - firstly as a reaction to the Depression of the 1930s and, after the Second World War, as part of a policy of import substitution. Despite being inward-looking, industrialisation has become a significant economic force in many Latin American countries, displacing agriculture in terms of productive importance in at least eleven countries (see Table 8.4). Although the processing of raw materials for export constitutes part of the manufacturing sector, more important in recent years has been the production of consumer goods. Not only has the emphasis of mid-twentieth century industrialisation world-wide shifted towards more consumer-good production, but the policy of import substitution has also promoted the production of a full range of consumer goods rather than a more specialised selection. Such an orientation of production has created a dependence on foreign technology. The production of consumer goods in Latin America had the effect of reducing imports of such goods but the corollary was the need to import the plant and inputs to produce these goods. As plant technique, machine capacity, inputs and product design changed overseas, there was continuous pressure on the industrialist to change. The national firm has tended to be very dependent on foreign technology, machinery and, at least initially, material inputs. The subsidiary of the multinational firm has been even more closely linked to sources of foreign technology and inputs (Cunningham, 1986). Returning to the question posed at the beginning of this section, the possibility for Latin American countries to reduce their external constraints by expanding

Table 8.4: Ratio of manufacturing to agricultural GDP in Latin America, 1965-84

Country	Ratio 1965	1979	1984
Chile	2.67	3.00	3.50
Mexico	1.50	2.90	2.67
Argentina	1.94	2.85	2.50
Venezuela	-	2.66	2.57
Peru	1.33	2.60	3.13
Brazil	1.37	2.55	2.08
Uruguay	-	2.39	-
Ecuador	0.67	1.27	1.36
Costa Rica	-	1.00	-
Nicaragua	0.72	0.83	1.04
Bolivia	0.76	0.77	0.80
Colombia	0.60	0.72	0.90
El Salvador	0.62	0.54	0.76
Honduras	0.30	0.53	0.56
Paraguay	0.43	0.52	0.65
Panama	0.66	-	0.65

Source: World Bank, World Development Reports

manufactured exports looks distinctly dubious. Brazil was able to change the orientation of its industry from inward- to outward-orientation mainly because of its large market, which had generally allowed imported machinery and plant to work with a high utilisation of capacity and therefore low per unit cost. Other countries in Latin America (with the possible exception of Mexico) have had to endure imported machinery and plant working at much lower levels of capacity utilisation and therefore much higher per unit cost (Gwynne, 1985).

Manufacturing production has, therefore, tended to be inward-looking and based on foreign labour-saving technology (ibid.). Such a pattern of industrialisation has managed to have an impact on indices of production (especially in relation to agriculture), but its impact on employment has been much less significant as Table 8.5 indicates. In 1980, 13 out of 17 Latin American countries still recorded employment totals in agriculture greater than those in industry. In the four middle-income countries in which this did not occur, employment totals in the tertiary

Table 8.5: Ratio of manufacturing to agricultural employment in Latin America, 1965-80; and percentage employment in services, 1980

Country	Employment in services (%) 1980	Manufacturing/ agricultural employment 1965	1980
Chile	58	1.07	1.56
Venezuela	56	0.80	1.75
Uruguay	55	1.45	1.81
Argentina	53	1.89	2.62
Panama	50	0.35	0.56
Costa Rica	46	0.40	0.74
Brazil	42	0.42	0.87
Colombia	42	0.46	0.71
Ecuador	42	0.34	0.51
Peru	42	0.38	0.45
Nicaragua	38	0.28	0.34
Mexico	34	0.44	0.78
Bolivia	34	0.37	0.43
Paraguay	31	0.36	0.43
El Salvador	30	0.27	0.25
Guatemala	26	0.23	0.46
Honduras	23	0.18	0.26

Source: World Bank, World Development Reports

sector were on average double those in manufacturing. Thus, the type of manufacturing developed in Latin America has been distinctly out of step with the demands of the labour market. Part of the problem is the rapidly increasing population (averaging 2.5 per cent *per annum* - See Table 8.6) and the even faster growth in the labour force (averaging 2.8 per cent *per annum*, but varying between countries - see Table 8.6). To give an idea of the scale of this problem, one need only refer to Brazil which had a working population of 77 million in 1984. However, with the labour force increasing by 3 per cent *per annum*, this means that the labour force will have increased by as much as 12 million in the last five years of the 1980s. The present mode of industrialisation will find it difficult to even provide 25 per cent of the jobs.

Industrialisation, then, has not provided the jobs for the rapidly increasing labour force. In Brazil, the most

Table 8.6: Recent growth of Latin America's population and labour force, 1973-84 (average annual growth of population/labour force, per cent)

	Population growth 1973-84 (%)	Labour force growth 1973-84 (%)
Brazil	2.3	3.0
Venezuela	3.3	3.9
Mexico	2.9	3.2
Ecuador	2.9	2.9
Bolivia	2.6	2.5
Peru	2.4	2.9
Colombia	2.0	2.8
Chile	1.7	2.5
Honduras	3.5	3.3
El Salvador	3.0	2.9
Nicaragua	3.0	3.2
Guatemala	2.8	2.8
Costa Rica	2.9	3.8
Panama	2.3	2.6
Argentina	1.6	1.1
Paraguay	2.5	3.3
Uruguay	0.5	0.5
Unweighted average	2.5	2.8

Source: World Bank, World Development Report, 1986. Oxford University Press, New York

industrialised of Latin American countries, industry accounts for only 27 per cent of the labour force. However, there is an interesting relationship between the urban hierarchy and industrial employment. Generally speaking, within each country, the larger the city, the larger is the proportion of industrial workers to total urban employment. Brazil's largest cities, Rio de Janeiro and Sao Paulo, for example, have an average of 36 per cent of their work-force operating in manufacturing, while in the smaller northern cities of Recife, Salvador, Belem and Fortaleza, the average is only 26 per cent. Part of the explanation lies in the extreme spatial concentration of manufacturing within most Latin American countries (Dickenson, 1978; Morris, 1981). The Sao Paulo agglomeration contains about 50 per cent of Brazil's manufacturing work-force, Santiago about 60 per cent of Chile's, and Lima/Callao about 70 per cent of the Peruvian. There is, then, an identifiable relationship between industrialisation and urbanisation in Latin America (Gwynne, 1985). It is precisely the large primate cities which have both the bulk of the national manufacturing work-force <u>and</u> a relatively high percentage of their own work-force in manufacturing that are the most rapidly expanding urban centres in Latin America. They constitute the major target for the steady flow of migrants from rural areas and from small towns with little economic growth.

Nevertheless, at all levels of the urban hierarchy, the growth in manufacturing production does not provide sufficient employment for the rapidly expanding work-force. This constitutes the fundamental economic problem of Latin American cities. Those without access to productive employment in either secondary or tertiary sectors inevitably become the poor of the city. Having made that basic point, however, problems of definition abound. The standard concept of the 1970s was of an urban duality, a capital-intensive, modern 'formal' sector juxtaposed with a labour-intensive, primitive 'informal' sector. The overriding problem of artificially dividing a functioning urban economy into two on the basis of simple typologies (Santos, 1979) has caused the partial demise of the concept in the 1980s (Bromley, 1985). The policy emphasis is now on how to increase the generation of employment in Latin American cities. The current emphasis, from all sides of the political spectrum, is to increase the number of small enterprises in a wide variety of manufacturing and service sectors (Harper, 1984; Bromley, 1985). In terms of future research direction, the recommendations of Schmitz (1982, p. 193), at the end of his classic book on small-scale manufacturing in Brazil should be noted:

> Hence the issue is not <u>whether</u> small enterprises have growth and employment potential but <u>under what conditions</u>. This question cannot be answered by a mere listing of factors ... The task ahead is more difficult as the relative importance of the various factors needs to be established ... this can best be done through branch-

> specific studies ... ultimately branch-specific studies must be situated in wider investigations which encompass the development in the national and international economy.

RURAL ISSUES

Before moving on to investigate the implications of employment problems on such urban issues as housing and amenities, it is worth considering the rural areas, from which the majority of migrants into the city come. In this context we may remind ourselves of Roberts's argument that 'the basic issue of internal migration is not why so many people leave the land but why more people do not do so' (Roberts, 1978).

The key issue of the 1960s, and to a lesser extent the 1970s, was that of agrarian reform, interpreted, in practice, as the break-up of the large semi-feudal estates that had characterised Latin America since colonial times. Present agrarian issues in each country are very much framed in terms of the legacy of agrarian reform. In some countries (Mexico, Bolivia), agrarian reform predated the 1960s. In other countries, it was non-existent for all practical purposes (Brazil, Argentina, Uruguay, Paraguay), but, in others, it has revolutionised the landscape (Cuba, Nicaragua). For many countries, however, agrarian reform has been selective and limited in spatial extent, but nevertheless heralded important changes in the countryside - Peru, Venezuela, Chile, El Salvador (Browning, 1983). Undoubtedly, where agrarian reform has occurred, it has brought greater social justice in the countryside, in particular the abolition of systems whereby peasant families were tied to outmoded contractual relationships with the landlord, such as days of work, service in the landlord's house and payment in kind. In some cases, rural employment and productivity also increased (Preston, 1978). However, at the national scale, agrarian reform has not been associated with major increases in food production - certainly not enough to feed the expanding urban populations. As a result, food imports have had to increase in many Latin American countries. The reasons for agrarian reform being associated with stagnant food production are complex and vary from one country to another. One factor that is pervasive throughout <u>all</u> Latin American countries, however, is the question of low prices and negligible support. In contrast to most Western governments, Latin American governments have either not given price supports to their farmers, or if they have, they have been set at low levels. Low market prices for food products have given little incentive for peasant farmers to increase farm productivity and tradable surpluses.

It could be argued that agrarian reform and low price-supports have operated to keep rural populations high. Firstly, agrarian reform only dealt with the problem of the latifundio, the large estate, and not the minifundio the small peasant plot. As a result, rural population densities are still high in areas of minifundio agriculture. In the Peruvian Andes, where it is considered that a holding of five hectares of arable land is necessary to provide an average peasant household with an adequate living, there were as many as 1,083,775 holdings of less than five hectares in 1972 (Smith, 1983); this represented a 50 per cent increase in these small holdings over figures for 1961. Furthermore, of the 1972 figure, 483,350 farms were less than one hectare in size. In not dealing with the minifundio problem, then, agrarian reform has effectively kept people on the land - normally at levels of production where tradable agricultural surpluses are meagre. Moreover, in those countries in which agrarian reform has operated to divide up the large estates between former tenant farmers, the absence of national price incentives for production has often meant static technology, low yields and low labour productivity. In this way, redistributed land may have kept people on the land but has not led to significant growth in agricultural production.

It is a feature of agricultural development that, within a framework of low prices, it tends to be only the larger farms that can risk investments in agricultural modernisation and significantly increase production. In those countries that have not experienced agrarian reform or have had only limited reform, significant economic changes have often occurred on the large estates. Goodman and Redclift (1981) pointed to two processes of change with particular reference to Brazil. Firstly, they identified the process of internal proletarianisation, characterised by the transformation of the tenants of large estates into wage labourers. They argued that as land prices increase with technical change in agriculture, it is more profitable for the landowner who has the resources to invest to pay labourers cash rather than to part with small parcels of land.

Furthermore, the greater emphasis on cash-cropping and mechanisation on large commercial estates has created more seasonal variations in the labour market; it has become less attractive for landlords to employ labour throughout the year and they have resorted to an increased use of casual labour. The second process of change was defined as that of external proletarianisation, and occurred when landowners of large estates did not have capital resources sufficient for radical commercial and technological change. In order to achieve these capital resources, the landlord would sell off part of his land and develop the rump of his land as a commercial enterprise, relying on wage labour. Those former tenants able to purchase their land from the landowner will also have to operate on a commercial basis to survive. Greater production has resulted from these processes, but at the expense of social problems - a dramatic increase in the number of landless labourers and the break-up of the tenanted peasantry.

Rural issues in each Latin American country depend on the legacy of agrarian reform. In countries that have had reform, issues of production tend to predominate. These issues have recently been highlighted by the difficult trading situation of most countries and their need to constrain imports, and particularly those of food which can be produced locally. The mid-1980s have begun to see higher support prices for farmers producing basic foods. Results have often been dramatic. Chile began to introduce support prices for basic food products in 1983 after food imports had reached $709 million in 1981. Domestic production has responded in such a way that wheat imports came down from 1.16 million tons in 1983 to 0.48 million tons in 1985, and refined sugar imports from 203,000 tons in 1983 to only 6,000 tons in 1985. The food import bill was less than one-third of its 1981 level by 1985 - at only $219 million. However, while economic problems of production can partly be solved by support prices, social problems such as that of the landless labourers have no such easy solution. Most rural people without direct access to land take another option and move to the towns.

URBAN ISSUES

The character of recent urbanisation in Latin America is

clearly related to the pattern of economic development. Rapid growth in production has occurred - but without an equally rapid growth in urban employment. Nevertheless, the image of rapidly expanding urban areas (Sao Paulo is the fastest growing city in the world, expanding by 600,000 per annum), images of urban affluence and the generation of considerable employment have fuelled one of the greatest migrations in history - from the Latin American countryside to the urban agglomerations. In 1940, Latin America was only 20 per cent urban; now it is 65 per cent. However, the mismatch of employment generation with labour force expansion in the urban agglomerations has caused large numbers of poor people seeking housing as best they can - and the problems of urban poverty have indeed been the major focus of recent geographical research into urban areas in Latin America. (Gilbert, 1981, 1983; Batley, 1982; Edwards, 1982; Gilbert and Ward, 1982, 1984; Ward, 1983, 1986; Chant, 1985a, 1985b; Varley, 1985; Gilbert and Healey, 1985; Ward and Melligan, 1985).

Research into housing for the urban poor in Latin America has been heavily influenced by the work of Turner in the early 1960s. His basic thesis was that squatter settlements offered the framework for being a solution to housing the poor in Latin American cities, rather than constituting a problem (Turner et al., 1963). He advocated a closer involvement of government in improving the housing conditions of the urban poor. In particular, he argued that if the government gave the land title to the squatter and if it provided basic infrastructural services for the settlement (water supply, electricity, paved roads, sanitation), then the individual squatter would gradually be able to build his own house to his own design with the basic services of urban life at his disposal. International agencies, national governments and city administrations all became attracted to the basic concept, partly because the provision of basic services also seemed a cheap solution to housing the urban poor. More than 20 years after its original appearance, it is worth reconsidering Turner's model in the light of subsequent research and the macroeconomic conditions of the 1980s.

Firstly, as has already been discussed, the macroeconomic conditions of the 1980s are very different from those of the 1960s. However, Turner's model has always had an important macroeconomic perspective. If one compares the relatively successful process of consolidation (in Turner's terms) of squatter settlements in the more prosperous countries and regions of Latin America in the 1960s and 1970s with the continued poverty of squatter settlements in the poorer countries and regions of Latin America, it becomes evident that it is only in what could be termed an intermediately developed country that also has a rapidly growing national economy that a sufficient surplus is created and urban employment generated for the urban poor to invest in construction materials and erect suitable dwellings on their own initiative over a period of time. The relationship between a successful process of urban consolidation in squatter or spontaneous settlements, on the one hand, and national levels of income, rates of economic growth and the distribution of national income, on the other, thus constitute an important issue for investigation, particularly in the light of the dramatic downturn in Latin American economic growth during the 1980s. Related to this at a more local level of analysis is the relationship between the process of self-help housing and the evolution of the petty-commodity sector providing building materials and services.

A second issue revolves around non-ownership accommodation for low-income groups in Latin American cities. It was implicit in Turner's hypothesis that each family would build their house and stay to live there. However, recent empirical research (Edwards, 1982; Gilbert, 1983; Gilbert and Ward, 1984) has shown alternative strategies, with families building a house and subsequently renting it out to a second generation of incomers. Gilbert and Ward found that in three barrios (neighbourhoods) in Bogota, an average of 43 per cent of families were renters rather than owners. However, in their comparative research across three countries, there was no consistent relationship. While renting was common in Bogota, it was less common in Mexico City (average of 12 per cent of families renting in six barrios) and virtually non-existent in Valencia. One important question, then, is who specifically are the renters? Furthermore, how does the land market shape non-owner housing opportunities and why do some cities such as Valencia still have very low levels of renting? As renting becomes more significant, are there implications for self-help schemes and might there be conflicts of interest between owners and renters? Should government policy be specifically geared to renters as it has undoubtedly been in the past to owners? The general point is that due to the Turner thesis, it became axiomatic that the consolidation of spontaneous settlements and the creation of self-help housing schemes based on sites and services were based around owner-occupiers and an enormous quantity of research and literature has been devoted to this theme. It is now necessary to counterbalance this vast literature on self-help housing that assumes ownership and investigate in depth the rented sector.

The Turner model also tended to put the inner city firmly into a context of intra-urban migration. It was commonly assumed that within the inner city, low-income accommodation would take the form of old colonial or post-colonial elite residences that have been sub-divided to provide one-room rental tenements for certain poor groups. The latter are frequently understood to be recently arrived migrants - the 'bridgeheaders' of Turner's model (1968). Later, these households move into 'owner-occupancy' in self-built housing at the periphery. This generalised pattern has been questioned by empirical research of migrant intra-city mobilities, though findings have been mainly drawn from populations living in spontaneous settlements on the periphery (Gilbert and Ward, 1982). Few people have examined the mobility patterns of inner-city residents. In fact, far from being a transient or upwardly-mobile population, it appears that there is considerable population stability among the poor in the city centre. This is a response to growing urban diseconomies associated with living at the periphery, or in certain cases to the effects of past rent control legislation that has reduced incentives to move from the inner city (Batley, 1982; Ward and Melligan, 1985).

A large number of squatter or spontaneous settlements has originated from invasions of land peripheral to the expanding city (Gilbert, 1981). These invasions have involved both public (state) and private land. Generally speaking, the invasion of state land by squatter groups has been better rewarded, as after the early 1960s, most national governments began to see the creation and improvement of squatter settlements as a solution to the problem of lack of housing for the urban poor. However, private land has also been invaded by squatters and here the issues have been much less clear. The invasion by squatters of private land is illegal but important questions, in these circumstances, include: what are the illegal processes of land development, and how do they relate to the state and national governments? In Gilbert and Healey's (1985, p. 126) case study of low-income housing in Valencia, they found that

> few of the residents have committed any crime in occupying the land. Rather than invading land they have merely bought a house ("improvement") from the invaders. In this sense, the invaders, aided and abetted by the authorities, are playing the role of urban developers in the poorer parts of the city.

They add that their evidence 'suggests that where land prices are high invasions are not permitted'. Venezuela presents a curious case where invasion of private land is common but where government involvement to regularise matters has been distinctly limited in practice. Under such conditions, how might national governments intervene to provide land legally for self-help schemes after an illegal invasion of private land has taken place (Ward, 1983; Varley, 1985)? How and why do Latin American societies differ so much in the manner of illegal land development and in the response of government?

Recent research (Gilbert and Ward, 1984) has shown that participation of households in community improvement tends to decrease with the time lapse after invasion and particularly after land regularisation and the installation of basic services such as water and electricity. However, such research has thrown up other interesting issues linking household structure with the housing process: for example, how do women and children participate in the self-help process? Gilbert and Ward (1984) have shown that in three contrasting Latin American cities (Bogota, Mexico City and Valencia), women have played the role of principal participant in community affairs in over 20 per cent of households, and in some individual communities the rate has been over 50 per cent of households. One crucial issue to be investigated in this context would be whether certain forms of household structure (such as those that are female-headed and extended) are better equipped to facilitate survival and home improvement (Chant, 1985a, 1985b). At the national scale, the related issue of social welfare provision offers important questions to be asked. How are government priorities for social welfare changing and why? What are the spatial and social delivery systems of social welfare? How far are conditions improving or deteriorating through the recession of the 1980s (Ward, 1986)? It should also be pointed out that the concentration of research activity on housing for the urban poor in Latin America has left significant research gaps in terms of understanding the urban fabric of those areas in which the more affluent groups of the Latin American city live.

SOME CONCLUDING REMARKS

This chapter has attempted to demonstrate that the major contemporary issues that are facing the development of Latin American societies in the last decade of the twentieth century are intimately linked to the peripheral and dependent nature of their economies in terms of the functioning of the world economy in general and that of the industrial market economies in particular. The major constraints on Latin American growth are now external, and particularly those related to international trade and capital flows. The 1980s have witnesssed economic stagnation in Latin America; but zero growth alongside a rapidly expanding population and labour force means lower average incomes and, in countries with a skewed distribution of income, increasing poverty.

It is difficult to envisage how these major external constraints on Latin American development can be significantly reduced. The combined problems of massive debt, high interest rates and stagnant exports mean that most Latin American countries cannot even pay off the annual interest on their debts, let alone pay back a portion of the original debt. Indebtedness is therefore increasing. Commercial banks reluctantly continue to increase lending to those Latin American countries whose economic policies have the IMF seal of approval, in the hope that economic restructuring will increase trade surpluses. They recognise that the long-term problem for Latin American countries is to find and develop export activities that expand in terms of both production and value. The policy of concentrating on primary exports in the 1980s may have increased export tonnage but, due to low world prices, the value of exports has invariably declined. Only Brazil, due to its successful shift from primary to manufacturing exports, has managed substantially to increase both its value and production of exports. Can such a trend be copied however? Some Latin American countries, such as Chile, preferred to deindustrialise in the 1970s (Gwynne, 1986); they argued that many national industrial firms were inefficient and should no longer be protected from the cold winds of world competition. Other countries did not provide sufficient incentives for industrialists to export. Complex but far-reaching restructuring of economies is required, but such restructuring requires fresh capital inputs that are difficult to obtain. The World Bank offers a programme of 'structural adjustment loans' (SALs) which is intended to combat this problem; but SALs only accounted for 7 per cent of total World Bank lending in 1984 and the programme is still confined in terms of both countries covered and funds available.

Without a reduction in external constraints and restructuring of Latin American economies, the long-term prospects for the continent's economies look grim. Stagnant economies meant stagnant employment, and the prospect of further increases in unemployment and underemployment in both the cities and the countryside. Due partly to the concentration of manufacturing in the major cities, the large cities appear the more dynamic locations within the national economy and, as a result, become the destination for massive migration from rural areas. Paradoxically, then, unemployment and underemployment are increasing more rapidly in the large cities than elsewhere, simply because of the exaggerated mismatch between labour supply and demand. As a result, poverty and the spatial extent of irregular settlements for the poor increase in the large cities. In this way, the links between the performance of the world economy, on the one hand, and the spatial spread of irregular settlements in large cities, on the other, can be appreciated.

ACKNOWLEDGEMENT

I would like to thank John Dickenson and Peter Ward for their valuable comments and advice.

REFERENCES

Batley, R. (1982) Urban renewal and expulsion in Sao Paulo. In A.G. Gilbert _et al._ (eds), _Urbanization in contemporary Latin America_, Wiley, Chichester, pp. 231-62

Becker, B.K. (1986) Geography in Brazil in the 1980s:

background and recent advances. Progress in Human Geography, 10 (2), 157-83

Bodini, H. (1985) Geografia de Chile: geografia urbana. Instituto Geografico, Santiago

Bromley, R. (ed.) (1985) Planning for small enterprises in Third World cities. Pergamon, Oxford

Browning, D. (1983) Agrarian reform in El Salvador. Journal of Latin American Studies, 15 (2), 399-426

Chant, S. (1985a) Family formation and female roles in Querétaro, Mexico. Bulletin of Latin American Research, 4 (1), 17-32

------ (1985b) Single-parent families: choice or constraint? The formation of female-headed households in Mexican shanty towns. Development and Change, 16, 635-56

Cunill, P. (1978) La America Andina. Editorial Ariel, Barcelona

Cunningham, S. (1986) Multinationals and restructuring in Latin America. In C.J. Dixon, D. Drakakis-Smith and H.D. Watts (eds), Multinational corporations and the Third World, Croom Helm, London

Dickenson, J. (1978) Brazil. Dawson, Folkestone

Edwards, M. (1982) Cities of tenants: renting among the urban poor in Latin America. In A.G. Gilbert et al. (eds), Urbanization in contemporary Latin America, John Wiley, Chichester

Gilbert, A. (1981) Pirates and invaders: land acquisition in urban Colombia and Venezuela. World Development, 9, 657-78

------ (1983) The tenants of self-help housing: choice and constraint in the housing markets of less developed countries. Development and Change, 24, 449-77

------ and Healey, P. (1985) The political economy of land: urban development in an oil economy. Gower, Aldershot

------ and Ward, P.M. (1982) Residential movement among the poor: the constraints on Latin American urban mobility. Transactions of the Institute of British Geographers, New Series, 7, 129-49

------ (1984a) Community action by the urban poor: democratic involvement, community self-help or a means of social control? World Development, 12 (8), 769-82

------ (1984b) Community participation in upgrading irregular settlements: the community response. World Development, 12 (9), 913-22

Goodman, D. and Redclift, M. (1981) From peasant to

proletarian: capitalist development and agrarian transition, Basil Blackwell, Oxford
Grenier, P. (1984) Chiloé et les Chilotes. Edisud, Aix-en-Provence
Gwynne, R.N. (1985) Industrialization and urbanization in Latin America. Croom Helm, London
------ (1986) The deindustrialization of Chile, 1974-1984. Bulletin of Latin American Research, 5, 1-24
Harper, M. (1984) Small business in the Third World. John Wiley, Chichester
James, P. (1969) Latin America. Cassell, New York
Morris, A. (1981) Latin America: economic development and regional differentiation. Hutchinson, London
Ortiz, J. (1983) Geografia de Chile: poblacion y sistema nacional de asentamientos urbanos. Instituto Geografico, Santiago
Palma, J.G. (1978) Dependency: a formal theory of underdevelopment on a methodology for the analysis of concrete situations of under-development. World Development, 6 (7-8), 881-924
Pinto, A. and Kakal, J. (1973) The centre-periphery system twenty years later. Social and Economic Studies, 22 (1), 34-81
Prebisch, R. (1962) The economic development of Latin America and its principal problems. Economic Review of Latin America, 7 (1)
Preston, D. (1978) Farmers and towns: rural-urban relations in highland Bolivia. Geo Books, Norwich
Roberts, B. (1978) Cities of peasants. Edward Arnold, London
Romero, H., Borgel, R. and Vio, D. (1983) Geografia de Chile: Fundamentos geograficos del territorio nacional. Instituto Geografico, Santiago
Santos, M. (1979) The shared space. Methuen, London
Schmitz, H. (1982) Manufacturing in the backyard. Frances Pinter, London
Smith, C.T. (1983) Central Andes. In H. Blakemore and C.T. Smith, Latin America: geographical perspectives. Methuen, London, pp. 253-324
Turner, J.F.C. (1968) Housing priorities, settlement patterns and urban development in modernizing countries. Journal of the American Institute of Planners, 39, 354-63
------, Turner, C. and Crooke, P. (1963) Dwelling resources in Latin America. Architectural Design, 33, 360-93

Varley, A. (1985) Urbanization and agrarian law: the case of Mexico City. Bulletin of Latin American Research, 4 (1), 1-16

Ward, P.M. (1983) Land for housing the poor: how can planners contribute? In S. Angel et al. (eds), Land for housing the poor, Select Books, Singapore, pp. 34-53

------ (1986) Welfare politics in Mexico: papering over the cracks. George Allen and Unwin, London

------ and Melligan, S. (1985) Urban renovation and the impact upon low income families in Mexico City. Urban Studies, 22, 199-207

World Bank (1985) World Development Report, 1985. Oxford University Press, New York

------ (1986) World Development Report, 1986. Oxford University Press, New York

Chapter Nine

CONTEMPORARY ISSUES IN TROPICAL AFRICA

A. O'Connor

This chapter will not provide happy reading: no honest appraisal of tropical Africa in the 1980s could do so. Of course, not all the people of the region have experienced famine; not every country has been torn apart by civil war; and the debt crisis facing national governments is not equally intense everywhere. However, all these circumstances are sufficiently widespread to be regarded as being among the key 'contemporary issues' of the region as a whole. This volume is sub-titled 'progress and prospect', but very little that has been happening in tropical Africa in recent years can be regarded by anyone as progress, and in many respects the prospect, at least for the rest of this century, is very bleak. This is not just the view of an academic outsider: it is one shared by most academics within Africa, and reflects the experience of most men and women in their daily struggle for survival. Many do manage to stay remarkably cheerful much of the time, but it is not only the Executive Secretary of the UN Economic Commission for Africa who sees the prospects for the next 20 years as 'almost a nightmare' (Adedeji and Shaw, 1985). Mercifully, only a small majority are on the verge of starvation, but the vast majority will remain in extreme poverty by the standards of the rest of the world throughout their lives.

Most were equally poor 20 years ago, but that was a time of hope for rapid improvement. Most African countries had just gained independence from colonial rule, and levels of both production and consumption were rising steadily. In teaching the geography of Africa, the key theme was most often seen to be 'development', and it seemed appropriate to make this the basis of a textbook written at that time (O'Connor, 1971). It would most certainly not be appropriate today. Among the writing of British geographers, a less optimistic view was presented by Hodder in 1978, and reinforced by Griffiths in 1984, while Grove (1986) has now painted an even gloomier picture. No-one from within Africa has been able to project a more favourable image, although the continent's most eminent geographer has provided a notable world-scale discussion of development processes (Mabogunje, 1980). (Indeed, 'progress' has been no more evident in most African universities than in other aspects of African life, hence the very few references in this chapter to writing from Africa). The World Bank has persisted in using the word 'development' in the titles of its influential reports on Africa (World Bank, 1981, 1984), partly for diplomatic reasons, but reality is better reflected in the titles of two of the most notable books to appear in 1985, Timberlake's Africa in crisis and Sandbrook's Politics of Africa's economic stagnation.

Perhaps this is just a temporary phase, and in due course the notions of progress and development will once again be applicable to most parts of tropical Africa - but on that we could only speculate. The concern here is with the current situation and the almost certain prospects for the rest of this century - as indicated, for example, in the Global 2000 report (Barney, 1982), which foresaw no improvements for this region with respect to any of the topics with which it was concerned. Over this time-scale, the question must usually be whether Africa faces decline and breakdown, or merely stagnation.

Before embarking upon this gloomy review, however, a few words might be said about this chapter in relation to this volume as a whole, and about the spatial framework adopted for the discussion.

TROPICAL AFRICA AND THE 'THIRD WORLD'

I hestitated before accepting the editor's kind invitation to contribute to this volume on 'The Third World', since I do not really believe that such an entity exists. It is largely a figment of the imagination of various academics and politicians. As a mental construct it has been of considerable value, assisting the identification and analysis of certain features shared by large parts of Asia, Africa and

Latin America, but over-use of the concept has led to much overgeneralisation, and even to the discussion of what amounts to the greater part of the world as 'a special case' deviating from some norm found elsewhere. From the perspective of tropical Africa, the contrasts among 'Third World' countries seemed more striking than the similarities even ten years ago (O'Connor, 1976), and this is all the more so today. Indeed, an awareness of the special problems and needs of Africa spread around the rest of the world in a quite remarkable way in the early 1980s, following media reporting of the appalling famine conditions in Ethiopia and elsewhere. Now perhaps there is as much need for geographers to query overgeneralisations about Africa.

The editor's invitation was accepted because Part II of this book provides an opportunity to demonstrate the distinctiveness of each major region, and also the diversity within each. It should be readily apparent that many of the features discussed here are <u>not</u> shared by all countries from Chile to China, and no further attempt will be made to labour this point. What is much more at issue within this chapter is the extent to which there are characteristics shared by all or most of tropical Africa. In many respects a proper understanding requires disaggregation not only to the national level, but beyond that. In many respects Nairobi and the Turkana District of northern Kenya are 'worlds apart'. However, this is not to argue that geographers should never generalise: their concern is equally with how far places differ and how far they are similar, and this chapter is much concerned with similarities in the context of tropical Africa.

Perhaps something should be said about the choice of tropical Africa as the context, rather than Africa as a whole. In purely physical terms, the Africa continent forms a very distinct entity, though Mazrui (1986) has recently asked some provocative questions about who decided that the Arabian peninsula was a part of Asia rather than Africa. This configuration of land and sea has come to have a considerable psychological impact, as a result of the diffusion of maps depicting it, leading to the establishment of institutions such as the Organisation of African Unity. However, do the people of Tunisia and Togo really feel that they have much in common? The Nile may provide a vital link between Egypt and Uganda, but is not Egypt in every way part of the 'Middle East' and at the heart of 'The Arab World'? At least in cultural terms, it seems most appropriate to consider Africa south of the Sahara in the same way that Mexico is normally viewed as part of Latin America rather than North America.

The case of South Africa is more problematic. In a straw poll of 100 of my students, 90 excluded it from the 'Third World'; yet the majority of its people live in circumstances that clearly fit in with that concept, and most data at national level suggest close similarities with countries such as Brazil. However, the fact that it is a middle-income country clearly sets it apart from most of Africa, while its political system most certainly makes it 'a special case'. It is also excluded here on the purely pragmatic grounds of my own ignorance of it, and the probability that anything written about it in 1986 will rapidly become out of date.

THE PEOPLE

Old-fashioned geography texts normally began with the physical environment before moving on to population; but in this chapter people are our primary concern, so let us start with four demographic matters.

Firstly, we have only a very rough idea of how many people there are in tropical Africa, and the situation is not improving. Some countries have conducted relatively reliable censuses, such as that which recorded 17.5 million people in Tanzania in 1978; but some, such as Somalia, have never had a national count, while in others, notably Nigeria, census figures have been so grossly distorted for political advantage that they have been officially declared null and void. The Nigerian case is of particular significance because whether it has 90 million or 120 million, this is certainly far more than in any other country. The population of the second largest, Ethiopia, is also not known with any accuracy, although the nearest thing to a national count yet made suggested a total of 42 million in 1984, 20 per cent more than had previously been estimated. For tropical Africa as a whole, we can only say that the population in 1986 was probably somewhere between 400 and 500 million (Table 9.1).

Secondly, the population is extremely unevenly distributed, with vast empty spaces and pockets of extremely high density. In national terms the range is from 2 per square km in Mauritania to over 200 in Rwanda.

Table 9.1: Tropical Africa: population estimates for the largest countries

	Population (millions) 1984	Annual % growth 1980-90
Nigeria	95	3.3
Ethiopia	42	2.6
Zaire	30	3.1
Sudan	21	2.8
Tanzania	21	3.4
Kenya	20	4.0
Uganda	15	3.3
Mozambique	14	2.9
Ghana	13	3.5

Sources: United Nations, Demographic Yearbook, 1984; World Bank, World Development Report, 1986

Similar contrasts are found over quite short distances within countries such as Kenya. The common view that most of Africa is sparsely settled is quite correct, but equally significant is the fact that most people are living in densely settled areas. The need for redistribution is quite apparent in many countries (Clarke and Kosinski, 1982; Clarke et al., 1985), though the manner in which this has recently been attempted in Ethiopia has caused world-wide concern.

Thirdly, sufficient data exist to say with confidence that the population is increasing faster here than in any other part of the world, and that in contrast to all other areas, the rate of increase has been accelerating. It is now around 3 per cent a year in most countries (Table 9.1), but is thought to have reached 4 per cent in Kenya (Ominde, 1984). A doubling of population over the next 25 years is thus a near certainty, with implications for every other aspect of life. Death rates have fallen sharply in recent decades, but there is still scope for them to fall further, and all must hope for this eventuality. Birth rates have not fallen at all, and there is no sign of their doing so in the near future. The 'demographic transition' that has taken place elsewhere has depended on increasing prosperity, of which there is no prospect; and most adults in Africa see a large family as highly desirable, with 7 or 8 children as the ideal.

Fourthly, a direct consequence of the rapid population

growth is that throughout tropical Africa, children account for almost half the total numbers (Green and Singer, 1984). This means a high dependency ratio, especially as more now spend at least some time in school rather than weeding crops and fetching water. The notion of Africa as a 'young continent' is absurd in relation to archaeological evidence of human origins here, but is wholly appropriate to the present and future population of every country.

A further key theme that must be stressed is that of ethnic identity, despite the dangers of entering this academic minefield. At one level there is a clear identity that spans Africa from Senegal to Swaziland - which has been greatly intensified by the shared transition from colonial rule to independence, and by the expansion of education in the post-independence era. People now think of themselves as African in a way that very few in India or China would think of themselves as Asian. At another level, there is the ethnic group with which most people still identify themselves, such as Hausa, Ibo or Yoruba in Nigeria. There is much dispute as to whether the term 'tribe' can be appropriately used for these entities, which vary enormously in size and have changed substantially over time, but there is no doubt that something frequently referred to as 'tribalism' is a widespread African phenomenon that has no real equivalent elsewhere. It involves the whole population, not just certain 'minorities', and cuts right across any emerging class divisions.

A third level of identification of profound importance in all parts of tropical Africa is the extended family (Bell, 1986). Numerous anthropological studies have shed light on the complex and diverse kinship structures that exist, but here we can only note the significance of a unit larger than the nuclear family for matters such as rural-urban migration. The extended family will have a part to play in many decisions as to who moves, and in enabling migrants to survive in the city without a home or a job. Many people would regard this as a very positive feature of African life, and certainly far more people will become totally destitute if it breaks down as rapidly as some fear.

THE POLITICAL MAP

A mention has already been made of the transition from colonial rule to independence as a shared experience which helps to give tropical Africa its own identity. At the same time, this consolidated the division of the whole area along lines drawn up by outsiders a hundred years ago, and there is no doubt that the present political map is one cause of Africa's desperate problems.

China, India, even Egypt, have existed in some form for thousands of years. Brazil, Argentina and Peru were independent nations 150 years ago. Tropical Africa also had its own political 'map' at that time, but there were no clearly defined boundaries on it, and the territories involved were those of the ethnic groups such as the Yoruba or the Kikuyu. Nigeria, Kenya, Cameroon and Zaire are wholly colonial creations that bear no relation to this earlier political map, and they still have little meaning for many of their citizens. They represented a new form of spatial organisation that was superimposed on earlier spatial structures without ever wholly taking their place (and geographical studies should not always be tied to this framework).

The great majority of African countries gained independence from Britain, France or Belgium in 1960 or soon after, though Portuguese rule was maintained in Angola and Mozambique until 1975. This process involved the creation of states in some respects literally overnight, and in every case the nature of the state has been of great importance. There are substantial differences from one to another, not least in the ideologies espoused. Angola, Benin, Congo and Mozambique are among those which now claim to be Marxist/Leninist, while Ivory Coast, Nigeria, Kenya and Zaire are among those which have consistently steered a course much further to the political right. However, there are at least three features sufficiently widespread to be seen as characteristic of tropical Africa as a whole. One is the frequency of coups leading to military rule and the extreme rarity of freely-elected governments. A second is the extent to which charismatic individuals have dominated political life, for good or ill. Few people can think of Tanzania or Uganda without immediately thinking of Julius Nyerere or Idi Amin. A third is the extreme weakness of the state, even where great efforts have been made to strengthen it, as in Tanzania (Hyden, 1980). In countries such as Zaire, it remains largely irrelevant to many people's lives.

The state in Africa is weak partly because it is so new, partly because it lacks resources, and partly because it does not represent a nation - which cannot in any sense be created overnight. Nation-building has been a major preoccupation of African governments, and where this has been relatively successful, as in Tanzania, it has often been at the expense of economic development. The lack of any real national integration in some of the largest countries has resulted in civil war, notably in Zaire, in Nigeria, and now for a second time in the Sudan, where most people in the south consider rule from a Muslim, Arabic-speaking north to be almost as alien to their interests as colonial rule.

The extent to which colonial rule left behind potentially helpful economic structures has been a matter of much debate, but it is at least arguable that Africa's economic crisis of today is largely due to the non-viability of the political structures partly inherited and in part newly created. Sadly, the political record has been no better in the two countries which did not share the usual colonial experience, Liberia and Ethiopia. An independent Ethiopian nation-state has existed for centuries, but recent years have witnessed not only a violent revolution but also bitter resistance in the northern provinces of Eritrea and Tigre to rule from Addis Ababa - which has greatly exacerbated widespread famine conditions. The scope for geographical research on national integration issues is certainly just as great in Ethiopia as in most other African countries (Wood, 1983).

The problems created by the political map are reflected in the fact that tropical Africa currently accounts for one-third of the world's refugees. The country most affected is Sudan, where the inflows from both Chad and Uganda have increased in the 1980s, and where the inflow from northern Ethiopia came to exceeed half a million. However, many other borders have also been crossed by thousands of refugees, such as that between Angola and Zaire. In some cases spontaneous rural settlement across the border has been possible, but in others, camps have had to be set up by bodies such as UNHCR. Of course, large quantities of outside assistance raise particular problems where the local population suffers extreme deprivation. The only geographer who has investigated the African refugee situation in any depth is Rogge (1985), and there is an urgent need for more work on both causes and consequences of these moves of desperation.

LIVELIHOOD

Twenty years ago most African countries were perceived as poor but 'developing'. Ten years ago that optimistic phrase was being used rather less often, and most of these countries were seen as being in a category designated by the United Nations and other bodies as the 'least developed'. Today, Africa is singled out in most surveys of the world economy as the one area in which standards of living are deteriorating, with no immediate prospect of reversing this process.

Most national economies were actually continuing to expand in the 1970s, but this was essentially a function of population growth, and in most cases gross national product (GNP) per capita had begun to fall (Table 9.2). Then, in each year from 1981 to 1984, both United Nations and World Bank sources indicate an absolute fall in GNP for Tropical Africa as a whole, and a fall of over 3½ per cent in per capita terms.

Table 9.2: Tropical Africa: per capita GNP of the largest countries

	US$ 1984	Average annual % change 1964-74	1974-84
Nigeria	730	5.3	-3.5
Ethiopia	110	1.8	-0.3
Zaire	140	0.4	-4.2
Sudan	360	-1.1	0.4
Tanzania	210	2.8	-1.2
Kenya	310	2.8	0.0
Uganda	230	0.7	-3.9
Ghana	350	0.6	-4.7
Mozambique	(220)	4.2	-4.9

Sources: World Bank, World Development Report, 1986; OECD (1986), Latest information on national accounts of developing countries

The most serious aspect of this deterioration relates to food production. Both malnutrition and seasonal hunger have long been all too widespread (Grigg, 1985, Chapter 8; UNICEF, 1985), and research on both these long-term issues by geographers as well as others is urgently needed.

Recently, however, they have turned into devastating famine (Giri, 1983; Gill, 1985; Hancock, 1985; ICIHI, 1985; Wolde-Mariam, 1985). This gravely afflicted both Ethiopia and the West African Sahel in the early 1970s, and then recurred in these areas, in Sudan, and also far to the south, especially in Mozambique, in the 1980s. At one stage more than 20 countries were seriously affected, and probably over a million people, mostly children, died in 1984, at least partly from starvation. Many more had to abandon their homes to seek food. The publicity given to the famine conditions has been such that they need not be described again here. Nor can they be properly explained here, although it should be said that much nonsense has been written both by those who suggest that drought and famine are synonymous, and by those who suggest that drought has been almost irrelevant. Famine has generally resulted from a lethal combination of drought and armed conflict superimposed on the more widespread phenomenon of failure to increase productivity in agriculture (Mlay, 1985; Hinderink and Sterkenburg, 1985; Commins et al., 1986).

Of course, the data base regarding agriculture in Africa is even weaker than that for population numbers. We know that in every country, the majority of people are rural dwellers, and that most men and women are engaged in farming; but they are not just farmers, and little is known about how much time is spent cultivating and how much in other activities such as house-building, fuel-gathering and water-fetching. We know that most people produce food crops for themselves and obtain cash either from a surplus of these or from other exports crops; but there are no records of either area planted or quantity produced in most cases. While there are important variations from one country to another, the general picture of declining agricultural exports per capita is quite clear, as is that of steeply rising food imports: it is just possible, however, that the estimates of crop production that are submitted to FAO by national governments (Table 9.3) previously included an element of exaggeration (or wishful thinking), and now sometimes indicate declines because this has become more fashionable. The low official prices paid for crops in many African countries, noted in most analyses of the agrarian crisis may have led some farmers to grow less for both export and the towns, but also to consume more themselves.

Table 9.3: Tropical Africa: index of per capita food production for the largest countries, 1982-4, (1974-6 = 100)

Nigeria	96
Ethiopia	100
Zaire	92
Sudan	93
Tanzania	100
Kenya	82
Uganda	98
Mozambique	73
Ghana	73

Source: World Bank, World Development Report, 1986

However, there is certainly no evidence of any general rise in recent years in total agricultural production per head in rural Africa to compensate for the loss of production arising from massive rural-urban migration. The food imports have been very largely for the rapidly expanding towns and cities. Until recently the level of urbanisation in tropical Africa was lower than in any other part of the world, but it is now overtaking that in both India and China. Migration probably accounts for about half the 7 or 8 per cent annual growth of most cities, resulting in a concentration of young adults sufficient to ensure that natural increase makes an equal contribution. At present, tropical Africa has only two cities with over 4 million inhabitants, Lagos and Kinshasa, but if present trends were to continue it would have one-fifth of the world's cities of that size by the year 2025. It is perhaps unlikely that present trends will in fact continue, but even so, city growth is among the most important changes currently occurring in the geography of the region (Adepoju, 1983; O'Connor, 1983; Peil and Sada, 1984).

The rapid rate of rural-urban migration is not easily explained. There is little to be gained from arguments about whether people are pushed from rural areas or pulled to cities, or about economic versus other motives. Motives are usually mixed, and people move because of perceived differences in prospects in two places. All over Africa rural-based families are deciding that income prospects are better if some members move to the city, and that there is just a possibility of someone doing really well there. At the same time, the city is seen as more exciting than the village by most of those who move. In eastern African countries such as Zambia and Tanzania, migration has been boosted by wives joining husbands who are holding on longer to their urban jobs, with particularly serious implications for agriculture in these countries where women do more than half the farm work.

One factor contributing to the especially rapid rate of migration throughout tropical Africa in comparison with, say, India, is the fact that of all Africa's 'development' efforts in recent years, the most successful has been in respect of school enrolments. The numbers of primary school leavers have been rising rapidly, and education has made them (and their parents) more aware of what the city has to offer them. Another favourable factor in every country is the kinship network mentioned earlier, which provides most migrants with a base for their first weeks or months in the city.

This support system means that many people can survive in the city for some time with no form of employment, while even more can remain with only very low earnings from casual work. However, up till now the move to the city has brought the majority a higher income than they would have obtained by staying in the rural area. Some have found wage employment, which has generally continued to expand, even if more slowly than the labour force, while others have found some form of self-employment which they consider preferable to peasant agriculture. The 'informal sector' was almost non-existent in the cities of eastern Africa 30 years ago, and there has been considerable scope for its growth: in western Africa, it has long been more important, but in Nigeria the oil boom permitted rapid expansion.

One point that must be emphasised is that contemporary urbanisation in tropical Africa is not based primarily upon industrialisation. As in most of the world through most of history, the cities and towns have grown primarily as centres for administration and commerce, with manufacturing playing a very subsidiary role. Some import-substitution industry developed in many countries in the early years of independence, but generally the opportunities for this have been largely exhausted, while there is no prospect of export-based manufacturing of the type that is growing so successfully in some Asian countries. By the mid-1980s stagnation was as much a characteristic of the small industrial sector of most African countries as of the large agricultural sector, and even the limited range of industries existing were generally functioning well below capacity.

The experience of individual countries has of course not always conformed to the general picture presented here. One important source of variation has been the very uneven distribution of mineral resources. Oil brought very substantial revenues to Nigeria, especially after the 1973 price rise, and permitted vast expenditures on roads and other public works. However, it also brought severe dislocation in many respects, fuelled corruption on a massive scale, and is now seen as a curse rather than a blessing by many observers (Watts, 1984), especially since oil prices have slumped. Copper helped Zambia to enjoy a per capita income well above average for many years, but now reduced prices have brought dire economic crisis there.

Political circumstances have had some economic impact in every country, but nowhere more so than in Uganda, where much of the 'formal' economy collapsed in the 1970s, giving way to a 'parallel' economy known locally as magendo. (In theory there is vast scope for geographical research on such activity over much of Africa, but this would be a hazardous undertaking). Conversely, political stability has contributed to healthier economic conditions in the Ivory Coast, Cameroon, Kenya and Malawi, though these are all countries in which the benefits have been very inequitably distributed. One of the most depressing experiences is that of Ghana, whose relative position has steadily deterioriated, without either drought or revolution as an excuse; while one of the most encouraging in the 1980s has been that of Zimbabwe, where there has even been success in expanding sales of food crops from small farms.

THE NATURAL ENVIRONMENT

Since so many people in tropical Africa depend directly on the land for their livelihood, the influence of the natural environment is profound. In a predominantly subsistence economy, drought does not just bring financial loss; it is often a matter of life and death. The most important single element in the natural environment for the human geography of this region is undoubtedly rainfall, and both its seasonality and its reliability are as significant as annual average totals. In all three respects, conditions in many areas have been much worse in the 1970s and 1980s than in the 1950s and 1960s, but it is not possible to say whether this deterioration is a long-term trend.

Disturbing long-term trends are far clearer with respect to relationships between people and environment, and the key theme of a notable recent book on Africa is 'environmental bankruptcy' (Timberlake, 1985). Various traditional systems of land use, by both cultivators and pastoralists, that were well adapted to environmental constraints, are becoming non-viable as densities of people and livestock increase. Similarly, systems well adapted to particular local environments are proving severely destructive as people have to move on to more marginal land. Some claims with respect to 'the advancing Sahara' have been exaggerated, but there is widespread evidence of desertification and other forms of environmental degradation, perhaps the worst cases being in Ethiopia. In recent years, efforts to mobilise the population there to build systems of terraces to combat soil erosion have brought some remarkable successes, but they do not begin to match the scale of the ongoing destruction.

One cause of accelerated soil erosion is removal of the natural vegetation, and this is proceeding at an ever faster pace - from the rain forests to the desert margins. Vast areas of forest remain in Zaire, but in countries such as Ivory Coast, they are shrinking rapidly through a combination of timber felling and clearing for cultivation. In some savanna areas such as parts of Burkina Faso and northern Nigeria, the cutting of trees and bushes has created a severe fuel-wood crisis. Both there and in Ethiopia, this has led to much increased use of animal dung for fuel, at the expense of soil fertility.

Meanwhile, the environment continues to provide hazards, such as the swarms of locusts which returned to many areas in 1986, and in the form of high rates of disease carried by insects or picked up from polluted water. Geographers have paid much attention to the tsetse-fly, which is not responsible for many human deaths but which prevents cattle-keeping over vast areas; but they have paid less attention to the mosquito and malaria, which kills millions of children and debilitates millions of adults every year. Bilharzia, transmitted through water-snails, is another widespread disease whose incidence is not decreasing. Opinions differ greatly with regard to the priority that should be given to efforts to reduce disease (whether mainly through Western science or with a greatly increased contribution from traditional health care). Such investment may not be seen as directly productive, yet it is sometimes suggested that other forms of investment will yield little if put into communities where most people are ill much of the time (Haswell, 1975).

Ever since the excesses of environmental determinism early this century many geographers have been reluctant to emphasise the direct importance of the natural environment for patterns of human activity. In tropical Africa it is profoundly significant not only with respect to 'what happens where' at present, but also with respect to how the expanding population can gain a livelihood in each locality in the future. This has been clearly demonstrated by the work of Grove over many years, though he emphasises human rather than physical factors in his recent review of the problems now afflicting Africa (Grove, 1986).

EXTERNAL RELATIONSHIPS

A central theme in studying the geography of poorer parts of the world must be how far their poverty results from internal factors, and how far it results from relationships with the rest of the world. Some would hold colonial rule very largely to blame for Africa's poverty, whereas others consider that there is very little evidence of pre-colonial prosperity, except for tiny minorities in some very unequal societies. Others again would say that there is no point in speculating on what might have been, and that efforts should be concentrated on investigating how far present-day relationships are helping or hindering (Shaw, 1983).

Even the latter question generally proves extremely difficult to answer. Most African countries conduct far more trade with richer countries elsewhere in the world than with each other, and do so because this trade is perceived to be beneficial. Many Kenyans do indeed gain from the export of coffee and the import of a wide range of manufactures. However, the emphasis on promoting exports has sometimes been at the expense of sections of the community not in a position to participate, and some of the imports benefit only a privileged few. Furthermore, the terms of trade seem to have moved in a direction unfavourable to Africa far more often than the reverse. Over the past 20 years, Africa's primary product exports have covered the cost of progressively fewer manufactured imports, and this intensified in the early 1980s.

Private investment by multi-national corporations has greatly contributed to economic growth in some parts of the world, but its role in most African countries has been very limited, and its developmental impact is much disputed. It has assisted the establishment of some plantations, some mines, and even a few factories. These have provided employment, government revenue, and in some cases desired products such as sugar or cement. However, the fact that profits flow out to rich countries overseas leads many to question whether the same activities might not be promoted in some other way, or other activities promoted instead.

While tropical Africa is of very small importance in the operations of most MNCs, it figures much more prominently in flows of official bilateral and multilateral aid. Does aid help? A recent study with that title (Cassen, 1986) argues that it generally does, but acknowledges that the record is poorer in Africa than elsewhere, and that it often helps far less than it might. There is much scope for geographical research on this subject, considering especially the distribution of benefits and losses from aid projects. The emergency aid from both governments and non-government organisations in response to the famines of the early 1980s, which undoubtedly overcame a vast amount of suffering, of course constitutes a very special case, though this is not to say that it will not be required once again quite soon. International bodies will probably become more effective in taking this type of essential remedial action, but whether they can come to operate in the right way and on the right scale somehow to prevent such situations occurring is much more doubtful.

A fundamental change in the external relationships of most African countries has occurred in the early 1980s in the form of a massive build-up of debt (Table 9.4) (Adedeji, 1985; Green and Griffith-Jones, 1985). Some of this results from aid which was in the form of loans rather than grants, but much results from commercial credit on which interest is now being paid at a far higher rate than was anticipated. The absolute volume of debt is small in comparison with that owed by some Latin American countries, but as a proportion of GNP or of export earnings, the African figures are now higher. It is not clear whether even the most severe austerity measures can resolve the problem in the most extreme cases. The body best able to assist is the IMF, but its terms for doing so are far from generous, and one result

Table 9.4: Tropical Africa: long-term public debt of the largest countries

	Debt in $ million		1984 debt service as % of exports	
	1970	1984	Scheduled	Actual
Nigeria	480	11,815	28	25
Ethiopia	169	1,384	20	19
Zaire	311	4,084	24	24
Sudan	307	5,659	96	25
Tanzania	250	2,594	47	22
Kenya	319	2,633	27	27
Uganda	138	675	36	35
Ghana	495	1,122	19	19

Source: World Bank (1986) Financing adjustment with growth in Sub-Saharan Africa 1986-90

of the debt crisis has been to give that institution great power over African states - whose dependency has been massively intensified.

CONCLUSIONS

The outstanding 'contemporary issue' particular to tropical Africa in the mid-1980s has surely been famine. The year 1986 brought some respite, but for how long? Many aspects of the famines, in Ethiopia, in Sudan, in the Sahel, and elsewhere, could hardly be more 'geographical', and thus provide us with an obvious and urgent research agenda.

In most areas, famine has reflected a disastrous breakdown in the relationships between people and their natural environment. Drought has generally been one of the prime causes, but all too little is known even about the extent and nature of the drought, much less its impact. What degree of rainfall deficiency, over what periods of time, has been critical for what activities in each locality? Furthermore, many environmental factors other than rainfall affect water availability, and all require investigation. In addition, how far is the issue in many areas a much broader one of environmental degradation, with people affecting the environment as much as the reverse?

An equally important contributory factor in some areas

has been the totally different issue of political conflict and even civil war. This brings in a second geographical theme, for the conflict has generally been between the people of distinct areas. In Ethiopia it is northerners who are resisting rule from Addis Ababa; in Sudan it is southerners who are resisting rule from Khartoum. Famine has often resulted from people having to move from their homes because of the fighting, in many cases becoming refugees in a foreign land.

Civil war not only contributes to famine but also greatly hinders relief efforts, and it profoundly affects the geography of such efforts. Thus, aid to starving people in much of Eritrea and Tigre has had to be channelled through Sudan, and people in southern Sudan have had to be reached via Kenya. At the same time, relief efforts have highlighted further consequences of the physical environment. Vast areas of Ethiopia are highly inaccessible because of rugged terrain, while both there and in Sudan the rains that are long-awaited by farmers often make the few roads that exist totally impassable. The logistics of famine relief as well as the causes of the spatial pattern of famine must be high priorities for geographical research.

However, it is clear that famine constitutes the tip of a massive iceberg in terms of the problems facing the people of tropical Africa. If periodic famine demands our attention, so also does the even more widespread hunger and malnutrition that persists year after year and the failure of many efforts to increase food production. Many other long-term dilemmas are presented throughout the region by the combination of political turmoil, economic stagnation and the world's highest rate of population growth. Not only are there numerous unresolved problems regarding appropriate policies: in many cases, even the basic facts of the current situation are not known.

Far too little is known, for instance, about the region's natural resources and about forms of environmental degradation such as soil erosion. There is a need to bring together the knowledge that does exist in local communities and the vast store of information now being provided by satellite imagery, but still mostly locked away in Washington and Moscow. Demographic trends and processes also need more attention, yet in Nigeria it has not even been possible to conduct a reliable census. The best indication of the relative sizes of Nigerian cities would probably be provided by data for primary school enrolments - if anyone could assemble the information. Transport is of critical importance in many ways apart from famine relief, and while the basic facts about rail or air traffic are known, we have no idea who carries how much of what, and how far, on their heads. Perhaps there could appropriately be far greater use of the bicycle; but where can anything be found on how much it is currently used in each part of Africa?

Even in the absence of data, policy decisions have to be made, and many of these are highly geographical. Thus, the perennial problem of whether to invest scarce funds where needs are greatest, or where returns are likely to be highest, is compounded throughout tropical Africa by ethnic attachment to territory. In development planning, the question 'where?' has sometimes been neglected not because it is unimportant but because it is too sensitive. Similarly, many African states are faced with decisions about whether to encourage or to resist redistribution of population, including rural-urban migration; but of course, 'the African predicament' is much broader than these policy issues of particular geographical interest.

Two books with quite different messages illustrate the types of response that academics may provide. Hyden (1983), in No shortcuts to progress, argues that most people across tropical Africa remain tied into an 'economy of affection' and remain quite largely independent of both the market and the state. His earlier work on Tanzania (Hyden, 1980) indicated that the state was incapable of mobilising the peasantry effectively, and so he feels that there is no alternative to the long and painful process of penetration by capitalism (which others, quite wrongly, assume happened long ago). Richards (1985), in Indigenous agricultural revolution, argues that there is a potential for progress within these peasant farming systems that still dominate the region; and whereas Hyden considers that the peasants must somehow be 'captured', Richards wants us to set them free. He recognises, however, that progress by this route may also be slow, with no 'short cuts'.

One possible geographical reaction to such writings is that there are no blanket solutions for the whole of tropical Africa, even on a long-term basis. Perhaps in some areas, indigenous solutions are about to emerge, while in others drastic changes will have to come from outside. More likely, a combination is required everywhere, but with considerable variation in the respective roles of the state, the market and the local people. This means almost infinite scope for investigating how both present circumstances and future prospects vary, both among and within African countries.

In many respects, Africa is in a mess; but it is not the same mess in Zaire as in Tanzania, in Nigeria as in Uganda, in southern Sudan as in northern Ethiopia. (The mess is totally different again, of course, in South Africa). There are various circumstances and problems shared by many areas and many people - hence the rationale for this chapter - but any search for common solutions, whether by Africans or by outsiders, must be supplemented by consideration of these variations from place to place.

REFERENCES

Adedeji, A. (1985) Foreign debt and prospects for growth in Africa. Journal of Modern African Studies, 23, 53-74

------ and Shaw, T.M. (eds) (1985) Economic crisis in Africa: African perspectives. Rienner, Boulder

Adepoju, A. (1983) Issues in the study of migration and urbanization in Africa. In P.A. Morrison (ed.), Population movements, Ordina, Liège, pp. 115-50

Barney, G.O. (ed.) (1982) The global 2000 report to the President. Penguin, London

Bell, M. (1986) Contemporary Africa. Longman, London

Cassen, R. (1986) Does aid work? Oxford University Press, Oxford

Clarke, J.I. and Kosinski, L.A. (eds) (1982) Redistribution of population in Africa. Heinemann, London

Clarke, J.I. et al. (eds) (1985) Population and development projects in Africa. Cambridge University Press, Cambridge

Commins, S.K. et al. (eds) (1986) Africa's agrarian crisis. Rienner, Boulder

Gill, P. (1986) A year in the death of Africa. Paladin, London

Giri, J. (1983) Le Sahel demain, catastrophe ou renaissance? Karthala, Paris

Green, R.H. and Griffith-Jones, S. (1985) External debt: Sub-Saharan Africa's emerging iceberg. In T. Rose (ed.), Crisis and recovery in Sub-Saharan Africa, OECD, Paris, pp. 211-21

Green, R.H. and Singer, H. (1984) Sub-Saharan Africa in depression: impact on the welfare of children. World Development, 12, 283-95

Griffiths, I.L. (1984) An atlas of African affairs. Methuen, London
Grigg, D. (1985) The world food problem 1950-1980. Blackwell, Oxford
Grove, A.T. (1986) The state of Africa in the 1980s. Geographical Journal. 152, 193-203
Hancock, G. (1985) Ethiopia: the challenge of hunger. Gollancz, London
Haswell, M. (1975) The nature of poverty. Macmillan, London (on The Gambia)
Hinderink, J. and Sterkenburg, J. (1983) Agricultural policies and production in Africa. Journal of Modern African Studies, 21, 1-23
Hodder, B.W. (1978) Africa today, Methuen, London
Hyden, G. (1980) Beyond Ujmaa in Tanzania. Heinemann, London
------ (1983) No shortcuts to progress: African development management in perspective. Heinemann, London
ICIHI (Independent Commission on International Humanitarian Issues) (1985) Famine, a man-made disaster? Pan, London
Mabogunje, A.L. (1980) The development process: a spatial perspective. Hutchinson, London
Mazrui, A. (1986) The Africans: a triple heritage. BBC, London
Mlay, W. (1985) Pitfalls in rural development: the case in Tanzania. In F.G. Kiros (ed.), Challenging rural poverty, Africa World Press, Trenton, pp. 81-97
O'Connor, A.M. (1971) The geography of tropical African development, 2nd edn. Pergamon, Oxford
------- (1976) Third World or one world? Area, 8, 269-71
------ (1983) The African city. Hutchinson, London
Ominde, S.H. (ed.) (1984) Population and development in Kenya. Heinemann, Nairobi
Peil, M. and Sada, P.O. (1984) African urban society. Wiley, Chichester
Richards, P. (1985) Indigenous agricultural revolution: ecology and food production in West Africa. Hutchinson, London
Rogge, J.R. (1985) Too many, Too long: Sudan's twenty-year refugee dilemma. Rowman and Allenheld, Totowa
Sandbrook, R. (1985) The politics of Africa's Economic Stagnation, Cambridge University Press, Cambridge
Shaw, T.M. (1983) Debates about Africa's future. Third

World Quarterly, 5, 330-44

Timberlake, L. (1985) Africa in crisis. Earthscan, London

UNICEF (1985) Within human reach: a future for Africa's children. UNICEF, New York

Watts, M. (1984) State, oil and accumulation: from boom to crisis. Environment and Planning D, Society and Space, 2, 403-28

Wolde-Mariam, M. (1985) The socioeconomic consequences of famine. In F.G. Kiros (ed.), Challenging rural poverty, Africa World Press, Trenton, pp. 11-20 (on Ethiopia)

Wood, A.P. (1983) Rural development and national integration in Ethiopia. African Affairs, 82, 509-39

World Bank (1981) Accelerated development in Sub-Saharan Africa. World Bank, Washington

------ (1984) Toward sustained development in Sub-Saharan Africa. World Bank, Washington

Chapter Ten

CONTEMPORARY ISSUES IN THE MIDDLE EAST

R.I. Lawless

INTRODUCTION

The 1970s inaugurated a period of dramatic and spectacular change in the Middle East. Egyptian troops crossed the Suez Canal in an offensive which marked the beginning of the October 1973 war, the first time the Arabs had taken the initiative against the Israelis; the Arab oil-producing states decreed a boycott of the world's most powerful state, the USA; and OPEC, the grouping of Third World oil producers, doubled the price of oil in a single afternoon and a few weeks later doubled it again. Within a matter of months, prices quadrupled from $3 to $11.65 a barrel. In 1979 following the Iranian Revolution, prices leapt up again, with OPEC oil selling at $37 a barrel by mid-1980. Huge sums accrued to the major exporters; oil revenues of the eight Middle Eastern members of OPEC rose from US $16,891 million in 1973 to US $219,107 million in 1980 (Table 10.1). Some observers spoke of the dawn of a new era of development and economic prosperity. OPEC countries would dictate the course of world energy development and dominate the world financial system; it was to be the Arabs' 'decade of destiny', while the Shah of Iran boasted that Iran would soon become an industrial power, the equal of Japan and Germany.

In March 1983, OPEC implemented the first price reduction in its history, and by mid-1986, oil prices had tumbled to below $10 a barrel (the price of Arabian light plunged to $8.25 in July 1986). Oil revenues of the major Middle Eastern producers, which had peaked in 1980, fell to US $119,100 million in 1984 (Table 10.1). Figures for the

Table 10.1: Middle East members of OPEC: revenues from oil, 1973-84 (millions of US dollars)

	1973	1974	1975	1976	1977	1978	1979	1980	1981	1982	1983	1984
Algeria	988	3,299	3,262	3,699	4,254	4,589	7,513	12,500	10,700	8,500	9,700	9,700
Iran	4,399	17,822	18,433	20,243	21,210	19,300	20,500	13,500	9,300	17,600	20,000	16,700
Iraq	1,843	5,700	7,500	8,500	9,631	10,200	21,291	26,100	10,400	9,500	8,400	10,400
Kuwait	1,735	6,543	6,393	6,870	7,615	7,952	16,863	17,900	14,900	9,477	9,900	10,800
Libya	2,223	5,999	5,101	7,500	8,850	8,400	15,223	22,600	15,600	14,000	11,200	10,400
Qatar	463	1,451	1,685	2,092	1,994	2,200	3,082	4,795	4,722	3,145	3,000	4,400
Saudi Arabia	4,340	22,574	25,676	30,755	36,540	32,234	57,522	102,212	113,200	76,000	46,100	43,700
UAE	900	5,536	6,000	7,000	9,030	8,200	12,862	19,500	18,700	16,000	12,800	13,000
Total Middle Eastern OPEC	16,891	68,924	74,050	86,659	99,124	93,075	154,856	219,107	197,522	154,222	121,100	119,100
Total OPEC	22,813	86,800	92,450	107,886	122,621	114,394	192,512	274,909	249,189	196,238	156,960	158,600

Source: <u>OPEC Annual Statistics Bulletin, 1984</u>. Organisation of Petroleum Exporting Countries, Vienna, 1985, p. 34

Table 10.2: Statistics on production and revenues for OPEC's Middle Eastern members, first half of 1985-6

Country	Crude production (000 b/d) 1H1986	1H1985	Oil revenue ($ million) 1H1986	1H1985	Revenue loss ($ million)	Revenue loss (%)
Algeria	662.40	668.70	1,453.25	2,538.44	1,085.19	42.75
Iran	1,935.70	2,159.10	2,896.11	6,225.00	3,328.89	52.47
Iraq	1,807.50	1,191.70	3,983,21	4,375.71	392.50	8.96
Kuwait	1,271.00	933.80	2,890.99	4,031.42	1,140.43	28.28
Libya	1,034.00	973.50	2,367.69	4,113.89	1,746.20	42.44
Qatar	291.70	280.20	716.49	1,301.30	584.81	44.94
Saudi Arabia	4,371.00	3,190.10	9,613.51	12,700.14	3,086.63	24.30
UAE	1,300.00	983.60	3,105.65	4,393.31	1,287.66	29.30
Total	12,673.30	10,380.70	27,026.90	39,679.21	12,652.31	

Source: Middle East Economic Survey. 29 (44), (11 August 1986), p. A9

first half of 1986 reveal a further fall compared with the same period in 1985 (Table 10.2). In the wake of the slide in oil prices, development budgets have been dramatically reduced and many large projects deferred, scaled down or simply cancelled. For Kuwait the dramatic fall in oil prices will result in a real deficit of KD 1,306 million in the 1986-7 budget, the first time in its modern history. Libya, for example, could face a $5,000 million current account deficit in 1986, whereas as recently as 1980 the country had a visible trade surplus of nearly $13,000 million. Saudi Arabia's deficit in 1984, at about $24,000 million, was the highest in the non-Communist world after the USA. The collapse of oil prices has exposed the vulnerability of the states which make up the region, their relative lack of economic weight in world economic affairs and their politically dangerous reliance on foreign supplies of food, weapons and technology. Japan, for example, has a gross domestic product six times that of the six major oil exporters (Saudi Arabia, Kuwait, UAE, Qatar, Iraq and Libya) combined (Owen, 1983).

Their political disunity and disarray is equally obvious. The political defection of Egypt, the most important and populous non-oil state following the Camp David Accords in 1978, increased the imbalance between rich and poor states in the region. Under the terms of the peace treaty between Israel and Egypt, Israel eventually relinquished control of Sinai but was able to strengthen its occupation of the West Bank and Gaza, and in 1982 embarked on a costly invasion of neighbouring Lebanon, which accelerated the collapse of that fragile and vulnerable political unit into a series of warring cantons. A solution of the Palestinian problem remains as remote and intractable as ever. For six years, two states, the new Islamic Republics of Iran and Iraq, have been locked in a bitter conflict which has soaked up oil revenues and destroyed or disrupted urgent development projects. Iraq's military expenditure is currently estimated at up to US $ 1,000 million a month. There has been a resurgence of protest movements based on sectarian and regional loyalties, of which the most extreme manifestations were the takeover of the Grand Mosque in Mecca, Saudi Arabia and the insurrection in the Syrian town of Hama and the subsequent levelling of the town by government forces. The Iranian Revolution and the Soviet military intervention in Afghanistan at the end of 1979 highlighted fears of superpower confrontation in the region.

Political violence close to the capitalist world's largest oil reserves placed the Middle East at the centre of the re-emergence of interventionist policies in the USA.

AN OIL DOMINATED ECONOMY: CONSTRAINTS AND CONTRADICTIONS

Owen (1981) has identified a number of key trends which characterised the region during the 1970s; the rapid accumulation of enormous financial resources by the major oil exporters; the transformation of the region into an oil-dominated economy in which the development of both agricultural and industrial sectors was inhibited by structural features connected with the lop-sided dependence on the export of one primary product; accelerated flows of labour and capital between oil-poor and oil-rich states; the growing integration of the whole region into the international economy; and the abandonment of economic policies which might be loosely defined as 'socialist' in favour of greater liberalisation for which the Arabic term infitah, or 'opening up', is often applied.

The massive scale of financial resources that were made available to the Middle East oil producers during the 1970s as a result of the oil boom can be easily illustrated. The sparsely populated Gulf states contributed 69 per cent to OAPEC's combined GNP by 1980, Algeria and Iraq 19.5 per cent and Egypt and Syria, with over half of OAPEC's population, a mere 11.5 per cent. Nine Arab oil producers, the Gulf states, Iraq, Libya and Algeria, allocated nearly US$ 275 billion to domestic investment under their development plans for the late 1970s, more than four times the total investment of the other twelve Arab states. Saudi Arabia alone received and spent twice as much as all the other OAPEC members put together (ibid.).

Most observers agree that much of this new wealth was wasted or spent on ambitious and expensive prestige projects of dubious long-term value. Owen (ibid.) argues that the oil money

> was accumulated much too fast, predominantly by underdeveloped desert states with rudimentary administrative systems. These states came under enormous internal and external pressures to create the infrastructures, the armies and the welfare services

> thought appropriate to a modern state. The cost of these projects was astronomical.

The disbursement of oil revenues was associated with high inflation and corruption. Sharabi (1983, p. 301) estimates that

> Between 1973 and 1980, 90 per cent of revenue from oil was spent on nonproductive goods and services and only 10 per cent was productively invested. Indeed, viewed from the standpoint of the ordinary citizen, daily life in the Arab world appears incoherent and mystifying. It is characterised by misery in wealth, by underdevelopment in development, by weakness in strength.

Renner (1984) reminds us that all major Third World oil producers face the same predicament - oil revenue is a form of rent and it must be transferred into forms of productive activity if these societies are to benefit from it after their oil reserves are exhausted. All the Middle East oil producers have failed to strengthen and develop the productive sectors of their economies in order to generate new wealth in the future and to diversify their economic structures in order to reduce dependence on the export of one primary commodity. Kubursi (1980) maintains that the last decade has seen wealth from oil in the ground converted to money in the treasury rather than the production of new wealth in the form of goods and services.

The high capital investment has been mainly devoted to infrastructure; a public works project can cost up to three times as much in Saudi Arabia as in California. Almost all the major industrial projects in the Gulf are geared to export and are in branches for which there is excess capacity in world markets. Furthermore, the capital cost of such projects can be 40 to 100 per cent higher in the Gulf than in OECD countries. The construction of major export refineries and petrochemical plants, for example, is taking place at a time when there is world-wide surplus capacity of 30 or 40 per cent in these sectors. It will be difficult for them to secure a big enough share of world markets to justify the high investment costs. Renner (1984) estimates that for most of the new Middle East export refineries to make even a minimum return on investment, product prices would be required to rise by about 30 per cent. To a large extent, the oil wealth has been recycled back, to the industrialised West through the high prices demanded by construction and engineering firms working in the region and massive imports of luxury goods. Sayigh (1982) deplores this 'leaking outwards' of the region's purchasing power, which stimulates the productive sectors of the supplying countries rather than those of the importing nations. Much of the funds that were invested rather than spent were held as liquid assets in Western banks where their value has been eroded steadily by inflation and the depreciation of the dollar.

The balance sheet is not entirely negative, and amid the inefficiency and waste there were some achievements. During the last decade illiteracy declined dramatically and enrolment in schools and universities rose rapidly. The number of doctors and hospital beds per capita more than doubled, infant mortality fell, and life expectancy rose dramatically. In the major oil-producing countries, per capita income rose sixfold between 1973 and 1981, but the extent to which inflation in the costs of food, housing and other basic commodities may have eroded away these gains is difficult to assess. Nor do such figures reveal income disparities within states which most observers agree have widened significantly during the last decade. Stork (1984) calculates that within the Arab states, 8.2 per cent of the population accounted for 30 per cent of GDP in 1970, while 72.5 per cent shared 50 per cent of GDP; while by 1981, 11.8 per cent of the population controlled 72.8 per cent of GDP, and 88.2 per cent were left with a mere 27.2 per cent.

Both those major oil producers with significant non-oil resources, such as Algeria, Iran and Iraq, and those without, such as Kuwait, UAE and Libya, have faced difficulties in using the new wealth from oil to diversify their economies. In Libya, after 1970, the revolutionary regime became committed to a higher rate of investment on development than under the previous regime in order to create viable productive sectors in agriculture and industry. Yet absorption of investment proved difficult both in agricultural and industrial sectors, and with a poor resource base abundant capital was no substitute for other factors of production such as skilled labour and management (Allan, 1981). In Algeria the rapid increase in the level of investment after 1974 proved too ambitious for the planning system, which was virtually overwhelmed by the scale of the tasks before it. The increased emphasis on non-industrial sectors and infrastructures envisaged under the 1974-7 Plan did not occur, and there was a further lengthening of the development lags between these sectors and industry which had appeared during the previous Plan. The distortions in the planning process which had begun to appear before 1974 were accentuated during the Second Four-Year Plan. Escalating costs and long delays in the completion of development projects, many of which had to be carried over from one planning period to the next, restricted the ability of the planning system to adjust to these processes (Lawless, 1984). The Algerian economist Benachenhou (1980) has described these years as 'la période du recul de la planification'. With reference to Iraq, McLachlan (1979) has shown that in spite of abundant water resources, the Iraqi environment is not easily controlled and represents a poor resource base. Consequently, it is not surprising that the government has decided to adopt the easy option of sustaining the country on an oil income. He concludes that Iraq, in spite of important non-oil resources, is becoming an oil economy on the model of that of Kuwait and the other Arab Gulf oil producers.

Abed (1986) is convinced that the oil producers believed that oil was synonymous with growth and that growth equalled development. Since 1980 a number of factors have demonstrated the limits of petrodollar power, reducing demand for OPEC oil and undermining world market prices. The world's economy went into its worst economic crisis since the 1930s and the deep recession resulted in a fall in demand for oil. The oil shocks of 1973 and 1979 encouraged industrial technologies that were energy-efficient while the use of alternative energy sources increased slightly. Above all, however, higher oil prices encouraged an increase in output in non-OPEC producers in both the industrialised countries and in the Third World. Keen to maximise their market share, these exporters have been ready to undercut OPEC prices, and as a result, OPEC's share of the capitalist world market for oil slumped from 67.4 per cent in 1976 to 44.6 per cent in 1982 (Renner, 1984). Mabro and Ait-Laoussine (1986) have pointed to the weaknesses in the policies adopted by OPEC in the face of this challenge. Between 1982 and 1985, its policies were defensive and, having decided to defend the price of the day, its members shouldered the burden of reduction in output. OPEC's policy of price maintenance failed because the volume of demand for OPEC oil fell below a critical level. In this situation some members were tempted to compete on price with non-OPEC producers in order to secure or augment their share of a declining market. By mid-1985, Saudi Arabia, which had resisted the temptation to compete, found itself in an impossible situation, with rapidly falling export revenues, and its actions to correct this situation triggered off a price war. In December 1985 OPEC endorsed the new Saudi strategy, announcing that its primary objective was no longer defence of oil prices through output reductions, but the securing of a market share. With both non-OPEC and OPEC states now seeking to maximise output, production quickly exceeded demand, leading to a price crash. OPEC members also embarked on a damaging price war between themselves. Mabro and Ait-Laoussine (ibid.) have warned of dangers inherent in this policy, and argue that a solution demands that the issues of prices and volumes be tackled together in a long-term perspective.

The transformation of the region into an oil-dominated economy has dictated a specific pattern of economic integration. Labour flows from the capital-poor states of the region to the oil-rich states have been virtually uncontrollable, and although some labour exporters have tried to regulate the movement, their efforts have met with little success. Capital flows have occurred in the oposite direction, from oil-rich to oil-poor states, mainly in the form of remittances from workers employed in the oil states of the Gulf and Libya, and transfers made by individual oil-rich states, by a large number of intra-Arab development organisations and by private institutions. At the end of 1980 the capital of the various intra-Arab institutions that make up the 'joint Arab economic sector', alone, amounted to more than US $23 billion (Sayigh, 1982). Unfortunately, these capital transfers, though considerable, have occurred with little co-ordination or control. Owen reminds us that the governments of the region have found it difficult to formulate their own development plans and consequently, 'They have no effective way of evaluating the impact of these massive shifts of money and labour or of managing them to promote their separate, never mind collective, interests ... The reasons are partly political involving narrow regime self interest' (Owen, 1981, p. 7).

In contrast with the exchange of capital and labour, trade among the countries of the region remains very restricted, accounting for a mere 6-7 per cent of both imports and exports of the Arab countries. However, one of the striking features of the 1970s has been the dramatic increase in the Middle East's share in total world trade and the growing integration of the whole region into the international economy. Owen (ibid.) calculates that between 1972 and 1979, the Arab countries' share in total world trade by value more than doubled from 3.6 to 8 per cent. This was achieved largely by an exchange of oil for growing imports of foodstuffs, armaments and consumer goods. A single commodity, oil, accounted for over 90 per cent of total Arab exports by the end of the 1970s and the proportions are higher if figures for only the major oil exporters are considered. Indeed, Sayigh (1982) observes a narrowing in the range of exports for most countries in the region. At the same time, some 93 per cent of all imports to the Arab countries come from outside the region, including about half of their food requirements. One country, Saudi Arabia, had become the seventh largest importer of American goods in the world by 1979. Sayigh points to the grave implications of the steep rise in exports, based almost entirely on one single commodity, and in imports. He sees this as one facet of an excessive dependence on the advanced industrial economies, other manifestations of which include a growing dependence on foreign sources of technology, foreign supplies of foodstuffs, foreign money markets, and the 'cultural invasion' of the region by foreign ideas, patterns of consumption and values. He concludes that

> The dependence of the Arab economies on the advanced industrial market economies is generations old, but it seems to have intensified and acquired more disturbing overtones since the early seventies when the Arab economies apparently acquired greater strength and a bright promise of independence (ibid, p. 162).

The intensification of the region's integration into the world capitalist economy was facilitated by the abandonment of 'socialist policies' in many countries during the 1970s and the 'opening up' or liberalisation of their economies. Economic liberalisation emphasised the need for greater foreign investment and accepted much greater income inequalities among the population. This policy shift was most dramatic in Egypt after 1974, but also occurred in Tunisia and to some extent in Algeria and Iraq. The Egyptian version of economic liberalisation aimed at restructuring the Egyptian economy by encouraging foreign investment, the reorganisation of the public sector and a revitalisation of the private sector (Owen, 1983). With its abundant cheap labour, Egypt would become an exporter of manufactured goods for the rest of the Arab world. However, foreign capital appeared reluctant to invest in Egyptian agriculture and industry, but the policy of infitah allowed Egypt to accumulate large amounts of foreign exchange by the end of the 1970s. It became the largest recipient of aid in the Third World, receiving some US $4 billion in non-military aid from the US alone. Remittances from Egyptians working abroad totalled some $3 billion a year, almost equal to the amount gained from oil exports. Large sums were also earned from Suez Canal dues and from tourism. The result was a consumption boom, and as local industry and agriculture could not supply the rapidly growing demand, there was a massive increase in imports of manufactured goods and foodstuffs. Hopes of transforming the Egyptian industrial sector in partnership with foreign capital were not fulfilled. Foreign companies were not interested in employing Egyptian labour or in encouraging Egyptian manufactured inputs, but merely sought access to the vast Egyptian market. Turkey has proved much more successful than Egypt at producing manufactured goods for Arab and Iranian markets. Progress in the agricultural sector failed to keep pace with the growth in population and demand for higher quality foodstuffs. By 1982 Egypt was spending nearly $3 billion a year on imported food. Reform of agricultural institutions and land tenure were avoided and government intervention limited to technical matters. Waterbury (quoted in ibid.) has pointed to Egypt's increased dependence on the United States during this period, while Hussein (ibid.) argues that infitah resulted in a weakening of Egyptians' control over their economy. The high price of increased reliance on external funds quickly became apparent when income from oil, tourism and remittances began to decline in 1982. Imports remained at a high level, giving rise to an acute balance of payments deficit.

OIL-INDUCED LABOUR MIGRATION: NEW RELATIONS OF EXPLOITATION

One of the most vivid patterns of change affecting the Middle East during the 1970s has been that of oil-induced labour migration. Migration, of course, is not a new phenomenon to the region. During the late nineteenth and early twentieth centuries, there was substantial emigration from Greater Syria to North and South America, while large numbers of Turks, Algerians, Moroccans and Tunisians joined the post-war migrant labour flow from the Mediterranean to Western Europe. Nevertheless, before the 1970s migrant flows from one Arab state to another were small. By 1980, however, at least 3 million Arabs (Halliday, 1984) had migrated to seek employment in the expanding economies of the Gulf states and Libya, where they were joined by an estimated 1.8 million migrants from outside the region, mainly from South Asia and the Far East. (1) Today, migrants account for over half of the total population in Kuwait, Qatar and the UAE, and over a third of the economically active population in Libya, Saudi Arabia, Bahrain and Oman. Denied the benefits of citizenship and political rights, they are segregated, often physically, but also socially, from the indigenous inhabitants and from other workers. Indeed, the increase in the participation of migrants from South-East Asia in Gulf labour markets since 1975 is linked to the introduction of large-scale 'enclave'-type industrial developments away from the major urban centres, where migrant workers have little or no contact with the local population. Examples of such enclave industrial areas include Umm Said in Qatar, Ruwais in Abu Dhabi and Yenbo and Jubail in Saudi Arabia.

If the rapidity of the demographic evolution of migrant communities, as workers have been joined by their dependents, has caused concern, another unforeseen aspect of labour importation identified by Serageldin *et al.* (1981) has been the large-scale withdrawal of nationals from the productive sectors of the economy and the proliferation of sinecure and luxury employment. The education system of these states reinforces this trend. Hostility to the growing migrant flows has become more widespread, and while many individual nations in the oil-rich states have benefited from the ambitious development programmes initiated by their governments, there are those who now question economic policies which inevitably require additional foreign workers. There are deeper issues involved, however, as Abu-Lughod reminds us:

> ... the countries of the Gulf/Peninsula are ... moving more and more towards a system that deals with labour as a rented commodity and treats labourers in the same

> impersonal way that one might treat a machine. One must ask what the effects of such a choice will be on the character of the host society. Can a culture that dehumanizes workers in this manner and that bases its own privilege on a caste system of this type endure, and can it train citizens who practise the ideals of Islam? (Abu-Lughod, 1983, p. 260).

For the labour-exporting states, there is no doubt that individual migrants have benefited financially from employment in the oil-rich states, but recent research indicates that migration for employment on the scale of the late 1970s has had a negative effect on the economies of the labour-supplying states. In 1975, for example, Jordan was exporting 40 per cent of employed Jordanians and the Yemen Arab Republic 24 per cent of her work-force (Serageldin et al. 1981). Because international migration is age- and skill-selective, it has often drained the capital-poor countries of their skilled workers and professional people, and in some cases has put at risk their own development programmes because critical skills have been lost or are in short supply. Often intended to absorb the unemployed and the unskilled in the case of Egypt, international migration has attracted skilled workers at all levels: the unskilled form only 20 per cent of Egyptian emigrants, but 40 per cent of the Egyptian work-force (Longuenesse, 1985). Furthermore, the capital-poor states pay for the education and training of the migrant workers which often represents a financial burden and opportunity cost in poor Arab countries. Because of manpower shortages, some labour-exporting states have themselves become importers of labour. In 1979 there were an estimated 50,000 foreign workers in the Yemen Arab Republic, while Jordan in the early 1980s had more foreign workers than some of the oil-rich states. Often described as 'replacement' migration, Seccombe (1986a) has shown that in the case of Jordan, labour demand has expanded beyond mere replacement, and foreign workers are to be found in sectors that are not suffering from labour shortages. The large inflow of mainly unskilled labour has contributed to rising unemployment among unskilled Jordanians.

Recorded workers' remittances in 1983 to the seven main Arab labour-exporting countries (Egypt, Jordan, Somalia, Sudan, Syria, Yemen Arab Republic and the People's Democratic Republic of Yemen) totalled $6.9

billion, compared with merchandise exports valued at only $8.6 billion. In 1982 earnings from remittances represented 31 per cent of GDP in Jordan, 34.8 per cent in Yemen Arab Republic and a staggering 65.2 per cent in the People's Democratic Republic of Yemen. Generally regarded as the most tangible benefit of labour migration, remittances are rarely invested in the productive sectors of the labour-exporting economies and most are used to purchase consumer goods. Changes in consumption patterns have encouraged the importation of foreign manufactured goods and foodstuffs, to the detriment of local agriculture and industry, and by stimulating demand, have provoked widespread inflation.

The official view in the oil-rich states during the mid-1970s was that the large inflow of migrant workers was a temporary expedient. The decline in the construction boom, together with the growth in the quantity and quality of the national labour force, would eventually reduce the demand for migrant workers and increase the share of the indigenous population in the labour force - a process known as 'localisation'. It was assumed by many observers that the decline in oil revenues since the early 1980s and the lower development expenditures that quickly followed would have a similar result. Since early 1985 there was much speculation that economic recession in the oil-producing states would lead to a rapid exodous of foreign workers. This view was reinforced when Libya expelled some 35-40,000 foreign workers, mainly Tunisians and Egyptians, in August-September 1985. At the same time, press reports suggested that 1.5 million Arabs employed in the Gulf might have to return home in 1986. Seccombe (1986b), however, argues that although the inflow of non-nationals to the Gulf states has declined and work permit cancellations have increased, there has been a relatively low level of net labour outflows and a high rate of retention of foreign labour. Recorded workers' remittances, for example, did not decline until 1984-5, and Bangladesh reported higher remittances in 1985-6 than 1984-5 (Table 10.3). He suggests that if demand for labour in the construction sector has declined, staff are still required to maintain and service the new infrastructure and industries. With more restrictive immigration procedures, employers have sought to retain existing labour, often at lower wage rates, while the economic recession appears to have had little impact on the rapid increase in employment of domestic services. In Kuwait, Bahrain and Oman, Asian

Table 10.3: Workers' remittances 1981-2-1985-6 ($ million)

	1985-6	1984-5	1983-4	1982-3	1981-2
Bangladesh	550	473	649	394	402
Egypt	na	3,781	3,931	3,166	1,935
India	na	2,600	2,650	2,599	2,281
Jordan	898	1,027	923	932	921
Pakistan	2,445	3,006	3,116	2,793	2,195
Philippines	na	624	944	810	791
South Korea	na	1,490	1,663	1,359	1,102
Syria	na	214	302	292	380
Yemen Arab Republic	1,059	1,276	1,176	975	1,323

Source: Seccombe, I.J. (1986) The myth of the expatriate exodus. Middle East Economic Digest, 26 July-1 August, p. 28

immigrants received an increasing share of new work permit issues. Seccombe argues that the economic recession will not necessarily make it easier for states like Kuwait to achieve their plan to balance expatriate/local population ratio by the year 2000. Evidence from work permits renewal and transfer data indicates that more and more migrants are staying on in Kuwait after their initial contracts have been completed. Seccombe (1986c, p. 51) concludes that '... the Gulf states may be repeating the experience of the labour-importing countries of Western Europe, where immigrant labour also proved resistant to the pressures of recession during the mid-1970s'.

FOOD AND FARMING: DEFICITS AND DEPENDENCY

In 1960 the Middle East was a net food exporter, but the 1970s saw rising food imports at a time when world food prices were climbing rapidly, provoking an acute 'food crisis'. The value of agricultural imports rose from about $4 billion in 1973 to over $30 billion in 1985 (Table 10.4). Today, the countries of the Middle East represent the world's most rapidly growing food deficit area. Nearly 60 per cent of the region's wheat requirements are now imported, as are between 15 and 20 per cent of meat supplies. Saudi Arabia spent $4.9 billion on food imports in

Table 10.4: The Middle East: total agricultural imports, 1983, 1984, 1985

Country	Total agricultural imports ($ million) 1983	1984	1985
Algeria	2,509	2,570	2,790
Bahrain	242	224	208
Cyprus	204	195	217
Egypt	3,887	4,084	4,257
Iran	3,440	3,670	3,490
Iraq	2,857	3,085	3,040
Israel	924	981	920
Jordan	668	700	685
Kuwait	1,510	1,440	1,285
Lebanon	573	601	640
Libya	1,515	1,525	1,495
Morocco	1,096	1,300	1,230
Oman	381	435	485
Qatar	224	227	218
Saudi Arabia	5,182	5,351	4,900
Syria	878	905	942
Tunisia	516	596	520
Turkey	285	713	550
UAE	1,300	1,380	1,240
Yemen (AR)	803	809	775
Yemen (PDR)	285	240	280
Total	29,279	31,031	30,167

Source: Middle East Economic Digest, 7 June 1986, p. 32

1985, Egypt $4.3 billion, and Iran $3.5 billion. Even Turkey, the only country which has not had a sizeable trade deficit in agricultural commodities, experienced a sharp rise in agricultural imports in 1984 and 1985? (Margulies, 1986). Conservative estimates project that the value of food imports to the Arab states alone could rise to a staggering $322 billion by the year 2000 (Bowen-Jones, 1986). Even the richest states in the region have become uneasy about their food dependence and potential economic vulnerability.

Increases in domestic food production during the 1970s only just kept pace with population growth in most Middle Eastern countries, and fell far short of the demand for better and more nutritious food. Agriculture continues to support a high proportion of the labour force, but its share of the employed population has fallen, often dramatically. Similarly, agriculture's contribution to GDP has contracted relative to other economic sectors. With justification, agriculture has been described as the 'Achilles heel' of Middle Eastern development efforts. Throughout the region, the agricultural sector has been accorded a low priority by planners and government elites. Weinbaum (1982) estimates that by the early 1980s, no more than 15 per cent of public investment in the region was granted to the agricultural sector. These sectoral priorities, he argues, must be interpreted as part of a political process. Public funds have gone disproportionately towards underwriting industry and subsidising food for city dwellers because influential elites are almost exclusively urban-based. Almost unanimously in their efforts to achieve modernisation, the region's policy-makers accepted the vision of an urban-industrial society as the key to national prosperity. The agricultural sector was too weak to bid successfully for investment funds - indeed, the dominant economic policies extracted capital and labour from the rural areas to sustain an expanding urban, industrial sector.

Resources have been transferred out of agriculture to the rest of the economy by price-fixing policies. By a system of depressed crop prices, the agricultural sector over much of the region has been forced to bear much of the burden of capital formation for the urban-industrial sector. Low procurement prices for agricultural products also ensure higher profits on the sale of export crops, profits which often accrue to governments where marketing is through state-controlled organisations. In Egypt, for example, transfers to Treasury from profits of the Cotton Organisation totalled LE 348 million between 1973 and 1976, whereas the direct producer-subsidies paid by the Agricultural Prices Stabilisation Fund on all crops for all purposes amounted to only LE 187 million over the same period. If the exchange rates gains are added in, the transfers out on cotton alone more than paid for all of the direct producer-subsidies, all public sector investment in agriculture and all of the current expenditures of both the Ministries of Agriculture and Irrigation.

Low agricultural prices, together with subsidised food in the cities, have acted as push and pull factors, respectively, contributing to the movement of peasants from the countryside to urban areas and to labour markets in other countries. Low profits from farming and the possibilities of more attractive employment in urban areas have created acute shortages of skilled and even unskilled farm workers in some parts of the region. Probably the most dramatic decline in the rural population has taken place in Iraq, where shortages of agricultural labour have become so acute that efforts have been made to encourage Egyptian, and more recently Moroccan, farmers to settle in rural areas.

In nearly all countries of the Middle East, governments have tried to reorganise the structure of land ownership by introducing programmes of agrarian reform. These strategies have been used by regimes with very different political and economic systems to achieve a variety of political, economic and social aims. Invariably, political considerations have taken precedence over economic factors. In Egypt and Iraq, land reform programmes were introduced following military coups and were intended to destroy the political power base of those large landowners who had dominated the deposed monarchical regimes. The reform in Iran was one of many land reforms carried out by conservative Third World governments to eliminate a real or possible revolutionary threat from a discredited peasantry and to create a new social grouping in the rural areas that would support government policies. Land expropriated from rural elites was distributed to peasant farmers and the prevailing mode of land distribution has been to extend small-scale private ownership. With too little land to go around, however, larger sectors of the rural population have been bypassed. Paradoxically the breakup of the large estates often succeeded in reducing employment opportunities for the large landless rural population - a substantial proportion of rural families throughout the Middle East.

Agrarian reform programmes were frequently accompanied by attempts to introduce new institutions and reorganise production relations. State-controlled co-operatives have been the most widespread mode of reorganisation, providing beneficiaries of reform land with credits and farm inputs and a marketing outlet for their crops. Intended to provide certain economies of scale and offset the land fragmentation of the reforms, too often the co-operatives have been starved of the necessary funds to provide adequate credits for farm inputs, while many farmers have resented their loss of control over decision-making. Peasant individualism has remained strongly entrenched. Some regions have been convinced that the centrally managed collective or state farm represents the only means of increasing marketable surpluses, providing more employment and extracting funds needed for industrial development. However, most have moved cautiously in introducing the collective or agri-business option, which in many cases challenges prevailing social and economic institutions.

All too often, new institutions have been given little time to develop fully and to resolve their problems before being replaced by new modes of production. Iran under the Shah's regime provides a classic example of abrupt policy changes dictated by political expediency. Under the Shah's land reform, membership of a village co-operative was obligatory. Many co-operatives existed in name only, however, and they were starved of funds because the government feared the consequences of strong peasant-run organisations. Dissatisfaction with the performance of the co-operatives from the mid-1960s led the Shah to favour the creation of farm co-operatives, professionally managed, collectively owned farm units modelled on the Israeli moshavim. However, impatient with the slow production increases, he turned increasingly to foreign agri-business companies to raise output through capital-intensive mechanisation. Yet dramatic increases in production were not realised, and the agri-business ventures were soon submerged by a host of problems. As the revolution approached, enthusiasm for agri-business began to wane and a new policy of encouraging medium-sized farms was introduced (Lawless, 1985).

It is estimated that by the year 2000, the Middle East will have a population of some 420 million. The question of how these extra people will obtain their food poses a serious problem. There are no large unexploited water resources, making it difficult to increase the cultivated area. Indeed, existing cultivated areas are under threat from urban expansion and a widespread process of ecological deterioration. During the 1970s, in Egypt nearly a sixth of cultivated land was lost to other purposes. Given the added problem of increasing yields rapidly. Beaumont and McLachlan (1985) argue that food imports will continue at a high level until the twenty-first century. They estimate that by the year 2000, the region will need to import 48 million tons of cereals alone, almost two-thirds of total world exports of cereals in the early 1980s. As it seems likely that the oil producers will be able to finance such imports, this will be at the expense of capital-poor states elsewhere in the world.

THE WEST BANK AND GAZA: ISRAEL'S BANTUSTANS?

At the very heart of political instability in the Middle East region lies the Arab-Israeli conflict and the unresolved plight of the Palestinian people. In 1967 Israel occupied the residual areas of Palestine, generally referred to as the West Bank and Gaza, which then contained over half the estimated 2,650,000 Palestinians in the world. Today, after 19 years of Israeli domination, the full absorption of the occupied territories into the Israeli system is well advanced, the economy of the territories continues to stagnate, and the Palestinians are denied permission to develop and expand their own industry, agriculture and social infrastructures (Roberts, 1986).

Whereas the Allon Plan of Israel's Alignment limited Israeli settlement in the West Bank to the arid and depopulated Jordan rift so as to guard Israel's eastern frontier, the pioneer zealots of Gush Emunim (Bloc of the Faithful) were committed to a Jewish presence in the heavily populated central regions of the West Bank. When the Likud government came to power in 1977, a number of Gush Emunim settlements were quickly legalised and the new government set out to implement a settlement policy that was almost identical to that proposed by Gush Emunim. At first Arab land was seized on security grounds, but more recently, the policy of declaring certain tracts as state lands has greatly extended the possibilities for settlement. By 1986 Israel had taken direct control of 41 per cent of the West Bank, and a further 11 per cent of the land was subject to severe restrictions that amounted in effect to direct control (Beaumont and McLachlan, 1985). Colonisation has fallen short of the 'Program of the 100,000' announced in 1981, but nevertheless some 52,000 Jewish settlers were living in about 100 new settlements in the West Bank and Gaza by 1986. The majority of these 'colonists' occupy dormitory settlements, taking advantage of relatively cheap housing and economic incentives such as tax deductions, and commute daily to jobs in Jerusalem and Tel Aviv (Demant, 1983).

The Israeli occupation of the West Bank also enabled Israel to secure control over its important water resources. Mekorot (the Israeli Water Company) immediately took responsibility for the West Bank's water resources, and no increase in the water supply to the Palestinian population was permitted on the grounds that this would threaten salination of supplies to pre-1967 Israel. Since the Israeli annexation, only seven new wells have been drilled to supply Arabs with drinking water. A number of irrigation wells supplying Palestinian farms have run dry, while others are experiencing a declining water table and increased salinity. No Palestinian village or individual farmer has been given permission to drill a new well for irrigation since the Israeli occupation, but between 1967 and 1982, Israeli settlements have drilled at least 17 new wells. The town of Ramallah had no choice but to accept connection to the Israeli national water system when its wells ran dry and the Israeli authorities refused drilling permission. The Yarkon and Crocodile basin, lying almost entirely under the northern West Bank, provides up to 40 per cent of the water consumed in Israel proper. By 1979 Israel boasted that it was exploiting on average 95 per cent of the annual renewable water resources between the Mediterranean and the Jordan river (Stork, 1983).

The demographic consequences of the occupation have been equally dramatic. After 19 years of Israeli rule and despite a rate of natural increase averaging 3.5 per cent a year, the number of Palestinians who remain is 1,350,000, approximately the same as in 1967. The West Bank and Gaza now constitute less than a third of the total Palestinian population. Abu-Lughod (1983) has estimated that between 1967 and 1983, some 700,000 Palestinians were displaced from their homeland. This has been achieved by massive expulsions of residents and by emigration. Especially after 1973, when it became clear that the occupation was to be long term, the emigration of skilled and professional workers and businessmen to the expanding economies of the oil states accelerated. For the peasantry, unable to make a living through farming because of land confiscation and lack of access to irrigation water, and for those without skills, day labouring jobs in Israel, many in black economy employment, have become the only means of survival. In 1983, 46.5 per cent of all employed persons in Gaza were working in Israel; the figure for the West Bank was 33 per cent. About half of these Palestinian workers were employed in construction. A once diversified society has been proletarianised and deliberately reduced to the status of a reserve labour army, dependent on the Israeli economy. (2)

Stagnation and decline in its own productive sectors and increased dependence on external sources (some 40 per cent of Palestinian disposable income came from abroad in 1984) have deepened the vulnerability of the Palestinian economy and its economic subservience to Israel. The acute economic crisis now affecting the Arab oil-producing states, rising unemployment in Jordan and continued recession in Israel, could bring about a steep decline in West Bank per capita income. A recent ILO report highlighted the rising unemployment rates in the West Bank, especially among young professional graduates, as opportunities for employment in Saudi Arabia and the Gulf diminish (Graz, 1986).

CONCLUSION: A SECOND CHANCE?

The non-OPEC producers possess a relatively small crude oil resource base and OPEC countries still control 67 per cent of the world's oil reserves. Beyond 1990, Abed (1986) estimates that OPEC may begin to re-establish some of its former power with five Middle Eastern producers - Iran, Iraq, Kuwait, Saudi Arabia, UAE - increasingly dominating OPEC's capacity and production. He believes that by the mid-1990s, the Middle East oil states may be in receipt of higher oil revenues and may therefore be given a second chance to set their economies on a course of self-sustaining economic and social development. However, will they have learnt the lesson of the first oil boom? Formulating and implementing new economic strategies will not be easy. In the smaller Gulf states, there is growing disillusionment with policies aimed at economic diversification which have produced numerous 'white elephants', like Dubai's massive Jebel Ali port complex and free-trade zone, where hopes of creating a thriving industrial zone from the desert wastes have remained largely unfulfilled. With tiny populations in which foreigners greatly outnumber nationals, and few natural resources other than oil, investment of the bulk of their oil wealth abroad could become increasingly attractive. For the more populous oil states, development of their non-oil sectors has proved equally problematic, in some case unleashing social forces that have threatened to undermine their regimes. Yet social and political changes are a prerequisite for the success of development programmes. Abed (ibid.) argues that economic development is not possible without political modernisation and the active participation and mobilisation of the majority of the population. He claims that major development decisions taken by autocratic regimes in a political environment filled with fear, indifference or corruption have resulted in disastrously bad programmes and projects. For the capital-poor states of the region, the future looks bleak. The generosity of their oil-rich neighbours is by no means a foregone conclusion, and if there is a second oil boom, there will be strong pressures on the oil producers to use the new wealth for the benefit of their own nationals. Already deeply affected by the first oil boom and their incorporation into the regional oil economy, their productive sectors have declined or stagnated, while important sections of their populations have acquired aspirations which their own governments cannot satisfy. The rhetoric of Arab unity and economic integration and harmonisation will no doubt continue to be voiced, but, as in the past, is unlikely to be translated into reality. Inequalities between states within the region and between different social classes seem likely to be accentuated rather than to diminish during the next decade. With large sections of their populations maintained just above the poverty line with the assistance of a range of government subsidies on basic commodities, and with a relatively low domestic savings ratio, the capital-poor states will remain heavily dependent upon foreign aid. Serious balance of payments problems will make it difficult to resist policy recommendations by creditors such as the World Bank and the International Monetary Fund, even though policies such as the reduction or elimination of subsidies could provoke internal unrest, threatening the political survival of their regimes.

NOTES

1. The proportion of migrant workers of Arab origin in the Gulf states declined from 77 per cent in 1970 to 57 per cent in 1980. The Asian share of the Gulf's non-national labour market has increased rapidly, reaching 34.5 per cent in 1980. Non-national Arabs still form a majority of the

expatriate work-force in Kuwait and Saudi Arabia, but a minority in Bahrain, Oman, Qatar and UAE. Following the closure of West European labour markets to new immigration in the early 1970s, Turkey achieved an impressive redirection of its emigrant work-force to the Arab world, notably to Saudi Arabia, Libya and Iraq. See I.J. Seccombe and R.I. Lawless 'Between Western Europe and the Middle East: changing patterns of Turkish labour migration', Revue Européenne des Migrations Internationales, vol. 2, no. 1 (1986), pp. 27-58.

2. It should be noted that Gaza was never as diversified occupationally as the West Bank. It had fewer skilled and professional people to lose and has been increasingly drawn upon as the source of cheap labour for the Israeli economy. Israeli policy towards the West Bank has been different, and has sought to reduce the number of inhabitants and their capacity to resist the occupation.

REFERENCES

Abed, G. (1986) Arab oil in the 1990s: a second chance? The Middle East, 139, 29

Abu-Lughod, J. (1983a) Social implications of labour migration in the Arab world. In I. Ibrahim (ed.) Arab resources: the transformation of a society, Centre for Contemporary Arab Studies, Washington, DC, Croom Helm, London

------ (1983b) Demographic consequences of the occupation. MERIP Reports, 115

Allan, J.A. (1981) Libya: the experience of oil. Croom Helm, London

Beaumont, P. and McLachlan, K. (1985) Agricultural development in the Middle East. John Wiley, Chichester, New York

Benachenhou, A. (1980) Planification et développement en Algérie 1962-1980. Presses de l'E.N. Imprimerie Commerciale, Algiers

Bowen-Jones, H. (1986) Import dependence is cause for concern. Middle East Economic Digest, 30 (7)

Demant, P. (1983) Israeli settlement policy today. MERIP Reports, 116

Graz, L. (1986) ILO report highlights West Bank unemployment. Middle East Economic Digest, 30 (24)

Halliday, F. (1984) Labour migration in the Arab world.

MERIP Reports, 123

Kubursi, A. (1980) Arab economic prospects in the 1980s. Institute for Palestine Studies Papers 8, Beirut

Lawless, R.I. (984) Algeria: the contradictions of rapid industrialisation. In Richard Lawless and Allan Findlay (eds), North Africa: contemporary politics and economic development, Croom Helm, London

------ (1985) The agricultural sector in development policy. In Peter Beaumont and Keith McLachlan (eds), Agricultural development in the Middle East, John Wiley, Chichester and New York

Longuenesse, E. (1985) Les migrations de travail dans les bouleversements de la société égyptienne. Peuples Méditerranéens, 31-32

Mabro, R. and Ait-Laoussine, N. (1986) We are worried. Middle East Economic Survey, 29 (16)

McLachlan, K. (1979) Iraq: problems of regional development. In Abbas Kelidar (ed.), The integration of modern Iraq, Croom Helm, London

Margulies, R. (1986) What future for Turkish agriculture? The Middle East, 137

Owen, R. (1981) The Arab economies in the 1970s. MERIP Reports, 100-101

------ (1983) Sadat's legacy, Mubarak's dilemma. MERIP Reports, 117

Renner, M. (1984) Restructuring the world energy industry. MERIP Reports, 120

Roberts, J. (1986) Palestine: Israel tightens its economic stranglehold. Middle East Economic Digest, 30 (15)

Sayigh, Y. (1982) The Arab economy: past performance and future prospects. Oxford University Press, Oxford

Seccombe, I.J. (1986a) Immigrant workers in an emigrant economy. International Migration, 24 (2)

------ (1986c) The myth of the expatriate exodus. Middle East Economic Digest, 26 July-1 August

------ (1986b) Economic recession and international labour migration in the Arab Gulf. The Arab Gulf Journal, 6 (1)

Serageldin, I. et al. (1981) Manpower and international labour migration in the Middle East and North Africa. The World Bank, Washington, DC

Sharabi, H. (1983) The poor rich Arabs. In Ibrahim Ibrahim (ed.), Arab resources: the transformation of a society, Centre for Contemporary Arab Studies, Washington, DC, Croom Helm, London

Stork, J. (1983) Water and Israel's occupation strategy.

MERIP Reports, 116
------ (1984) Ten years after. MERIP Reports, 120
Weinbaum, M.G. (1982) Food, development and politics in the Middle East. Westview Press, Croom Helm, London

Chapter Eleven

CONTEMPORARY ISSUES IN SOUTH ASIA

J. Soussan

The South Asian sub-continent presents many paradoxes to the student of development issues. It contains a relatively small number of states (India, Pakistan, Nepal, Bangladesh, the Maldives, Bhutan and Sri Lanka) but a vast number of people (around one billion, or over 20 per cent of the world's population in 1985). The region is an area of enormous diversity, both environmentally (with massive mountain ranges, vast deserts, densely populated areas of high productivity) and socially (containing many distinct languages, ethnic groups and religions, remote rural areas and huge cities), and yet has a coherence and sense of identity born of a monsoon climate and (more importantly) a common political and cultural heritage which permeates all aspects of life. These make South Asia a distinct and identifiable region. To generalise about such a huge and diverse area is a thankless task, but generalise we must if the development experiences of this region are to be understood and their implications drawn out.

For many people, one image of South Asia dominates - that of the starving child. The region is seen to represent the poverty of the Third World - poverty which is frequently presented as being born of over-population. How true is this image? Without doubt South Asia contains many extremely poor people: for example, it is estimated that 273 million Indians (37 per cent of the total) live below the official poverty line, a standard which represents an income from all sources which provides the barest minimum of food necessary to sustain life and no more. Large regions of India and Pakistan and most of Nepal and Bangladesh are poor by any standards and have stagnant, or even declining, economies. Yet South Asia contains areas, both urban and rural, of relative prosperity and dynamic economic development, many of which produce significant food surpluses. Similarly, even within the poor regions, significant sectors of the population are far from destitute. In South Asia, then, the key issue (if one can be so reductionist) is less one of poverty than of inequality - inequality which is both spatial and social.

There are many other, more specific, issues which could be discussed here: environmental maintenance and the threat of degradation from forces such as deforestation and soil erosion (World Resources Institute, 1986); international trade relations and financial flows (UN, 1985); population trends and health issues (World Resources Institute, 1986); migration - both international (to the Middle East and the developed countries) and intra-national; the nature and causes of social change (Bardhan, 1984); political instability (in particular, the rise of regionalism and, in places, separatist conflicts) (Farmer, 1983); rapid urbanisation and the accompanying growth of urban problems (Bose, 1978); and energy supplies - both commercial fuels such as oil, coal and nuclear power, and non-commercial fuels, such as wood and agricultural residues (Soussan et al., 1984). All of these issues and more are of importance in contemporary South Asia, and each could be discussed at length. There is just not the space to do so here, however, and in consequence, the remainder of this chapter will concentrate on contemporary developments in the two sectors of the economy which are most crucial for South Asia's hope for generating economic development and alleviating poverty and inequality - agriculture and industry.

AGRICULTURE

Agriculture remains the most important sector of the economy of all of the countries in South Asia. For most, it is the largest sector as a percentage of GNP and the dominant export earner, and for all it is the largest employer. Agriculture feeds the people and provides inputs and capital to other sectors of the economy. In many parts of South Asia, agriculture is characterised by low and stagnant productivity, rural people by extreme poverty, and rural societies by restrictive structures which produce great inequalities. Rural indebtedness, fragmentation of landholdings, unequal land distribution and usurous tenure relations combine with factors such as archane technology, poor infrastructure and restricted access to inputs, credit and markets to depress productivity and preserve inequalities. To these factors can be added the vagaries of the environment, with in particular South Asia's monsoon climate producing highly seasonal and irregular rainfall which can seriously affect output if controlled irrigation is not available: a gloomy picture indeed.

Yet South Asia has over recent decades increased food output to more than keep pace with population growth, with India in particular moving from being a major grain importer (with 24 million tonnes of emergency supplies in the drought years of 1965 and 1966 staving off mass starvation) to being a net grain exporter, with the Indian government now faced with problems of storing massive buffer stocks. Much the same is true of Pakistan and Sri Lanka. The story is less optimistic for the very poor countries such as Bangladesh, but even those have on the whole managed to increase food production to outpace population growth. That this rapid expansion of production has been possible can be summed up in two words: Green Revolution.

How can these contradictory pictures of South Asia's agriculture be reconciled? Which is true? The answer is simple - they both are. The Green Revolution, which is responsible for so much of the increased output, is highly concentrated in a few privileged regions of South Asia, with other areas receiving few benefits and even (it is argued) suffering adverse consequences from the 'revolution'. This high concentration is no accident; it stems from the very nature of the Green Revolution itself. Few subjects arouse such polarised opinions as the Green Revolution: commentators tend to be either all for it or totally against it, and few hold a balanced view. More objective authors, such as Baker (1984) and Chambers (1984), argue convincingly that the Green Revolution in South Asia has had many benefits, but is not and cannot be the universal panacea claimed by its proponents. This simple statement is a key to understanding one of the main issues for agriculture in contemporary South Asia.

What is the Green Revolution? It is a development strategy, based on a package of technological and institutional innovations, whose aim is to solve a perceived problem in Third World countries where a growing population is seen to be pressing on food supplies and available land. The package generally consists of the introduction of new, higher-yielding varieties of seeds (HYVs, usually foodgrains), a new set of cropping practices and greatly increased inputs of fertilizers, pesticides, mechanisation and water. These are essential to provide the controlled environment required by HYVs if they are to grow. As important as these technological changes are the establishment of extension services, to provide technical support; credit programmes, to permit farmers to purchase the vital inputs; marketing organisations, so that the increased output can be sold; and infrastructural development, so that inputs can be distributed and crops marketed. In other words, the Green Revolution is a sophisticated (and expensive) package of changes to the agrarian economy, which are introduced as a conscious state policy.

Why should the state wish to introduce such policies? Pearse (1980) identifies two distinct reasons:

(1) Freedom from national food dependence by increased food production leading towards self-sufficiency at a national level; and
(2) Freedom from hunger for all sections of the community.

In South Asia, the first policy goal has dominated the concern of the governments introducing the Green Revolution, and has largely been achieved. Some progress has also been made towards the second goal, but this has been highly selective, and much of the progress has had little to do with Green Revolution policies. The main effect of the Green Revolution has been to increase dramatically the already wide and widening disparities in agricultural production and rural welfare between different regions and between different sections of the community within some regions.

The characteristics of South Asia's agricultural development can be illustrated by looking in more detail at some of the contemporary trends in the region. Table 11.1 shows the area, yields and total production for each of the major crops in the countries of South Asia. These figures reveal much about the agricultural geography of the region. The main crops are grains, with rice and wheat particularly dominant, but cash crops are regionally significant, and pulses, an important component of South Asian diet, are also

Table 11.1: South Asia: production and yields of principal crops, 1984

	Area (1000 ha)	Yield (hg/ha) (1974-6)	1984	Production (1000 M.T.)
Bangladesh				
Rice	10500	(1790)	2048	21500
Jute	537		1365	733
Pulses	315	(734)	691	218
India				
Wheat	24395	(1310)	1851	45148
Rice	42800	(1692)	2126	9000
Millet	18500	(493)	638	11800
Pulses	23414	(474)	539	12620
Jute	1075		1365	733
Tea	410		1573	645
Cotton	8000		469	3750
Nepal				
Wheat	472	(1148)	1343	634
Rice	1335	(1981)	2067	2760
Millet	124	(1130)	924	114
Pulses	164	(397)	427	70
Bhutan				
Rice	31	(2000)	2001	61
Pakistan				
Wheat	7322	(1330)	1510	11053
Rice	1999	(2272)	2507	5009
Millet	553	(486)	463	256
Pulses	1435	(522)	518	743
Cotton	2360		1258	2970
Sri Lanka				
Rice	750	(1975)	3027	2270
Tea	310		742	230

Source: FAO (1985) FAO production yearbook, 1984. FAO, Rome

notable. The figures also reveal a distinct distribution of the main crops. Rice dominates in the wetter high productivity areas in Sri Lanka, Bangladesh, the coastal plains of

Table 11.2: Grain yields for selected Indian states (kg/ha)

	Punjab		Tamil Nadu	Bihar		All India	
Year	Rice	Wheat	Rice	Rice	Wheat	Rice	Wheat
1966-7	1186	1544	1551	366	451	863	887
1970-1	1765	2238	1974	788	957	1123	1307
1974-5	2072	2360	1855	872	1353	1045	1338
1977-8	3362	2537	2210	987	1261	1308	1480

Source: Joshi, P. and Kaneda H. (1982), p. A-3

Pakistan and India, and the lower reaches of the Gangetic Plain. Wheat is the main crop in the upper Indo-Gangetic Plain (which is again an area of high potential) and on the better land in much of the Indian peninsula. Of the cash crops, tea is largely a highland crop, jute is confined to the lower Ganges and cotton is found mainly in the north-western part of the sub-continent and the western parts of the Indian peninsula. Pulses are widely distributed and millett is the main grain crop on low potential land. Many of the rice areas have been cultivated intensively for many centuries and have extremely high population densities. Bangladesh, for example, contains over 1,000 people per square kilometre of cultivated land, a staggering figure in a predominantly rural country. Some of the wheat areas also have long histories, but have experienced rapid agricultural growth in the last century or so. The lower potential lands of central India, Pakistan, Nepal, north-east Sri Lanka, and so on, are far less densely settled, and are frequently characterised by extreme poverty and economic stagnation.

Baker (1984) summarises this spatial pattern into those arbitrary but useful categories for India: the old rice lands, the dry lands and the 'intermediate' areas of rapid development in the last century. The Green Revolution has been most successful in the intermediate areas, where the conditions for the adoption of the new technologies and forms of organisation were by far the most favourable. As has been suggested, these areas are regions which were already relatively prosperous and dynamic. This is revealed in Table 11.2, which demonstrates the disparities in grain yields between more prosperous (e.g. Punjab) and poorer (e.g. Bihar) states in India, both at the beginning of the Green Revolution era and after a decade of adoption of the technologies.

Table 11.3: The Green Revolution in Punjab State, India

	Production (Million tonnes)	
	Wheat	Rice
1966	1.9	0.3
1972	5.6	1.5
1980	7.9	3.1

	Area under HYV (% total)	
	Wheat	Rice
1967	35%	5%
1975	90%	91%

Farm size distribution (1971)	
Size (ha)	% total
below 1	37.6
1-3	31.3
3-5	13.8
5-10	12.3
10+	5.0

Land owner-cultivated (% total cropped area) 81% in 1970

Irrigation: 76.8% cropped area in 1974

Tubewells	1973:	362,000
	1980:	565,000

Fertilizers: total consumption	1972:	290,000 tonnes
	1980:	680,000 tonnes

Sources: Kahlon et al. (1972) and Gill (1983)

Punjab State in India is the classic example of a successful Green Revolution area. As Table 11.3 indicates, HYV seeds are almost universal, as are the associated inputs of fertilizers and secure irrigation. Mechanisation is the norm rather than the exception, land distribution is relatively even and most farmers own the land they farm (in contrast with many Indian states, where tenancy is common and serves to restrict agricultural innovation). In the late 1960s, the Punjab State government established and vigorously pushed a Green Revolution strategy. New seeds were provided, credit made available, an extension service to provide expert advice established, and an infrastructure to provide inputs and market the grain developed. The State government's policy found willing partners in Punjabi farmers, who adopted the Green Revolution package with enthusiasm. The results were dramatic. Wheat yields, already high by Indian standards, shot up, and the Punjabi wheat crop increased from 1.9 million tonnes in 1965-6 to 5.6 million tonnes in 1971-2. By the early 1970s, Punjabi wheat yields exceeded those of the USA - an amazing achievement. Progress during the 1970s was less dramatic, but it did continue, and the early gains were consolidated. Total food-grain production (including rice, the production of which grew rapidly with the extension of secure tubewell irrigation) increased from 7.3 million tonnes in 1971 to 12 million tonnes in 1980.

By the late 1970s the Punjab was the breadbasket of India, contributing 3.2 million tonnes of wheat (63 per cent of the total) and 3.4 million tonnes of rice (56 per cent of the total) to the Indian government's strategic grain reserve in 1978. During the 1970s, the Punjab was using around 30 per cent of India's total consumption of chemical fertilizers, and the use of these and other inputs (in particular water) enabled Punjabi farmers to overcome the vagaries of the monsoon climate. These achievements were remarkable for a state with 2.5 per cent of India's farmland and less than 3 per cent of her population.

In contrast to other Green Revolution areas, such as those in Latin America, the success did not lead to dramatically increasing inequalities within the Punjab. This partly reflects the state's land distribution, but the Green Revolution led to a rapid rise in wage levels for agricultural workers (attracting migrant labour from poorer states such as Bihar), and created many new opportunities in agro-processing and agricultural service industries, which grew as quickly as agricultural output. Thus, for example, many members of the 'blacksmith' caste (traditionally a 'Harijan' or 'untouchable' caste) have developed prosperous businesses servicing agricultural machinery.

The Punjab's story, then, is a remarkable one. It is one of the successful development of the agrarian economy of an area which has resulted in a general increase in prosperity and the production of food surpluses which are of major importance to the rest of India.

The success of other Green Revolution areas in South Asia is perhaps not as dramatic, but is frequently comparable. In particular, many have had rapid increases in per hectare yields and total output. The general distribution of prosperity is perhaps less typical, as a number of areas in South Asia which have had sustained agricultural development have also experienced some widening disparity of welfare. This issue has frequently been overstated, however. The real impact of the Green Revolution has been to widen disparities between, as much as within, agricultural regions.

This leads on to the key question for South Asia's agriculture: can the dynamic development of the agriculture of a few privileged regions such as the Punjab be reproduced elsewhere? Time alone will answer this question. The trends emerging are not encouraging, however. There has without doubt been some progress in all areas of South Asia, and Green Revolution technologies have been adopted successfully by farmers in most regions. In many areas, however, increasing yields and technical innovations have been concentrated in the hands of big farmers, who have access to the capital resources and the contacts in the bureaucracy to take avantage of the new technologies. Byres (1972) argued this case a number of years ago, and his argument has been supported by the experience of regions of South Asia such as Bangladesh (Jones, 1982), Bihar (Corbridge, 1984) and Tamil Nadu (Farmer, 1977). In such areas, small farmers, and in particular tenant farmers and share-croppers, have been disadvantaged by the Green Revolution. Grain prices locally tend to fall, but their output remains static, the new demands of the big farmers place pressure on local supplies of vital inputs such as water and fertilizers, and so on. In such areas the general trend is towards stagnant or declining production for all but the big farmers, worsening the already acute poverty of the poor. It is doubtful whether any policy based on the traditional Green Revolution style will be able to overcome the poverty of such areas. In consequence, in South Asian agriculture the signs are that the prosperous people and regions will develop further while the poor will continue to get poorer.

INDUSTRY

The development of South Asia's industrial sector is even more concentrated than that in agriculture, as it is inevitably centred on the region's cities. Taken over time, the track record of industry in India, Pakistan and Sri Lanka has been one of slow but steady progress. There have been periods of more rapid growth, but South Asia has never experienced the sustained high growth rates in industry that a number of Latin American and East and South-East Asian countries have had since the mid-1960s. Nor has South Asia incurred the problems - excessive foreign debt, dependence on foreign capital and markets, and so on - that this rapid growth has entailed. Industrial policy and patterns of industrial development differ for each of the South Asian countries, and consequently, it makes sense to look at each of the countries in turn. Industrial success is largely confined to India, Pakistan and Sri Lanka. Bangladesh has very little manufacturing industry, and Nepal, Bhutan and the Maldives virtually none.

In Nepal, industry contributed just 14 per cent of GNP in 1983, with what manufacturing there is largely confined to industries such as bricks and food products to serve local consumption. The only significant export activity is the processing of forestry and agricultural products, but even this is limited in extent, and Nepal is totally dependent upon India, which buys over two-thirds of Nepal's total exports. Tourism is the only 'industry' which has significant development prospects in Nepal, but the scope for the expansion of tourism is limited by a chronic lack of facilities and capital internally and the fickle nature of tourism internationally.

All the points made above for Nepal hold for Bhutan and the Maldives. Both are very small, very poor and very remote. Manufacturing industry is non-existent, except for limited small-scale production for local needs. The Maldives has been vigorously promoting tourism, with some success, in recent years, but the high costs and limited market constrain the possibilities of this activity. Tourism apart, industry is unlikely to play any part in the development of the small, poor South Asian countries in the foreseeable future.

Industry in Bangladesh accounts for less than 15 per cent of GDP and is largely confined to the capital Dhaka. Despite efforts to promote industry, the growth rate of the manufacturing sector has been lower in the first half of the 1980s than in the second half of the 1970s. This in part reflects a serious recession in 1982-3, when output actually declined, but is largely due to the structural weakness of Bangladesh's industrial sector. Recovery in the mid-1980s (industrial output grew by 5.1 per cent in 1985) cannot hide these weaknesses, and the prospects for the sustained industrial development Bangladesh requires so urgently are slight.

Following independence from Pakistan, the main industrial activities were nationalised. In consequence, public sector enterprises dominate the main sectors, such as jute manufacturing, cotton textiles and iron and steel. The performance of these enterprises has been particularly disappointing, with inefficiency, capital starvation and serious over-staffing, combined with an unfavourable international climate for these sectors. These problems have been recognised, and the government announced a significant liberalisation under the New Industrial Policy in 1982. The effects of this policy have not been dramatic, but some improvement has occurred.

The private manufacturing sector in Bangladesh is limited in size, but has shown a better growth rate. Those serving the domestic market have been constrained by limited demand, however. A recent success story is the production of ready-made garments, exports of which grew from $31 million in 1983-4 to $116 million in 1984-5. Overall, the potential for expansion of manufacturing in Bangladesh is mixed. A massive pool of cheap labour and potentially large internal demand are favourable, but a lack of resources, very poor infrastructure and capital shortages mitigate against the development of this sector.

Sri Lanka has fared better. Industry contributed 27 per cent of GNP in 1983, and grew by 5.8 per cent per annum between 1977 and 1983. This is a great improvement over the earlier growth rate of 2.1 per cent between 1970 and 1977. In recent years, the internal conflict between the Tamil minority and the government has damaged the economy, but despite this the industrial sector grew by 11.4 per cent in 1984 and 7 per cent in 1985.

Sri Lanka's industrial structure is a mix of public (40 per cent of value added) and private (60 per cent) ownership, and contains a wide range of activities. Traditional industry concentrated on serving internal demand and processing agricultural outputs. These have been augmented by activities such as textiles and garments, chemicals, metal products and electronics, all of which are aimed at least in part at the export market. Sri Lanka adopted a policy package in 1977 which attempted to emulate the Newly-Industrialising Countries (NICs) of East Asia, with features such as an orientation towards exports, favourable tax and incentive measures, infrastructure development and the establishment of an Export Processing Zone on the Singapore model. This policy package has met with some success. This is particularly true in the export sector, which has grown more rapidly. Sri Lanka is a long way from becoming another NIC, however, and the extent to which the success can be sustained in the face of escalating internal strife and an unfavourable, protectionist-minded, international setting must be questioned. The late 1980s will tell whether Sri Lanka can build on early successes and develop into a vigorous economy with an established industrial base. As yet, the signs are mixed.

Pakistan has an important industrial sector. At 27 per cent, industry's share of GNP rivals that of agriculture. Manufacturing has grown dynamically over the last decade; at 7.8 per cent per annum from 1976 to 1980 and 9.9 per cent per annum from 1981 to 1985. Manufactured goods provided 61.4 per cent of merchandise exports in 1983, with textiles and clothing the dominant sector at 50 per cent of total export earnings. Pakistan's other main source of foreign exchange is remittances from migrant workers in the Middle East, a source of earnings which is under great pressure because of declining oil prices in the mid-1980s. This further increases the importance of manufactured exports for Pakistan's economy.

Successive governments have vigorously promoted industrial development, with the package of policies changing over time, but with the prominence of industry in planned development paths remaining. The industrial development which has occurred reflects this. There is a mix of public and private ownership, with an earlier emphasis on state-owned activities (many of which were in the capital goods and textile sectors) replaced by a greater emphasis on private capital and a wider range of consumer goods in recent years. This change of emphasis was in part forced on Pakistan by the increasing pressure upon traditional industries (in particular textiles, which is facing increasing international competition and encountering increased protectionism).

As is true for most of South Asia, Pakistan has achieved notable industrial development without building up massive overseas debts. The growth of state-owned industries in particular relied on internal capital and overseas aid, which Pakistan has traditionally had great success in attracting.

Commercial borrowing (especially by the vigorous private sector) has grown in recent years under the policies which have encouraged liberalisation and emphasised the private sector. Pakistan has not, however, adopted the full range of NIC-style policies. This is realistic, given the nature of her economy, and the prospects for continued industrial growth (though at perhaps a slower rate than in the early 1980s) appear to be good.

By any criteria, India is a major industrial power. She ranks in the world's top ten nations in terms of industrial output and has a range of manufacturing industries which is perhaps more complete than that of any Third World country except, perhaps, Brazil and China. Industry provides 26 per cent of GNP, employs over 10 per cent of India's massive work-force, and in 1983, provided 58 per cent of total exports.

The scale of India's manufacturing industry reflects the size of the country, and in particular, that of the internal market. Government policy has emphasised industrial development since independence in 1947. The form that this policy took was until recently an import-substitution approach, in which imports were restricted or taxed heavily and indigenous industry was encouraged to provide for the internal market. These types of policy were very popular in the 1960s, but failed in many parts of the Third World. India experienced mixed success. The size of her markets and diversity of resources permitted a wide range of large-scale industry to develop, again with local resources providing most of the capital. However, the consequence of this has frequently been extreme inefficiencies, outdated technologies and acute bottlenecks. The rate of growth of India's industry has been steady but slow (4.3 per cent per annum from 1976-80 and 3.3 per cent per annum from 1981 to 1983).

The industrial policies pursued until the mid-1980s emphasised the state sector, which was expected to be the leading edge of industrial development. This was particularly true for large-scale, capital-goods industries. These tended to be concentrated in a small number of major industrial centres such as Bombay, Khanpur and Ahmedabad. The important textile industry is similarly concentrated in a limited number of major cities. Bangalore has emerged as the centre of the electronics and aeronautics industry, and cities such as Delhi, Madras and many others contain a range of industries which mainly supply regional markets.

Many sorts of consumer goods are provided by tens of thousands of small factories and workshops, and growth in the small-scale sector has frequently been more dynamic than that of the large-scale sector.

Under the import-substitution policies, the licensing system, trade restrictions (including restrictions on importing technologies), restrictions on foreign ownership and the movement of capital and a range of other bureaucratic controls protected Indian industry, but fostered gross inefficiency. Indian industrialists did not have to try: whatever they made was bought anyway. Nevertheless, industry did grow, and India avoided crippling debts and foreign exchange problems. By the 1980s, however, India's industrial economy was in danger of fossilisation.

This has been recognised by the new government of Rajiv Gandhi, which has pursued a policy of dramatic liberalisation since 1984. A whole range of bureaucratic controls and import restrictions has been removed, the acquisition of modern technology encouraged (frequently via joint ventures with foreign companies), and private-sector industry strongly emphasised. The changes are a very long way from total de-regulation, however, a point emphasised by the seventh five-year plan (1985-90). The early results of the new policies in India have been dramatic. Industrial production grew by over 5 per cent in 1984 and 1985, a whole range of more modern technologies have been adopted, and many bottlenecks have been removed (the government has facilitated this by concentrating on infrastructure improvements and power development). The rapid growth of India's electronics industry (which produces a range of consumer electronics, computers and sophisticated inputs into the aeronautics and space programmes, the nuclear industry, and so on) and vehicle manufacture typify what is heralded as the success of the liberalisation policy.

Whether this initial success will be maintained is difficult to assess. The early euphoria is tempered by the growth of external debt and the deterioration of India's balance of payments. It has also been suggested that the initial burst of growth is in part due to the take-up of slack demand, and is unlikely to be sustained. What is clear is that India's industry will continue to develop (whether rapidly or slowly) and India will continue to develop as a major industrial power. It is also clear that this development will not provide employment for anything but a small proportion of India's huge work-force, and that industry will tend to widen the gap between the prosperous and the poor. This is true for the rest of South Asia, as throughout the region the already wide rural-urban gulf continues to grow.

In conclusion, the story of contemporary South Asia is one of progress and stagnation, increasing prosperity for some and deepening poverty for others. These disparities are as much along spatial as along social lines. The trends identified in the discussion of agriculture and industry suggest that these gaps will continue to grow. Whether, as is widely hoped, the success of the 'haves' will, in time, drag up the 'have-nots' is open to doubt. As such, until the question of inequality is tackled, poverty will not disappear - whatever the success stories which emerge.

The issues discussed in this chapter have centred on agriculture and industry, two sectors which dominate the economy of South Asia. It was mentioned earlier that concentration on this set of issues is inevitably done at the expense of a range of other issues. These issues are, of course, all closely interrelated to form the overall context of contemporary South Asian development. The environmental threats facing the region are closely associated with the agrarian sector: deforestation, erosion and land degradation reflect agricultural colonisation and the cultivation of fragile environments. The pattern of international economic relations is changing with industrial growth. Population trends, and in particular migration and urbanisation, are strongly influenced by emerging spatial inequalities of poverty and prosperity in both industrial and agrarian sectors. The same is true of social and political change and stability, with in particular regional conflicts reflecting real or perceived disadvantage and inequalities. Patterns of and opportunities for the development and use of energy and other resources is closely related to the spatial structure of development and stagnation. All these and other issues are influenced by development in agriculture and industry, just as these two sectors are influenced by many aspects of the wider economic, social and physical environment. To understand and account for these complex patterns of interaction is a daunting challenge, but is one which must be confronted before a full understanding of contemporary trends in a region such as South Asia can be understood.

REFERENCES

Baker, C. (1984) Frogs and farmers: the Green Revolution in India. In T. Baylis-Smith and S. Wanmali, (eds), Understanding Green Revolutions, Cambridge University Press, Cambridge

Bardhan, P. (1984) The political economy of development in India. Basil Blackwell, Oxford

Baylis-Smith, T. and Wanmali, S. (eds) (1984) Understanding Green Revolutions. Cambridge University Press, Cambridge

Bose, A. (1978) India's urbanisation 1901-2001. Tata McGraw-Hill, Bombay

Byres, T. (1972) The dialectic of India's Green Revolution. South Asian Review, 5 (2), 99-116

Chambers, R. (1984) Beyond the Green Revolution. In T. Baylis-Smith and S. Wanmali (eds), Understanding Green Revolutions, Cambridge University Press, Cambridge

Corbridge, S. (1984) Agrarian policy and agrarian change in tribal India. In T. Baylis-Smith and S. Wanmali (eds), Understanding Green Revolutions, Cambridge University Press, Cambridge

Day, R. and Singh, I. (1977) Economic development as an adaptive process. Cambridge University Press, Cambridge

Farmer, B.H. (ed.) (1977) Green Revolution? Macmillan, London

------ (1983) An introduction to South Asia. Methuen, London

Financial Times (1986) India. 12 May

Food and Agricultural Organisation (1985) FAO production yearbook, 1984. FAO, Rome

Gill, M.S. (1983) The development of Punjab agriculture 1977-80. Asian Survey, 23 (7), 830-44

Harriss, B. (1984) Agrarian change and the merchant state in Tamil Nadu. In T. Baylis-Smith and S. Wanmali (eds), Understanding Green Revolutions, Cambridge University Press, Cambridge

Jones, S. (1982) A critical evaluation of rural development policy in Bangladesh. In S. Jones et al. (eds), Rural poverty and agrarian reform, Allied Publishers, New Delhi

Joshi, P. and Kaneda, H. (1982) Variability of yields in foodgrain production since the mid-sixties. Economic and Political Weekly, 17 (13), A2-A8

Kahlon, A.S. et al. (1972) The dynamics of Punjab agriculture. Punjab Agricultural University, Ludhiana

National Westminster Bank (1984) Pakistan. Natwest, London

------ (1985) India. Natwest, London

Pearse, A. (1980) Seeds of plenty, seeds of want. Clarendon Press, Oxford

Soussan, J., Ferf, A. and O'Keefe, P. (1984) Fuelwood strategies and action programmes in Asia. AIT, Bangkok

United Nations (1984) Economic and social survey of Asia and the Pacific 1984. UN, Bangkok

------ (1985) Economic and social survey of Asia and the Pacific 1985. UN, Bangkok

World Bank (1984a) Bangladesh: economic trends and development administration. World Bank, Washington, DC

------ (1984b) Sri Lanka: recent economic developments, prospects and policies. World Bank, Washington, DC

World Resources Institute (1986) World resources 1986. Basic Books, New York

Chapter Twelve

CONTEMPORARY ISSUES IN EAST ASIA

J. Gray

It is with hope and satisfaction that one turns from the rest of the developing world to the countries of East Asia, whose economic growth in the last two or three decades has been so remarkable. Between 1960 and 1978, real per capita growth in South Korea was 9.9 per cent per annum; in Hong Kong, 9 per cent; in Taiwan, 6.2 per cent and 6 per cent in Singapore and in Japan; and Japan's high rate of growth during this period actually represented a diminution, in industrial maturity, from its double-digit peak a decade earlier. This growth, moreover, has been accompanied by a decrease in inequality of incomes which has been so sharp that income distribution in Taiwan and South Korea is now comparable to that of Holland, Denmark or Sweden.

The success of the Far Eastern NICs (Newly Industrialised Countries) has left in disarray much of the conventional wisdom concerning economic development in poor countries. It gives no comfort either to the theorists of the left nor to those of the right. East Asian experience challenges the left-wing assumption that the industrialisation of surplus-labour peasant societies in capitalist conditions must inevitably increase inequality. It also challenges the idea that poor countries on the periphery will inevitably find themselves at an increasing disadvantage as their dependence on the international market increases.

The right can also expect only cold comfort from the East Asian experience, which strikes just as sharp a blow at the assumption that only individual enterprise free from government intervention can secure effective growth; for of the 'four tiger cubs' (South Korea, Taiwan, Hong Kong and Singapore) plus the mature economic tiger, Japan, four operate a system of vigorous government intervention in the economy. Nor does the East Asian case lend support to those in the liberal tradition who want to believe that successful development (in terms of both growth and social equity) requires a democratic political system, and one in which trade unions play a full part. Taiwan has hitherto been a one-party state. South Korea, although formally a plural state, is in fact ruled by one dominant party against which only assassination has so far proved effective. In Singapore the ruling party has marginalised opposition. Hong Kong is virtually without democratic institutions. Japan continues to be ruled by one party representing an unshaken coalition of bureaucrats, businessmen and peasants. Although trade unions exist in all of the countries of East Asia, their influence on wage bargaining is negligible and they have contributed little to the rapid rise in real wages which in so short a time has raised the standard of living of the peoples of East Asia from Asian to European levels.

The East Asian experience may be exceptional but it cannot be said to be marginal. If one includes Japan (as historically and logically one must), then this experience involves a population approaching 150 million people. It has led to a new situation in the world economy in which the flow of trade goods across the Pacific is now greater than that across the Atlantic. It has created a new centre of advanced technology able to put up an alarming challenge to the older centres of industrial innovation in the West. The development of the East Asian countries is all the more remarkable when one considers the poverty of most of them in terms of natural resources. Except for China, they have few basic ores or fossil fuels. They have people:land ratios among the worst in the world. (Hong Kong and Singapore, of course, have virtually no agricultural hinterland.) The average size of farms in South Korea and Taiwan is only approximately one hectare. Their population densities are among the very highest in the world.

Japan and the East Asian NICs, however, are not the whole of the region, nor can they entirely monopolise the limelight of economic triumph. The Chinese People's Republic and the Democratic Republic of Korea have been as distinguished in the Communist world as the others in the capitalist world. From 1960 to 1978, both achieved a *per capita* rate of growth of over 5 per cent per annum. Of the countries of the socialist bloc, only oil-rich Romania did better. Their rate of growth has put them well ahead not only of their Communist rivals, but also ahead of India and the countries of South-East Asia, except for Singapore. Moreover, economic growth in China since the new reforms began in 1978 has considerably increased in pace, bringing China much closer to the exceptional speed of development of her East Asian capitalist rivals. Thus, the study of East Asia's success must now include China. The old contrast between efficient free enterprise and inefficient socialism is found to require revision, not only because China's modified socialism is doing so well, but because capitalist East Asia is far from representing 'free enterprise' *tout court*. China's participation in the regional economic miracle suggests that the explanation of that miracle may lie deeper than ideological distinctions.

At the other extreme, the success of Hong Kong, which is perhaps the only totally free-enterprise, free-trade economy left in the world, casts doubts on any explanation of East Asian achievements which depends too heavily on the assumption that government intervention, socialist or non-socialist, has played a necessary role in these achievements. Further problems arise for an interpretation of East Asian success in terms of institutions when we compare the institutional arrangements of the four economies in which there is obviously a close working relationship between government and private enterprise, i.e. Japan, Taiwan, South Korea and Singapore, for their institutional arrangements are very different at points crucial to the argument.

The most obvious common factor is bureaucratic participation in economic decision-making. Hong Kong is an exception to this, but it is perhaps one that proves the rule. The success of East Asian economic planning is the result of a peculiarly fruitful relationship between the bureaucracy on the one hand and self-regulated private enterprise on the other. Self-regulation is just as characteristic of the Hong Kong economy, while the benevolent neutrality (economically speaking) of the British administration in Hong Kong does not preclude a very positive role in, for example, the creation of an adequate infrastructure and the negotiation of access to markets: the administration has even been known on occasion to exercise somewhat shamefaced Keynesian policies in order to manipulate demand.

At other points East Asian institutions can often be contrasted. In Japan, autonomous trade organisations play a powerful role, while the successors to the zaibatsu, whose operations are as diversified as those of the zaibatsu were specialised, are almost microcosms of the national economy - fiefs in a sort of industrial feudalism. In South Korea and Taiwan, however, trade organisations are as feeble as they are strong in Japan. In Taiwan, they are replaced by government-initiated para-statal organisations for the major industries. In South Korea, there is less of either type of organisation: the same tasks are performed through the informal network (largely of ex-soldiers) which penetrates both government and business.

In Japan and in South Korea, business finance is bank-dominated, debt-equity ratios are very high, and little capital is raised through the stock exchange. In Taiwan, the sources of industrial capital are rather more diffused. In Hong Kong and Singapore, on the other hand, the stock exchange plays a major role.

In the field of planning, the governments of Taiwan, South Korea and Singapore (and of course China) formally plan the economy; in Japan decisions on future development are made informally. There is no planning (except for infrastructural investment) in Hong Kong.

In financial policy South Korea and Taiwan stand in sharp contrast. Korea has followed a policy of high growth and high foreign borrowing, accompanied until recently by the tolerance of high inflation. Taiwan, on the other hand, has in this respect followed policies which would gladden the hearts of the IMF: strict control of inflation by restrictive fiscal policies and no foreign debt.

South Korea and Japan have been very sparing in their admission of foreign direct investment and the employment of multinationals. Taiwan has made much greater use of such international facilities. The industrialisation of Singapore has been largely based on inviting in the multinationals. Hong Kong is, of course, wide open, although the strength of its economy nevertheless lies in its small local businesses.

The Taiwanese economy has a large public sector - larger than that of India or of Tanzania - which comprises all the commanding heights of the economy. The public sector in South Korea is considerably smaller, and it is very small in Japan and negligible in Hong Kong. It is of course largest of all in China, but in conditions in which its relationship to the state is in the process of reform.

In sum, then, there is as much contrast in the

institutional structure of the East Asian economies as in any other region of the world.

II

One must ask the question whether there have been special conditions operating in the East Asian area to account at least in part for its extraordinary success. Firstly, all the countries concerned suffered massive devastation as the result of war. It might therefore be argued that these countries were a tabula rasa, on which modernisation could proceed unhampered by unsatisfactory existing economic structures. Yet the growth figures of the countries of East Asia during the years of their recovery were not exceptional. Their great leap was in fact only taken after the possibilities of recovery were exhausted, and in the case of Taiwan and South Korea, when their home markets were saturated and growth consequently beginning to falter.

A second special factor which may have acted as a stimulus to the East Asian economies was their political insecurity. South Korea lives under the threat from across the 38th Parallel: the Democratic Republic of Korea, more heavily armed than the South, recently made its continuing intentions clear by attempting to dig under the frontier tunnels capable of being used by military transport, and Seoul is extremely vulnerable to the sort of blitzkrieg for which the North Korean forces are ominously armed and trained. The Taiwanese regime is an émigré government, facing across the narrow Straits a Chinese government which has made no secret of its intention to recover the island. China, too, since her breach with the Soviet Union, is acutely conscious of her vulnerability and painfully aware that if the confrontation of the Soviet Union and the United States were to mutate here and there into collusion, China's international position might be very weak. It is this consciousness which has led China to make drastic ideological compromises in the hope of speedier growth. In the case of Singapore, the need to assert independence against Malaysia and Indonesia has undoubtedly provided a strong political motive for its government's determined and successful effort to industrialise. In Hong Kong a population mainly of refugees and the descendants of refugees has struggled to find personal security in profits and property, as refugees are wont to do everywhere.

A strong sense of national insecurity has assisted the governments of South Korea, Taiwan, Singapore and China to assert greater pressure on their peoples, in the interests of rapid economic growth, than might have been tolerable in other circumstances. The same sense of insecurity may also account to some extent for policies which aimed to reduce disruptive social inequalities.

Another factor to which great importance has been attached is the patronage which, in varying degrees, the East Asian countries have enjoyed from allies for political reasons. During the cold war years the United States poured aid into Taiwan and South Korea: Taiwan was receiving $6 per capita in civil aid and South Korea $10. On the other side, the Communist countries of the region were receiving considerable aid from the Soviet Union. China's First Five Year Plan, the foundation of her subsequent growth, was dependent on the provision on credit by the Soviet Union of almost 200 turn-key industrial plants.

In South Korea and Taiwan the main contributions of US aid was to enable both countries to spend massively on defence without incurring hyper-inflation. Beyond this, aid was not of great significance. Many other countries were in receipt of substantial aid and yet achieved no economic miracles; Singapore and Hong Kong did just as well without aid; and the real burst of growth in Taiwan and South Korea came after American aid had ended, and was in fact stimulated by the prospect of having to stand alone. It was then that both countries gave up their conventional strategy of import substitution and went for broke in the world market.

Trade favours were more important than aid. The tolerance of Western governments towards the increasing competition of East Asian products in their home markets has been remarkable. It is to be explained only by the importance of these countries as part of the Western bastion against the perceived threat of Communist expansion. Western tolerance has been all the more striking considering the determination and the deviousness with which Japan, Taiwan and South Korea protect their own industries from reciprocal competition behind a façade of free trade. The exports of Hong Kong and Singapore, owing to the historical connection with Britain, have also enjoyed privileged access to the Western economies.

The political connection between the capitalist states of East Asia and the advanced countries of the West may have had another important economic consequence. There is no reason to suppose that the economic system adopted by Taiwan and South Korea, i.e. the development of free enterprise within a framework of indicative planning and manipulated incentives, was their natural choice. Both governments during the 1950s showed a strong disposition to exert a high degree of direct economic control, in contrast with their more recent, more liberal and dramatically more successful policies. Involved in an international ideological struggle in which their allies looked to them to provide proof of the superior efficiency of the capitalist system, they accepted the role, and became much more capitalist than they had been, although considerably less than they pretend to be. Yet aid, trade and good advice cannot alone explain success in East Asia. These have not always been in short supply elsewhere.

Japan has also assisted in the transformation of its neighbours, both through its former influence as a colonial power and its present relationships with them. Japan ruled Taiwan from 1895 to 1946 and Korea from 1905 to 1946. While Japanese rule was not noted for its benevolence, its economic results were not all negative. Japan's economic interest in Taiwan as a supplier of rice and sugar gave her a strong motive for encouraging Taiwanese agriculture and she did so with typical efficiency. One result was a highly efficient system of irrigation. At the same time, on the uncompromisingly imperialist theory that Taiwan was not a colony but an integral part of Japan, the highly successful Japanese system of elementary education was imposed on the Taiwanese, who, as a result, were by 1937 probably the most educated peasants in Asia outside Japan itself. Thus, the Japanese conquerors removed from Taiwan the two biggest constraints on the transformation of pre-modern agriculture - drought and ignorance. By 1937 Taiwanese rice yields per acre were twice those of China. Japanese colonial policy in Korea was much the same as in Taiwan, but as far at least as the southern part of the peninsula is concerned, the incentives were less and the policies therefore applied with less vigour and success, but with similar effects.

Japan has also been, and remains, a major source of both technological and institutional change. East Asia is an area historically dominated by one culture, derived from China. If we can speak (historically) of Europe as Christendom, we can with equal validity speak of East Asia as the Confucian realm. Moreover, just as Latin provided the lingua franca of European culture, so the Chinese ideographic script (being independent of the pronunciation of local languages) provided a means of communication from Turkestan to Japan and from Korea to Vietnam. It still works. Although Japan and Korea created their own syllabaries in mediaeval times, the Chinese ideographs are still used, at least for serious and polite writing, and are still taught in the schools. Communication is therefore still easy. Ideas, values, techniques and institutional forms can travel relatively unimpaired throughout the area. For most of history, the flow of ideas was from China to the rest, but in the twentieth century it has been from Japan. Japanese administrative methods, production techniques, management systems and marketing organisation have had an immense influence.

In Taiwan and South Korea, planners refer systematically to the course of Japanese development, and to its changing priorities and changing strategies at successive stages of growth, and consciously (though not slavishly) follow the Japanese precedents.

One particular Japanese institution has played a key role in the area: the general trading companies or the sogo shosha. These are trading enterprises: they do not themselves manufacture, but they are in close relations with the great Japanese industrial firms and financial institutions and they play a great part in the encouragement of manufactures. They possess marketing expertise incomparable in range and quality and the heart of their operations is the constant search for new commercial opportunities and the constant updating of market predictions. Their operations are not tied to particular products or to particular producers. They do not generally trade on their own account, but act as agents earning a commission on sales and purchases. Thus, they have a direct incentive to increase trade flows of all kinds.

For developing countries, marketing skills are perhaps more difficult to master than production skills, yet we still tend to see the capacity to produce, rather than the capacity to sell, as the key to success. East Asian success, however, has been based largely on the ability to anticipate changing market opportunities, to get prices and specifications right, and to pursue a strategy based on 'dynamic comparative advantage'.

In Taiwan and South Korea the Japanese sogo shosha, which handle a substantial part of the foreign trade of both countries, have provided much of the necessary marketing skills. Perhaps they have done more. It has often been noted that government intervention in the Taiwanese and Korean economies is quite extraordinarily detailed and specific, so much so that many people are sceptical that governments could ever achieve such remarkable success in the precise long-range forecasting of supply and demand. Yet they have seldom put a foot wrong. Further research might well show that the sogo shosha have been closely involved in the planning of Taiwanese and Korean success.

The rapid industrialisation of Hong Kong and Singapore was made possible by the fact that as the principal entrepôts of East Asia, they possessed high levels of marketing skills. The traditional agency houses such as Jardine Mathieson and Coy. played, and continue to play, a role analagous to that of the sogo shosha, which themselves are active in both places, but the main element in the export-orientated industrialisation of Singapore has been the multinationals, which have been freely admitted to the island explicitly in order to exploit their capacity to sell in world markets; while in Hong Kong, the main stimulus to industrialisation has come from commercial firms offering specific contracts which have made possible the establishment or expansion of manufacturing firms at a minimum of risk.

Thus, in all four cases highly successful entry into the world market, and rapid industrialisation on that basis, have largely been made possible by the mediation of international marketing organisations. If this is 'dependency', the Third World could do with more of it.

III

THE STRATEGY OF GROWTH IN EAST ASIA

In Taiwan until 1958, development policy was of the conventional import-substitution type. Domestic industry was highly protected by tariffs and severe quantitative restrictions on imports. By 1958, however, the possibilities of import substitution of consumer goods was exhausted, the industrialisation flagged. The first reaction to this situation was conventional: it was to attempt to deepen import substitution by encouraging the domestic production of such products as chemicals, chemical fertilisers, rayon and motor vehicles; but it quickly became obvious that the home market was too small to sustain such developments at that stage.

The strategy was changed to one of concentration on exports. Taiwan was fortunate in having the capacity and (thanks to Japanese colonial enterprise) the experience to export processed agricultural products, and such exports had already afforded shelter from the sort of exchange problems which have plagued so many other developing countries: to this extent the Taiwanese economy was already export-orientated. The new export drive was based on labour-intensive production, and policy favoured this against more capital-intensive enterprise. Commercial policy was liberalised to maximise trade flows. Tariffs on imported raw materials and capital goods were drastically reduced, as were quantitative restrictions on imports. The currency was sharply devalued and tied to the American dollar and the new exchange rate made uniform. Backward linkages were encouraged by tax relief and the selective easing of quotas. These changes have been interpreted as representing a conversion to free trade, but they were nothing of the kind. They did not represent a lessening of government economic management but simply a change of emphasis. The liberalisation of trade was mainly limited to the encouragement of exports. There was a hidden agenda, represented by unpublished restrictions on trade which ensured that protection was systematically maintained where it would not impair the export drive. At the same time - and this is a vital point - as Taiwan and South Korea moved up the technological scale, the less capital-intensive industries which had at the earlier stage been protected were progressively deprived of their protection: import-substitution privileges in both countries have been, and have been seen to be, strictly temporary.

While the emphasis was almost exclusively on labour-intensive development, by the mid-1960s the government was already preparing the next stage of its strategy. Aware of the possibility of growing protectionism in world markets, faced with rapidly increasing real wages, and with competition in labour-intensive products from other developing countries, Taiwan laid plans well ahead to move up the technology scale, to increase the value-added content of exports and, in the light of the rapid growth of the domestic market, to deepen import substitution. In the implementation of these plans, the government was the initiator.

In South Korea, war continued until 1953 and economic development lagged behind that of Taiwan. As a result of the devastation of war, output fell by 17 per cent and GNP by 16 per cent. In 1951, prices rose fivefold, and doubled again in 1952, and inflation proved difficult to control thereafter. Until 1958, South Korea, like Taiwan, pursued conventional import-substitution policies, and found by then that the domestic market was saturated. The Korean government did not respond as quickly to this situation as the government of Taiwan: until Park Cheng Hee came to power, the administration took little interest in economic development. Park, however, gave priority to growth, setting up a powerful planning board and introducing civilians with technical and economic expertise into a government which had been essentially military. Powerful incentives were offered to export industries. Import-substitution enterprises lost the comforts of prohibitive levels of protection and negative interest rates. The new real interest rate was 20 per cent, which both stimulated savings and encouraged labour-intensive development. In general, much the same liberalisation measures were introduced as in Taiwan, but within the same carefully planned limits. There was an extraordinarily rapid 35 per cent per annum growth of exports, almost all in labour-intensive products, with GNP rising at 11 per cent per annum. As in Taiwan, this phase was accompanied by preparations to upgrade technology and value-added content at a later but already anticipated stage.

As far as agriculture is concerned, Taiwan and South Korea offer both comparisons and contrasts. The withdrawal of Japanese power, followed by land reform, had created in both countries a rural society of peasant proprietors farming an average of about one hectare, with very few farms over three hectares. Policy towards agriculture differed sharply, however. In Taiwan, dependence on agricultural exports induced the government to adopt positive policies towards agriculture. Korea's agricultural prospects being less hopeful, she was content to depend on subsidised American grain, the import of which depressed domestic agricultural prices; and by this oblique means, agricultural incomes were squeezed for the benefit of industrialisation. In Taiwan the exaction of a surplus from agriculture was carried out in a more systematic way, by state monopoly purchase of grain at fixed prices, by a system of barter of chemical fertilisers for grain and by relatively high taxation of the rural community, and these measures were enforced by one-party control of the peasant co-operatives.

Thus, in both countries agriculture was squeezed to provide capital for industrialisation. Yet one must not rush to conclude that their agricultural policies were quasi-Stalinist and inevitably counter-productive, or that they had the consequence so obvious in many other developing countries (especially in Africa), that of depressing the purchasing power of the peasant majority and thus severely limiting the domestic market. The squeeze was not so hard as to prevent either a brisk rise in agricultural production of about 3.7 per cent per annum or a steady rise in peasant incomes and peasant purchasing-power. In spite of relatively low farm-gate prices, farmers in both countries had both the means and the incentives to improve through increasing production, diversification into high-quality foodstuffs as both urban and rural incomes grew, and the investment of part of the increase in income accruing from land reform. In addition, the incomes of farming households were increased (though much less in Korea than in Taiwan) by the growth of industrial employment in the rural areas. It is also important to note that the policy of squeezing agriculture was maintained for no longer than was necessary, and after about 1970, it rapidly changed in both countries to one of encouraging agricultural production and maintaining peasant incomes by subsidising agriculture.

In Korea, by 1970, as a result of rapid industrialisation, rural incomes which had hitherto increased in step with urban wages had dropped relatively to 67 per cent; and as a result, in the 1971 elections many peasants deserted the ruling party and for the first time brought the regime within sight of possible electoral defeat. President Park responded swiftly by raising rural development to the highest priority; in his Saemaul movement for the regeneration of the village, rural investment increased from 4 to 38 per cent of government expenditure by 1978.

In Taiwan, the reasons for the reversal of previous agricultural policies were more complex and more obviously economic in nature, although the political consequences of a threatened relative decline in agricultural incomes was not a negligible factor, the more so as the government was dominated by Mainlanders while rural interests were mostly native Taiwanese. This threatened decline, however, was in itself an economic factor resulting from the very success of economic growth: the standard of living in Taiwan had reached the stage when farm incomes, relative to urban wages, began to be prejudiced by income inelasticity of demand for grain - a problem of advanced countries. Grain production in Taiwan is now highly subsidised, as it is in Korea and Japan.

In sum, although in the early years agriculture was squeezed in both Korea and Taiwan, policy did not fit the 'primitive-accumulation' paradigm. Agriculture incomes rose rapidly and buoyant peasant purchasing power played an important part in the rapid expansion of the domestic market without which the export drive would have been unsustainable.

IV

THE CHINESE PEOPLE'S REPUBLIC (CPR)

The economic strategy of the CPR, with its Soviet precedents, would be expected to fit the 'primitive-accumulation' paradigm. During the First Five Year Plan of 1953-7, it at least appeared to do so, but in fact even then, Chinese policy towards agriculture was nearer to that of Taiwan than to that of the Soviet Union. While massive resources were transferred from agriculture to industry, the regime was careful to ensure that taking one year with another, peasant incomes on average rose. Then, in 1958, Mao Zedong in his trenchant critique of Stalinism, repudiated the 'primitive-accumulation' thesis and insisted that the true motive force of development in China must be the increase of peasant purchasing power. This he sought through the policies represented by the Great Leap Forward and the Communes of 1958: the decentralisation of economic decision-making to the peasant communities, the use of surplus rural labour to transform the infrastructure, and the industrialisation of the countryside on the basis of appropriate technology.

His first effort to change the parameters of socialist development was a gross failure. This was not due to the deficiencies of his economic assumptions, but to the fact that the overwhelming political power of the Communist Party prevented any real decentralisation: the Party simply preempted economic decision-making at the grassroots, and the commune, so far from being an autonomous and democratically organised social unit, simply became a new organ of state and party dictatorship. While it is true that the commune system did succeed in vastly improving the rural infrastructure and did eventually lay the foundations of rural industrialisation, it succeeded largely by coercive methods, which in the end proved politically intolerable and economically self-limiting.

After the death of Mao in September 1976 and the rise of Deng Xiaoping to power in 1978, rural China was offered a new deal. The agricultural collectives were permitted to dissolve themselves. Family-scale farming returned, and the state sought to secure the fulfilment of its plans by offering tenure of farmland in return for a production contract which incorporated adequate economic incentives and which, after its fulfilment, left the farming family with unused resources which could be used in production for the free market. Peasants were encouraged and assisted to supplement the industrial and other enterprises owned by the communes and brigades by creating 'new economic combinations', which might be co-operatively or privately owned. The state price offered in the grain contracts was increased until it became in effect a support price. The results in agriculture were startling: agricultural production rose at a rate approaching between 7 and 8 per cent per annum.

In the field of industrial development, a strenuous though not yet entirely successful effort has been made to distance the state from direct industrial management, to give state sector firms autonomy, and to encourage them to enter the market. At the same time, the Chinese economy has been to a remarkable extent opened to the world, although strictly on China's own terms. Every effort is being made to maximise exports, beginning with labour-intensive products but with the firm intention of gradually raising their technology content to world standards.

Superficially the great change in China came in 1978 under Deng Xiaoping. In fact, the decisive breach was Mao's repudiation of Stalinism in 1958. In certain essential respects - insistence on the primacy of peasant purchasing power, on the industrialisation of the countryside, and on the view that the state's role in socialist society should not be to command but to initiate, enable and supervise - policy has been continuous since then. What has been shed from Maoism are those assumptions which actually militated against the success of such policy - excessive egalitarianism and the idea that collectivised living had some intrinsic moral value above and beyond its economic rationale.

It is clear that there has been a parallel development in China and in Taiwan and South Korea. It is significant too that the common characteristics of this Confucian world are quite specifically Confucian. They can be restated in Confucian terms as follows: a prosperous peasantry is the foundation of the state; the good magistrate does everything he can to encourage the development of local resources; the state should supervise the economy but civil society should conduct it. In the modern application of these Confucian (but still operative) economic assumptions, the encouragement of local enterprise is the fulcrum. It is a major instrument in creating a prosperous peasantry, directly by the employment opportunities created, and indirectly by providing resources for investment in agriculture. In this respect it can compensate for the need, in the earlier stages of growth, to keep farm-gate prices relatively low. It encourages the wide diffusion of decision-making power, and it promotes the urbanisation of the countryside and therefore of urban attitudes to family limitation.

The establishment of industry at village level was a main plank in the platform of the Great Leap Forward of 1958, and although the initial attempt to do so was a failure, the policy was re-applied in 1970 after the victory of the left wing in the Cultural Revolution. Commune and brigade industry became by far the most rapidly developing sector of the Chinese economy, with rates of growth of gross output value nearing 30 per cent per annum. The policy was heartily endorsed by Mao's successors. Opportunities to establish local enterprises have now been offered not only to township and village governments (the successors in this respect of the Commune), but to the rural population at large, subject to local government approval.

While rural industry (including transport and agricultural services as well as manufactures) in China has been developed as state policy, in Taiwan its growth has been at least assisted, if not determined, by historical and geographical factors. Rapidly growing industry was forced to spread, faut de mieux, up and down the narrow western coastal plain from the three major western ports. Early dependence on food-processing industries, which by their nature tended to be located at the rural points of supply, assisted this diffusion. Concentration on labour-intensive industries drew enterprises towards rural supplies of labour.

A developed infrastructure of rural roads and of electric power assisted the movement. Finally, the rapid increase in rural purchasing power, which resulted largely from the employment opportunities thus generated, gave rural industrialisation, as well as service trades, a local momentum of their own. In the end, this diffusion of industry has made Taiwan almost unique in the world in that, with total industry rapidly developing, the proportion situated in the cities is actually falling, and more than half (perhaps now as much as 70 per cent) of farm household incomes is derived from non-agricultural sources.

In this respect, South Korea stands in contrast to China and Taiwan. Industry is exceptionally concentrated in three great cities, and little was done until recently to encourage its dispersal. This, however, was not entirely a matter of policy: indeed, every South Korean politician knows that, faced with possible blitzkrieg from the north, the country should be doing everything possible to avoid concentrating industrial production in a few vulnerable spots. Otherwise, however, circumstances did not facilitate or favour diffusion. The inherited rural infrastucture was poor, the processing of agricultural crops was of no significance, the purchasing power of the rural population was much lower than in Taiwan, and a population of which 25 per cent were refugees from the north provided ample labour in and around the cities.

Yet in fact, given the rapidity with which urban industry developed, the rural spin-off effects were by no means negligible: rural factory employment had increased three-fold by the 1960s as compared with pre-war. At the same time, the healthy development of agriculture, growing and diversifying in spite of gross neglect, helped to raise rural standards and provide better rural markets for simple local industries and services. Finally, in the present Sixth Five Year Plan, the diffusion of industry has been given high priority: plans include 100 rural industrial estates, and a general switch (assisted by special incentives) of new industrial growth to the less urbanised south.

V

THE CONFUCIAN ECONOMY

This brings us to the point of considering the fact that this

East Asian area of dramatic economic success is definable only in cultural terms, and that the only common factor in this success is culture. Are we faced here with the phenomenon of a definable 'economic culture'? Is there a creature identifiable as homo economicus confucianensis? If so, what are its characteristics?

The first characteristic, universally recognised, is the achievement ethic, expressed in willingness to work hard and to work to high standards, in unrivalled receptivity to training and education, and in the habits of frugality which have given Taiwan a domestic savings rate as a proportion of national income of 33.4 per cent, Singapore 28.4 per cent, Japan 18.1 per cent, South Korea 16.5 per cent and Hong Kong 13 per cent, figures all well above the international average. Low or even stationary birth rates also reflect the tradition of frugality.

A second characteristic, related to the first, is the capacity for entrepreneurship, already proved in pre-modern history. One need only refer to the high levels of pre-modern commercial entrepreneurship in China and Japan, the remarkable success of Chinese emigrant enterprise throughout South-East Asia, and the extraordinary and well-attested ingenuity and adaptability of Hong Kong's small manufacturers. The apparent slowness of China in the early twentieth century to take advantage of modern economic opportunities may seem to cast doubt on this, but this slowness is something of a myth. The coastal provinces of China in the 1920s and 1930s put up as high a rate of industrial growth as Japan: her new native industrialists thrived and held their own, even in the face of privileged foreign competition; and her native handicraftsmen adapted their technologies and not only survived but flourished as never before. In fact, Chinese coastal enterprise under imperialism provides an a fortiori argument for the vigour of Chinese entrepreneurship.

A third characteristic is that while fiercely competitive, there is nevertheless a willingness that competition should be subjected to rules laid down by the self-regulating organs of trade or by government which, as a Confucian, s/he assumes represents and has a right to represent the economic interests of the whole of which s/he is a part. This assumption has been strengthened rather than weakened by the nature of modern East Asian economic development. Western capitalism grew up fighting against government, but Eastern capitalism has been largely the creation of government: bureaucrat and entrepreneur have lived in symbiosis. An illustration of this difference in assumptions can be given from the experience of Western investors in, for example, South Korea. They are appalled at the strictness of the rules under which the Korean government obliges them to operate, at the government's assumption that it can lay down the product mix of a foreign company and the maximum market share it can expect, demand sensitive information and examine accounts, and unilaterally change contracts. Foreign firms assume that they are being subjected to discrimination, but they are not: this is how the Korean bureaucracy deals with its own capitalists, and the capitalists expect nothing else. Tradition and modern history alike have led them to expect that such subservience best serves their own interests in the long run, and they will no doubt put up with it as long as it continues to pay such handsome dividends.

A further cultural characteristic, the importance of which has been observed in industrial management, is the lack of polarisation between higher and lower levels. A traditional combination of patron-client relationships and concensus decision-making has been passed on into modern industrial organisation - as if ICI were to be run on the lines of a feudal fief, by a consensus of the knights, sitting as a court chaired by the lord as primus inter pares. This creates an ethic of mutual obligation and group commitment scarcely known in the West; and it applies at all levels within industry.

For their part, the bureaucracies, having relinquished the conservative paternalism of their pre-modern social ethic and embraced the promotion of growth as the greatest national good, have put their formidable administrative ability, their high sense of public service and their traditional moral authority at the service of enterprise; but it is at the service of enterprise viewed nationally, and not of any particular sectional interest. They are able to do this the more effectively because Japanese colonial control, followed by the devastation of war, followed by land reform, wiped out existing vested interests.

In stressing the cultural elements which the East Asian countries share, the intention is not to propose a one-factor explanation. The other, often special, factors in their situations have been described; of these, perhaps the most important was access to powerful international market organisations such as the sogo shosha, able to offer opportunities for export development which in fact actually implied a particular strategy of growth. The question remains, however, why the sogo shosha's implicit (if it was merely implicit) strategy of dynamic comparative advantage was so successful in these countries and not in others - and why Communist China's strategy is now rapidly converging with that of the rest. The answer must lie largely in the cultural terms here suggested.

A question of considerable practical relevance for the rest of the world is whether the East Asian experience is transferable. To assume not would be to take too deterministic a view of culture. Expectations, values and behaviour can be changed. Many of the characteristics of East Asia's fortunate cultural inheritance might well be replicated elsewhere by judicious and constantly maintained policies.

REFERENCES

Adelman, Irma and Robinson, Sherman (1978) Income distribution policy in developing countries: a case study of Korea. Oxford University Press, London

Allen, G.C. (1981) The Japanese Economy. Weidenfeld and Nicholson, London

Asian Development Bank (1971) South East Asia's economy in the 1970s. Longman, London

Fei, John C.H., Ranis, Gustva and Kuo, Shirley W.Y. (1979) Growth with equity: the Taiwan case. Oxford University Press

Hofheinz, R. and Calder, Kent E. (1982) The East Asia edge. Basic Books, New York

Hopkins, Keith (ed.) (1971) Hong Kong: the industrial colony. Oxford University Press, Hong Kong

Kuznets, Paul W. (1977) Economic growth and structure in the Republic of Korea. Yale University Press, London

Wade, Robert and White, Gordon (eds) (1984) Developmental states in East Asia. IDS Bulletin, 15 (2)

Chapter Thirteen

CONTEMPORARY ISSUES IN ISLAND MICRO-STATES

J. Connell

It was widely assumed that the post-war era of decolonisation would draw to a halt long before all colonies became independent since many were simply too small (Fieldhouse, 1982, p. 411). However, size has never been a criterion for statehood, and as decolonisation proceeded, smaller and smaller states became independent, culminating in the independence of Tuvalu in 1978, then with a population of about 7,500. Although the demand for independence was couched in cultural terms, there were clear economic benefits from a separate independence.

Few remaining colonies have significant independence movements. With the exception of Namibia and separatist movements within states (as in Eritrea and Irian Jaya), what may be the last remaining independence struggle is currently being waged in New Caledonia (Connell, 1987a). Elsewhere, colonies have either not challenged their dependent status, or have negotiated firm ties of association with colonial powers, such as the ties between New Zealand, the Cook Islands and Niue, between the USA and Guam, the Marshall Islands, and Palau. Such choices demonstrate that independence is not the only option in asserting self-determination, but that free association or closer integration is also a possibility. Consequently, although this chapter primarily examines development issues in island micro-states (Table 13.1), it inevitably makes passing reference to similar sized territories that have chosen other options, and also to some larger states; its conclusions are often also relevant to small land-locked states with whom island states are often compared (Selwyn, 1978; Shaw, 1982). Nonetheless, despite structural similarities, diversity

Table 13.1: Island micro-states in the Third World: selected characteristics

	Population (mid-1984)	Area (km)	Population density	GNP per capita US $ (1984)	Life expectancy (1984)
Caribbean					
Antigua and Barbuda	78,000	442	176	1860	73
Bahamas	229,000	13,940	16	6690	69
Barbados	253,000	430	59	4370	73
Dominica	77,000	750	103	1010	75
Grenada	84,000	345	272	860	68
St Kitts-Nevis	55,000	269	204	1150	64
St Lucia	134,000	616	218	1130	70
St Vincent	117,000	388	302	840	69
Indian Ocean					
Comoros	283,000	1,865	205	330 a	55
Maldives	173,000	298	581	430 a	53
Seychelles	65,000	404	161	2700 a	69
Atlantic Ocean					
Cape Verde	320,000	4,033	79	320	64
Sao Tome and Principe	105,000	826	127	330	64

Table 13.1: continued

	Population (mid-1984)	Area (km)	Population density	GNP *per capita* US $ (1984)	Life expectancy (1984)
South Pacific					
Fiji	686,000	18,272	37	1840	65
Kiribati	63,000	690	91	460	65
Solomon Islands	259,000	28,530	9	640	57
Tonga	99,000	700	141	833 b	63
Tuvalu	8,700	26	335	471 b	60
Western Samoa	161,000	2,935	55	723 b	65
Vanuatu	130,000	11,880	11	784 b	55

Notes:
a. Approximate
b. 1982 data

Sources: World Bank (1986) *Pacific Economic Bulletin*, 1, p. 243

characterises micro-states even within a single ocean; cultural and linguistic differences have been enhanced by divisive colonial and sometimes post-colonial policies.

Underlying the assumption that tiny colonies could not become states were hazy notions of viability, querying what meaning independence might have where any semblance of self-reliance was impossible. Such notions remain valid for island micro-states:

> by reason of their small human and environmental resource base micro-states are of necessity rather more sensitive to external pressure than many other states and are ill-equipped to cope with the rigours of life within the international community (Harden, 1985, p. 51).

Attempts to define micro-states have been fraught with problems (cf. Dommen, 1985; Hein, 1985): hence, island micro-states (IMS) are here arbitrarily defined as those with less than a million population and with relatively low per capita incomes (cf. Dolman, 1985a, 1985b). The exclusion of states such as Trinidad and Tobago, Mauritius and Papua New Guinea, which share many characteristics of smaller nearby states and play important regional roles, and the IMS of the Mediterranean, demonstrates the limitations of any classification.

DEVELOPMENT CONSTRAINTS

Recognition of the critical constraints to conventional economic development in IMS has been long-standing. Two decades ago, these were summarised as, firstly, reliance on very few primary export products; secondly, a small domestic market, therefore limited industrialisation and a heavy reliance on imports; thirdly, problems of maintaining a wide range of machinery and the experts to repair them; fourthly, dependence on foreign capital; and fifthly, a high and disproportionate expenditure on administration, including education and health services (Benedict, 1967, p.2), and high transport costs. Crudely, IMS have no advantages of economies of scale, a limited range of resources, a narrowly specialised economy, primarily based on agricultural commodities, minimal ability to influence terms of trade, dependence for key services on external institutions, a narrow range of local skills and problems of matching local skills and jobs, a small gross domestic product (hence, problems of establishing import-substitution industries), yet alongside considerable overseas economic investment in key sectors of the economy, and especially commerce. The size and isolation of IMS has also contributed to limited ecological and biological diversity and a tendency towards instability when isolation ends: in the case of atoll states, and other low islands, even the most basic resources, such as water, are limited. Several states are prone to natural disasters and monoculture, permanent construction, coastal development and migration have all increased hazard vulnerability. IMS are thus unusually dependent on external relations, of trade, aid, migration and investment, yet are largely unable to influence those international events which affect them most critically.

IMS economies are very open to trade and capital movements. The combination of a large share of export goods in the GNP and the small size of the economy tends to produce an external concentration of production in a few export products, in which world prices are subject to considerable fluctuation. This concentration of domestic production 'exposes the economies of ministates to real shocks of an intensity unparalleled in larger countries' (Galbis, 1984, p. 37). Geographical constraints (soils and climate) have often limited the range of agricultural products though colonial policy has most effectively induced the trend towards monoculture (especially sugar). Nowhere is this more apparent than in atoll IMS such as the Maldives or Kiribati, where domestic development options are naturally constrained by limited land areas and the simplicity of atoll environments (so that natural ecosystems may easily be disrupted). Limited domestic resources have produced heavy dependence on imports and inflationary pressure has often been high: fossil fuels are conspicuously absent, energy is a substantial proportion of import bills and the first two global oil-shocks were severe economic blows.

Inadequate overseas diplomatic representation (Diggines, 1985, p. 199-201), limited financial expertise and weak control of the monetary system, hamper access to international capital markets (Commonwealth Consultative Group, 1985). Social ties are so powerful and pervasive that anonymity, impersonal role-relationships and impartiality are difficult to maintain - hence, the public service can rarely be politically neutral (Benedict, 1967, p. 7-8;

Simmonds, 1985) and corruption is almost inescapable. Distinctive philosophies rarely distinguish political parties and governments are often shifting, unstable coalitions, 'soft-states', seeking short-term solutions rather than committed to long-term development (Diggines, 1985, p. 197-8), vulnerable to external political pressures, and hence likely to enter into a 'special relationship' as a client state of a metropolitan power. Administrative authority has shifted from public servants to politicians, reducing administrative capacity to constrain more rapacious political and economic activities.

Post-independence development strategies have been evolutionary rather than revolutionary, usually stressing the limited objectives of a more self-reliant approach to development though the development of domestic resources: agriculture, fisheries, and where possible, minerals. Subsequently, as domestic development stagnated, greater stress has been attached to tertiary sector activities, primarily tourism. Transnational corporations, once feared, are more widely welcomed, because of their promise of economic growth and employment. Diversity has increased, but IMS have become more dependent on overseas resources (especially aid, loans and remittances), and hence more vulnerable to international recessions. Though seabed resources in EEZs suggest future economic potential, such resources are largely unproven, beyond contemporary exploitative skills, and are likely to require expensive imported technology (Dolman, 1985b). The isolation of several states has given them particular strategic importance - hence, the IMS derive some income and employment from hosting the strategic installations of metropolitan powers. The few advantages of remoteness include military and strategic uses, alongside fuel and chemical storage, quarantine, incarceration of offenders and gene pool conservation (Wace, 1980), though for whom these are advantages starkly highlights the problems of IMS powerlessness. The IMS, even in the remote Pacific, are part of an international economy, and economies once oriented almost solely to agriculture are now vastly more complex and externally oriented. In these economic changes, the IMS of the Caribbean and the Indian Ocean have broadly established trends later followed in the Pacific; in every case the post-independence era has witnessed greater attempts to achieve diversification.

Incorporation into a wider arena is symbolised in the

extending domination of island media by the media of the metropolitan world, nowhere more marked than in the rapid emergence of videos, and, increasingly, television. The cultural transformation of television has tended to be rapid, and widely regarded as negative (Thomas, 1984; Gould and Lyew-Awee, 1985), alongside the proliferation of other modern media, including cinema, video, records and radio (Peet, 1980). The various media have raised expectations yet, through commodity advertising, diverted the resources available to satisfy them. So pervasive is television in Guam that, 'not only does it make us feel homesick for places we have never been, it gives us the uneasy feeling that what we experience daily is abnormal' (Underwood, 1985, p. 171); and emigration has followed. Island cultures are as fragile as their economies and ecosystems.

The past decade, especially in the Pacific, has been a period of unprecedented political and economic change, the latter especially concentrated in capital cities, with localisation and the emergence of indigenous elites and bureaucracies. Although there is no inherent reason for this to be so, many IMS, especially in the Caribbean, are characterised by rapid population growth, high population densities, low agricultural productivity and unequal land distribution, high unemployment and underemployment, lack of real economic growth, high rates of inflation and relatively large foreign debts (Marshall, 1982, p. 452). Prospects of a more self-reliant form of economic development are minimal. Increased self-reliance is more likely to be effectively imposed from outside. A significant proportion of IMS have 'least developed country' status and all are heavily dependent on aid. There is substantial aid bias in favour of both colonies and IMS (de Vries, 1975; Connell, 1986b; Knapman, 1986). This primarily reflects the strategic objectives of donors, and the UN membership of IMS, since needs are usually better satisfied and strategies to achieve economic growth less well defined than elsewhere in the Third World. Though aid levels may decline they are likely to remain substantially in excess of those of larger Third World states. Islands, once remote, are now in thrall to the world economic system.

POPULATION AND MIGRATION

With few exceptions, populations of IMS are now as large as

they have ever been, and despite recent declines in fertility, growth rates usually remain at high levels (Caldwell et al., 1980). A small number of countries, which are just entering into the demographic transition, such as those in Melanesia (Vanuatu and the Solomon Islands) and the Comoros, have annual growth rates of over 3 per cent. However in most IMS high rates of natural increase are effectively siphoned off by emigration, which is widely regarded as a safety-valve. IMS are thus different from much of the Third World in having low death rates, lower birth rates and much higher emigration rates (Caldwell et al., 1980). Most IMS have responded to high fertility levels with family planning policies and programmes, but in recent years, these have lost their impetus (e.g. Connell, 1987c), and in some countries (e.g. Vanuatu and the Solomon Islands) are widely regarded as unacceptable, though a small-family notion is more common in the Caribbean than in the Pacific (Cleland and Singh, 1980; Brookfield, 1984, p. 155).

Higher life expectancies have followed a drop in infant mortality associated with an epidemiological transition from infectious and parasitic diseases (acute respiratory infections, malaria, diarrhoea) to chronic non-communicable diseases (cardiovascular diseases and cancer): this transition is most developed in the Caribbean (e.g. Halberstein and Davies, 1979) and least developed in IMS like Vanuatu, the Solomon Islands and the Comoros. Increasingly, nutrition problems have also become those of affluence and unequal distribution, rather than scarcity, especially where there is high dependence on imports. The major nutritional problems in IMS are protein-energy malnutrition, anaemia and obesity: there is rarely starvation of the kind found in parts of sub-Saharan Africa, though Cape Verde and Comoros have experienced severe food shortages and Cape Verde is one of the few IMS to have received significant food aid.

Movement of peoples within and between islands and from the IMS has intensified in volume, increased in distance and become more complex in pattern and purpose over the past century. International population flows are now the major regulators of demographic change in most IMS. In the past, these movements tended to be circular or repetitive but permanent, and relatively long-distance migration is becoming more general. There are a number of trends in population movement, although not all occur in every country: international migration to metropolitan states remains important; small islands are being depopulated as people move to large islands; and thirdly, urban populations are continuing to grow. Glaring economic and social differentials, both between the IMS and metropolitan countries and between remote islands and urban centres, have an important role to play in influencing migration, as expectations rise and relative deprivation becomes more acute.

High rates of emigration have characterised the Caribbean since the early nineteenth century. In Cape Verde migration was also massive by the end of the nineteenth century and probably no independent state has more of its population (perhaps a half or even two-thirds) living overseas (Meintel, 1983; Gomes, 1986). In the Indian Ocean migration effectively began around the start of the present century (Newitt, 1984, p. 78-9); and by contrast, emigration from the Pacific was almost unknown until the 1960s. However in every region, international migration has dramatically increased in the post-war years.

International migration is greatest from the smallest states and the principal destinations are New Zealand, the USA, Canada and the United Kingdom. This kind of movement is resulting in urban concentrations of islanders on a scale that is scarcely paralleled in the home countries. This is a recent phenomenon which has necessarily had a significant effect not only on the welfare of the migrants, but also on their attitudes and contribution to development in their home countries. Its major influence is in the return migration of those who have been overseas who have become urban men, women and children and are unwilling to step back into a rural environment and culture.

Migration is invariably selective and results in the loss of the more energetic, skilled and innovative individuals, and this loss, which may reduce rural political bargaining power, business expertise and so on, may not be compensated by remittances or by some other form of trickle-down effect from urban and national development. The brain-drain is widespread, as in the Comoros, where 'there has been a natural tendency for qualified Comorians to seek employment abroad and not return to the poverty and lack of prospects in the islands' (ibid., p. 91).

Emigration is counteracted and compensated by a return flow of remittances. Over time the amount and regularity of remittances usually falls, although the decline is more rapid for rural-urban migration than for international migration, where the probability of return migration is considered to be higher. International migration, involving higher costs, but also higher incomes and greatest uncertainty, records the highest rates of remittances in most Third World contexts (Connell et al., 1976, p. 91; Lipton, 1980). In many IMS, remittances are the largest single source of foreign exchange, greater than exports or aid. In the Caribbean this pattern has been long-established (Rubenstein, 1983; Brierley, 1985a), and the proportion of households depending on them is very great. In the Pacific this occurred later, though in Tonga, remittances were the largest single foreign currency earner by 1973. In Western Samoa, where remittances by 1973 were around the same level as the national agricultural income, migrants, for more than a decade, have been 'the most valuable export' (Shankman, 1976, p. 28; cf. Yusuf and Peters, 1985, p. 14). Remittances in IMS, where international migration is important, thus represent a very substantial component of cash incomes, much more than the contributions made by migrants within Europe.

Globally the overwhelming weight of evidence suggests that the use of remittances reflects the poverty and lack of investment opportunities from which the migrant came (Connell et al., 1976, p. 98). The vast majority are used for everyday household needs or in conspicuous consumption, especially of food and housing (Connell, 1980; Marshall, 1982; Rubenstein, 1983, p. 297). This has contributed to food dependency, and in northern Kiribati, where remittances represent the largest portion of cash income, a situation typical of smaller islands:

> the people of Butaritari and Makin are becoming increasingly dependent on remittances to pay their taxes and their children's school fees, to buy corned beef and rice for feasts and to purchase even moderately expensive items at the store. The export of labor has become the principal means of maintaining the local standard of living (Lambert, 1975, pp. 299-31; emphasis added).

The dominance of consumption expenditure reflects the belief that 'money is needed to buy the cultural symbols of social importance sought by Islanders' (Watters, 1970, p. 135). The remaining remittances are mainly invested in the agricultural sector. Although there are widespread assumptions that remittances emphasise dependency and produce rural stagnation (Connell, 1980; Rubenstein, 1983), their contribution to the generation of foreign exchange earnings and employment, especially in the service sector, in small islands where there are few other income-earning opportunities, ensures that there are many exceptions (e.g. Marshall, 1982, p. 459).

On balance, domestic perceptions of both overpopulation and substantial remittance flows have strongly discouraged IMS from formulating policies directed towards reducing migration and establishing more self-reliant development strategies. In the Pacific, only Western Samoa has even contemplated such policies, whereas other states have sought out new opportunities: both Tuvalu and Kiribati have Marine Training schools to train workers for overseas employment (Connell, 1986a). In the Caribbean the same is true, with Barbados granting financial assistance to those who wished to emigrate, and other governments actively seeking new outlets (Benedict and Benedict, 1982, p. 158; Marshall, 1982; Momsen, 1986). In many respects it is those countries without real permanent emigration outlets, such as Kiribati and Tuvalu, where they have been lost (Connell, 1986a, p. 468), that are increasingly the poorest. It has even been suggested that 'the inhabitants of some of the very small and isolated Pacific communities might be well advised to accumulate the aid expenditures now being expended on them as cash grants to help them move to larger neighbouring countries' (Blazic-Metzner and Hughes, 1982, p. 94), a suggestion directed at Kiribati and Tuvalu in a recent influential review of Australian overseas aid (Connell, 1986b), but not acted upon.

Crucial to the role of emigration in development is its future viability, which necessarily depends on the economic situation and political decisions in host nations. Continued colonial status enables a steady migration flow from small states and, in colonies such as American Samoa, Wallis and Futuna, and the Cook Islands, is a powerful deterrent to independence sentiments. A high proportion of the population of some colonies, like American Samoa and the US Virgin Islands, are overseas-born: these and other islands have become 'transhipment stations' (Maingot, 1983, p. 15) or stepping stones, sometimes illegally, for islanders *en route* to metropolitan nations. The future for intra-regional migration is exceptionally poor, and the future of emigration to metropolitan states is uncertain. Though, *in extremis*, 'no community can function as a giant incubator for male workers, who once they reach maturity, are exported' (Yusuf and Peters, 1985, p. 22), many IMS now approximate to this situation, and are not always discouraged.

URBANISATION AND UNEVEN DEVELOPMENT

It is no longer possible to regard islanders as solely rural people: only in the very smallest states (all of which have large urban populations elsewhere) are less than a quarter of the population living in towns or cities. In most cases, urbanisation is a result of the rapid post-war and post-independence expansion in government activity and spending, the consequent boom in well-paid, secure bureaucratic job opportunities, primarily for the educated elite and skilled workers, and the resultant growth in service employment. Migration initially played the most important part in contributing to urban growth (Cross, 1979), though natural increase is of growing significance (e.g. Hope, 1986). Although economic motives are dominant in migration, access to services, especially education and health, have also been critical determinants: services are overwhelmingly concentrated at the centre, and especially in the primate city. Only in the larger, fragmented IMS is there any alternative to extreme urban primacy, emphasised and exacerbated by the peripheral, coastal location of colonial cities.

Much rural-urban migration is permanent or at least long-term. As children are born in town, this permanence is enhanced, whilst migrants are increasingly preferring urban to rural life. This is significant, not merely for the breakdown of traditional social organisation that it implies, but because it effectively ensures that these second-generation migrants are essentially destined to remain (and raise families) in urban areas. The implication of this is growing unemployment amongst those who are effectively permanent urban residents; and the most visible evidence is the growth of shanty towns. As their employment and welfare needs are so much more visible than those of a dispersed rural population, and their ability to organise and influence politicians so much greater, policies have increasingly focused on the urban poor rather than the rural poor. This is part of the nature of emerging urban bias.

Uneven development is apparent even in the smallest

states, accentuated by widespread urban primacy, minimal decentralisation and urban and industrial sectoral bias in the allocation of resources, such as Development Bank Loans (Connell, 1986c), avowedly designed to foster rural and regional development, which thus exacerbate the existing industrial bias of commercial bank loans (Watson, 1985). A widespread problem is the limited development of the more remote islands; thus, the 1985 National Plan of the Maldives stressed that

> the emergence of a dual economy has created a modern sector which employs a minority of the workforce but concentrates incomes, resources and investment. It has accelerated the drift towards Male, the capital, undermined the natural economy and created a social imbalance between Male and Raajjethere (the outlying islands), between the rich and the poor.

Transport costs and frequencies, especially in the multi island states of the Indian and Pacific oceans, emphasise regional inequality and core-periphery problems. There have been secessionist threats in a number of IMS, notably the Solomon Islands (Premdas and Steeves, 1984), Vanuatu (Shears, 1980; Beasant, 1984) and St Vincent (Nanton, 1983, p. 241), though only in the Comoros has secession succeeded in the case of Mayotte (Gaspart, 1983; Newitt, 1984). Elsewhere, nationless nationalism has often integrated poorly remote, ethnically distinct regions, such as Anuta in the Solomon Islands (Feinberg, 1986). Only single-island IMS are without some fissiparous friction.

The predominantly urban location of industrial and tertiary sector employment, and thus the high-wage sector, has contributed to urban-rural migration, and often unmanageable urbanisation, epitomised in the extreme levels of deprivation crime and violence of the slums of Kingston, Jamaica (Eyre, 1986), but slowly becoming more apparent elsewhere. Shanty towns visibly demonstrated inequalities and the environmental aspects of urbanisation have increased. In Kiribati and Truk (Federated States of Micronesia), the disposal of sewerage in densely populated lagoon areas has led to recent severe cholera outbreaks. Urban growth has increased the distance to urban gardens, increased erosion in nearby areas, depleted firewood and contributed to the 'poor man's fuel crisis' (Connell, 1984b, p. 310). Industrialisation and tourism, especially in urban areas, have emphasised spatial inequality, even in small IMS such as Barbados, where the capital, Bridgetown, has remained a centre for the extractions of surplus value, while the economy has evolved from sugar plantations to tourism and manufacturing (Potter, 1985, p. 77; 1986). This kind of uneven development is typical of most IMS: the rare exceptions, where population growth is not intensifying in one area, such as Western Samoa, are those where international migration (and remittances) are vastly more important than domestic economic growth. Inequalities thus appear to have worsened, except in contexts in which international migration is of great importance.

AGRICULTURE, FISHERIES AND FOOD

Without exception IMS were historically characterised by their agricultural economies, a situation which remained until well into the twentieth century. Virtually nowhere is this still true, except in terms of employment, where it usually remains the most important sector. However there are vast historic differences between the largely autonomous Pacific islands and those of the Indian and Atlantic oceans and the Caribbean, where islands were populated by European settlers and their slaves, developing plantation agriculture, supported by external funds and thus never having to adjust to a life of subsistence broadly within the econological limits to the island environment (Benedict and Benedict, 1982; Newitt, 1984, p. 95; Dommen and Hein, 1985, pp. 155-6). In Sao Tome and Principe, growing food crops was actually forbidden in the 500-year colonial period (ibid., 1985, p. 176). The structure of plantation economies limited the contribution of agriculture to development, because of primarily foreign ownership, the high import content of plantation investment, the high consumer import propensity, their export orientation, and their minimal contribution to skill formation (Beckford, 1972). In every case, surpluses were not invested in the IMS, domestic savings were inadequate to generate industrial growth and colonial policy-makers were uninterested in such a transformation. In the post-independence era, national demands for industrial development coincided with stagnating agriculture, rising imports, but equally inadequate savings.

The post-war expansion of cash cropping in the South

Pacific IMS has brought these states closer to the situation of most other IMS, where the plantation capitalism of the nineteenth century had frozen land ownership, and resulted in the bulk of income being dependent on global markets over which they had no control. Large-scale estate and plantation agriculture continues to characterise most IMS: the attendant problems of monoculture, management, environmental degradation and the marginalisation of a poorly paid work-force have all contributed to post-independence decay and destruction of many plantation systems, and resulted in most IMS governments struggling either to manage nationalised systems (e.g. Brierley, 1985a), achieve land reform, or attempt to diversify cash crops and stimulate food production. Few have had much success.

In the Caribbean, both food crop and export crop production have declined, a result of various factors, including droughts, floods, fires (especially in sugar) and hurricanes, crop diseases and declining soil fertility, industrial disputes and labour shortages, foreign exchange problems, shortages of chemicals and machinery, loss of agricultural land, political instability, advancing age of farmers, lack of physical planning and technical services, poor technology (Gumbs, 1981; Axline, 1986, p. 57; Hope, 1986) and urban bias in commodity prices. Higher wages in the industrial and tourist sectors, combined with an increasing disdain for agricultural work, have produced rising unemployment alongside agricultural labour shortages (Axline, 1986, p. 52), a function of increasing relative deprivation in the agricultural sector. Broadly the same is true in the Pacific. Food and beverage imports are proportionately high, crudely summarised as a situation of 'food dependency' (McGee, 1975), representing more than a third of all imports by value in Cape Verde, Comoros, Dominica, Grenada, Kiribati, St Kitts-Nevis, St Vincent, Sao Tome, Tuvalu and perhaps elsewhere. Food dependency or 'dietary colonialism' has both encouraged substantial trade imbalances and resulted in the decline of subsistence agricultural systems. The simultaneous decline in subsistence agricultural production and growing consumption of imported foods (especially tinned meat and fish, rice, biscuits and flour), both by reducing the regularity of food consumption (as cash flows are variable) and increasing the consumption of sugar especially, has substantially increased the incidence of diet-related diseases such as diabetes (Ward and Hau'ofa, 1980, pp. 39-48).

Subsistence agriculture had universally declined, and there has been a transition away from labour-intensive crops towards crops like cassava, often at the expense of rural nutrition (Thaman and Thomas, 1985). The transition has been slower in the more densely populated Caribbean (Brookfield, 1984, pp. 153-4) and the Indian Ocean. In some circumstances, where emigration has not acted as a safety-valve, as in the Seychelles, 'there is simply not enough land to provide either subsistence or the cash crops to buy subsistence' (Benedict and Benedict, 1982, p. 109) or in the Comoros where very high population pressure on resources has resulted in erosion and declining per capita food production levels (Newitt, 1984, p. 104). Land tenure has slowed the development of some forms of agriculture, and increasing population pressure on resources and the devotion to cash crops have contributed to increasing inequality in land ownership (see for example, Rodman, 1984; Thompson, 1985). Even on Taveuni island (Fiji), where population densities are low, inequalities in land ownership are argued to be of Latin American proportions (Brookfield, 1979, p. 45). Such inequalities and land shortages occur even in situations of high emigration rates.

Migration and the movement into local wage labour have nonetheless contributed to a general decline in the use of marginal and distant land. Several Caribbean islands, such as Antigua, St Lucia and Grenada, have as much as a third of their potential agricultural land lying idle; and other factors that have contributed to this include patterns of inheritance (and 'parcelisation'), price fluctuations, 'modern' education and resultant labour shortages (Rubenstein, 1975; Brierley, 1985a). In classic plantation economies, such as those of the two Atlantic states, landlessness is extensive. The processes of agricultural decline that have gone on earlier elsewhere are being replicated on the small islands of the Indian and Pacific Oceans, and especially on those where international migration has been common. Migration and remittances have emphasised the trend towards the disintensification of the traditional agricultural system (Brookfield, 1972; Lea, 1972) that essentially followed the expansion of cash cropping. These trends are being maintained, and even exacerbated, by rapid urbanisation, whereas increased urban demand might have been expected to stimulate rural production.

Broadly the same kind of cumulative downward spiral has also affected artisanal fisheries in IMS, though less dramatically than in agriculture. The expanded EEZs offer potential for large-scale development, especially in the vast Indian and Pacific Ocean area. However, the massive cost of modern fisheries vessels, technical skills, fuel costs and market access have all hampered growth. International restructuring, incorporating more purse-seiners at a time of global market saturation, and lack of onshore facilities, have further restricted the attempts of IMS to participate in this sector (Kearney, 1980). IMS have yet to gain significantly from the new EEZ legislation; moreover, the countries of the eastern Caribbean have acquired exclusive rights 'to some of the most biologically unproductive waters in the region' (Dolman, 1985a, p. 58). In the Indian and Pacific Oceans, the potential is greater, yet ocean space gains, because of non-existent capacity to exploit them, are more theoretical than real, and even policing these waters is extremely difficult.

The fishing industry consequently remains primarily based outside the IMS. IMS, even with enlarged EEZs where stocks are largely unknown, are seeking to enter a depressed and highly protected market in competition with metropolitan producers; hence, even licensing fees have failed to live up to expectations (Dolman, 1985b, p. 137), though IMS who do gain much income from fisheries do it primarily through leasing their waters rather than through owning fishing fleets.

INDUSTRY AND INVESTMENT

With rare exceptions, IMS have no mining industry, so the whole primary production sector is experiencing something of a downturn, influenced by domestic socio-economic change, and in part due to recession and protection in the rich world. This has led to and followed from migration, emphasised uneven development and underlined the desire of IMS governments to diversify out of the primary sector. Far from adhering to Lipton's dictum, 'if you wish for industrialisation, prepare to develop agriculture' (Lipton, 1977, p. 24), there has been a withdrawal from agriculture and self-reliance in favour of a more rapid modernisation of the economy in pursuit of the chimera of sustained economic growth.

IMS have generally made some transition from the export of agricultural staples towards service and light manufacturing industries: in cases like those of Bahamas, Bermuda and Antigua (Henry, 1985), the agricultural system has virtually collapsed and been replaced by what are increasingly the 'quasi-staples' of light manufacturing and tourism. However, in contrast to larger world regions, industrialisation in IMS is largely absent, other than of basic import substitution and food-processing industries, in part because of the limited resources, small domestic market, inaccessibility and hence the impossibility of economies of scale (Selwyn, 1975), and in the Pacific, high wages. Open economies enabling cheap untaxed imports, and consumer tastes that discouraged investment and much 'inefficient' industrialisation is hidden behind tariff barriers (Kaplinsky, 1983). The domestic market is so small that in many respects 'the foreign sector is the economy' (Dommen and Hein, 1985, p. 152). Consequently, seemingly paradoxically, a significant part of industrialisation in IMS is oriented to export, where market access is possible, either because of linkages with global corporations or because of tax and trade concessions. Nevertheless, most overseas investment in IMS is towards manufacturing, or service provision, for local (or regional) markets: in the Pacific, foreign investment is aimed at securing existing export markets and at 'serving local markets and not at generating the export led growth which the island governments so earnestly desire' (Taylor, 1984, p. 347) or stimulating corporate economic growth as a result of 'corporate ephemeralism' (Taylor, 1986, p. 64). Exported capital does not have as its primary goal the industrialisation of the remote periphery.

The growth of manufacturing is most true of the Caribbean, where high-technology 'screwdriver' industries are increasingly common, as in Barbados, attracted by access, political stability, low wages and weak unions (Sunshine, 1985, p. 141). Most governments offer favourable terms to potential foreign investors: indeed, in the Caribbean, industrial development is characterised as 'industrialisation by invitation' (Barry et al., 1984, p. 73). Several IMS, such as Tonga, are virtually a single export-processing (rather than producing) zone (ESCAP, 1985; Howard, 1986), yet little investment has been stimulated. After more than half a decade of operation, Tonga's small-industrial centre had 14 firms, the total employment created was less than 300 people (less than 1 per cent of the work-force), and manufactured exports were valued at about 5 per cent of all exports. Both in Tonga and St Lucia, factory workers see the principal goal of factory employment as that of obtaining skills for subsequent emigration. There, in St Lucia and elsewhere in the Caribbean (Kelly, 1986) and in most IMS EPZs, the majority of factory workers are women, earning low wages in difficult conditions. Light manufacturing in EPZs has also been successfully established in some Indian Ocean states, but they often remain enclaves (de Vries, 1984) and infant industries face the chilling influence of protectionism in the developed industrial economies (Legarda, 1984), even when their impact on these markets is slight.

One manufacturing activity that has contributed to substantial exports throughout IMS is postage stamps, and in most IMS this is a major focus of economic activity, although no IMS actually prints the stamps. In Tuvalu and in other fragmented IMS, such as Antigua and the Cook Islands, stamps are now issued for several different islands as well as for the state as a whole. The task of maintaining sales in a fluctuating global market without saturating it remains a challenge, and the market is exceptionally volatile. Nonetheless, the attention given to this area of export is indicative of the crisis in IMS exports.

Limited industrial production in IMS is also a function of preferential trade and non-discriminatory tariffs. In general, trade preferences extended to LDCs have had only a marginal effect on increasing trade flows (Moss and Ravenhill, 1983): preferences are often extended to goods not imported by industrialised countries, there are restrictions on 'sensitive goods' such as textiles, as in the case of textiles from Mauritius and Fiji (Harden 1985, p. 91; Sutherland, 1986). Where relatively favourable trade agreements have been signed with regional groups, as in the case of the South Pacific Area Regional Trade and Economic Commission Agreement (SPARTECA), benefits have principally gone to the larger, or more favoured states. Indeed, SPARTECA, despite its primary aim of contributing to industrialisation in independent Pacific states, has largely failed (Robertson, 1986; Sutherland, 1986) - hence, in practice, exports of manufactured goods from IMS have primarily occurred only under competitive free trade conditions.

A substantial proportion of foreign investment in IMS is increasingly in the service sector, in general commerce (especially in the Pacific) and in more sophisticated financial and insurance ventures. In general, transnational corporations (TNCs) in IMS have attempted to rationalise by moving away from resource-based activities into the tertiary sector, where skills are in greater demand and linkages to rich-world countries are crucial. The prime example of this is tourism and more recently, finance. Financial service industries have gained in their attractions to IMS, as other sources of earnings have dwindled, and have certain advantages:

> precisely because their economies lack internal linkages, there is little difficulty in designing a set of tax advantages which not only do not weaken the domestic tax base but actually widen it beyond what the local economy itself could achieve (Dommen and Hein, 1985, p. 166).

Most island tax havens are in the Caribbean (Cayman Islands, Bahamas) or nearby (Bermuda) and around Great Britain, though they have more recently been established in Vanuatu and the Cook Islands. More generally, problems of distance and strong overseas competition have restricted the growth of even this extremely flexible and mobile activity. Finance centres have increasingly become 'functional', rather than 'paper', more directly contributing to employment, whilst the costs to the local economy may be minimal (Francis, 1985), though in small colonies, such as the Cayman Islands, extreme dependence on tax havens and associated tourism (Caulfield, 1978) has wholly destroyed the historic social ecology and economy. Elsewhere, this extremely footloose industry has been largely beneficial.

Transnational involvement in raw material exploitation is most marked in the Pacific, where timber, fish and minerals are more plentiful. There has been widespread criticism of timber exploitation, based on ecological devastation (clear-felling without replanting), minimal benefits (if any) to villages owning the resources, capital-intensive operation, minimal local processing and exemption from legislation (Shankman, 1978; Waddell, 1982; Marjoram and Fleming, 1983). Conventional gains from rents, formal sector employment and skill gain, technology transfer, infrastructure and taxation, and especially market access, in circumstances where domestic capability is limited, are more apparent in other areas of transnational operations. Outside the natural resources sector, where concessions can be won, TNCs can close unprofitable enterprises and move elsewhere, as in the Caribbean, in a process of 'island-hopping' (Kelly, 1986) as part of the flight from regulation. This mobility distinguishes the minimal integration of TNCs into national IMS economies, and stresses the enclave structure that marks integration into a global economy.

Regional trade is conspicuous by its absence in the Indian, Atlantic and Pacific Oceans, though more substantial in the Caribbean (Axline, 1986, p. 34-7) where, despite regional similarities in the production structures, diversity is greater and distance less than elsewhere. In the Pacific, production is competitive, rather than complementary, communications follow colonial lines rather than integrate the region, and there is some consumer bias against local products. Consequently, 'unless there is political will to cooperate the potential, limited as it is, for greater trade within the region will not be fully realised. Regional cooperation demands some sacrificing of national interests' (Sevele, 1983, p. 27). In a sense, then, not only are most enterprises within enclaves in the IMS, but the IMS themselves are enclaves, often better linked to metropolitan economies than to island neighbours.

The two principal regions of island states, the Caribbean and the Pacific, have a number of regional organisations, such as CARICOM, and the South Pacific Commission (SPC) which have contributed to greater expressions of regional solidarity and recognition of common interest, but such organisations have rarely contributed either to sustained economic growth, social development or regional trade, and have largely been overshadowed by updated bilateral historic colonial ties. Only the SPC loosely unites all independent states and colonies in a single ocean. Attempts to develop more formal, federal structures have been shortlived.

> In theory federal structures are able to provide many of the advantages of a larger economic unit, if factors of production, including labour, are allowed to flow freely within the component parts of the federation. In practice, political strains develop, with the outlying territories tending to resent what they see as the balance of advantages flowing to the centre and people at the centre tending to resent the influx of migrants from the periphery. Recent history does not give us great cause to believe that independent countries will succeed in recreating those federal structures that

came apart at the end of the colonial era (Faber, 1984, p. 375).

The diversity of IMS, rather than encouraging regional economic ties, has tended to lead to separate development: in the Caribbean, according to former Trinidad and Tobago Prime Minister, Eric Williams, there is no 'tradition of political cooperation and consultation. It is a tradition of divisiveness and separation' (quoted by Dolman, 1985a, p. 50) and particularism (Clarke, 1976; Lowenthal and Clarke, 1980), where political fragmentation has suggested that 'small is not beautiful' (Sanguin, 1981). The rhetoric of regional unity is unlikely to overcome the reality of dependent development. Fission is more obvious than fusion.

TOURISM

The growing orientation towards external sources of income is emphasised by the development of tourism, especially in the Caribbean, and more recently in the Indian and Pacific oceans. In countries like Antigua, Barbados, Bahamas and Seychelles, tourism is much the largest contributor to the GDP, the highest contributor to employment, with annual tourist numbers substantially greater than the local population. The status of economic development in IMS is symbolised by the increasing influence of tourism, and the massive dependence upon it. The few countries in which tourism is unimportant, in the Atlantic, Comoros and the central Pacific (Kiribati and Tuvalu) are amongst the poorest of all LDCs. Tourism generates more income for IMS than do commodity exports (Legarda, 1984) and has a proportionately greater economic role than in any other world region.

Tourism generates jobs, provides foreign exchange, taxes and other indirect revenues and generates activity in other sectors of the economy. However tourism is largely externally controlled and subject to cyclical and irregular fluctuations. The goods (especially food and drinks) consumed by tourists are often imported; hence, the most positive role of tourism is in employment in services, handicraft and construction industries, rather than in direct income benefits. Much of the best coastal agricultural land has been lost to tourism (Bryden, 1973; Benedict and Benedict, 1982, p. 153), and agricultural production has often declined, through the withdrawal of labour, rather than been stimulated by tourist demand. There are similar distortions in other price structures, including wage rates, and other economic disadvantages have been attributed to tourism, including seasonality, the subsidisation of infrastructure without alternative uses, excessive tax concessions, extreme concentration of hotels and landownership amongst a few (and hence increases in income inequality), and the difficulty of regulating an industry largely controlled by foreign technology, tastes and economic change - hence, success tends to be measured primarily by easily measurable growing tourist numbers rather than by more diverse criteria. A range of ecological and social problems, such as the destruction of traditional values, new consumption patterns and leisure habits, prostitution and crime are also attributed to tourism (e.g. Potter, 1983), though such problems and processes were usually already in existence.

Tourism, by its nature an almost entirely international activity in the Third World, is largely foreign-owned, and the foreign content, through the crucial roles of advertising and air (and sea) transport, is tending to increase rather than decrease. Tourism is very much an enclave economic activity that has, in some areas, replaced other enclave activities, and, in places such as Fiji and St Vincent, resulted in the foreign ownership of particular islands where the indigenous population have limited rights (Nanton, 1983). The inevitability of foreign ownership, the uneven development that has followed tourist development, and above all the cultural conflict argued inevitably to ensue, initially discouraged Pacific states from orienting development strategies towards tourism. However, limited economic growth and external interest have resulted in what was once regarded as the 'last resort' (Lea, 1980) now being central to development. Dependent on foreign investment, fluctuating tastes, metropolitan economic change and domestic stability (Connell, 1987b), tourism is the most vulnerable and competitive of all industries. Nonetheless, for all the economic, ecological and social problems, there are substantial benefits that have enabled IMS largely to negotiate successfully another strand in the network of external economic relations.

SMALL IS VULNERABLE

Remote islands are now firmly part of the global political economy. These extremes of the Third World are more closely incorporated into the world system than larger states where a semblance of self-reliant development is possible. Though 'small is dangerous' (Harden, 1985) and hence vulnerable, small size and a paucity of resources ensure a primarily economic vulnerability. Isolation has given several IMS unusual strategic importance, massively contributing to the economies of dependent territories, like the Marshall Islands and French Polynesia, that are used for missile and nuclear testing respectively, but subverting nascent nationalist social and political systems. In a crude economic sense, the global nature of superpower rivalry has transformed some remote islands and colonies.

Minimal domestic military strength has increased metropolitan fears over security and 'instability' in regions of perceived strategic significance rather than concern over the aspirations and legitimate diplomatic and trading ties of small states. Thus, in the Caribbean, 'American considerations about a security policy for the region have originated primarily in reaction to perceived threats; they have so far failed to concentrate on the needs and aspirations of the governments of the area' (ibid., p. 149), and policies have been discriminating and divisive. Similar perceptions have posed problems in the Indian Ocean (Kapur, 1986) and most recently in the Pacific. In almost every case, IMS have been pawns in superpower rivalries, sometimes as beneficiaries, sometimes as victims, but never as principal actors.

Small size and belated independence, though delaying the evolution of democracy and occasionally stifling tiny nationalist movements, have also hindered the emergence of radical political movements. IMS have been characteristically conservative, nowhere more so than in the Pacific, where Melanesian 'socialism' has proved more illusory than real (Howard, 1983); and similarly, Cape Verde and Sao Tome have long had rhetorically socialist governments that have practised little. The single important exception was the existence of the People's Revolutionary Government (PRG) in Grenada from 1979 to 1983, hence the variety of interpretations of the achievements and failures of the PRG are enormous (Maingot, 1986). Though Brierley suggests that 'perhaps no nation in Caribbean history advanced so rapidly and effectively on such a variety of social and economic issues in less time than Grenada did under Bishop and the PRG' (Brierley, 1986b, p. 51), the brevity and incompleteness of the revolutionary transformation prevents widespread acceptance of this conclusion. Though the conclusion that 'in the heightened cold-war tension of today's Caribbean some countries are too small for successful revolutions' (Momsen, 1984, p. 150), may be too peremptory, the general conservatism of IMS, in a global system where negotiation but not control is possible, emphasises the relative powerlessness of small states. Small size is a powerful constraint on national policies.

Grenada demonstrates both the extraordinary difficulty of breaking away from convention trajectories of dependent development and the extent to which achieving a greater degree of self-reliance is fraught with problems (cf. Thompson, 1985, pp. 123-5). Rapidly rising expectations have fostered uncertainties, even where economic growth has occurred. In Seychelles, 'people had participated in higher standards, but they have been shown standards that were higher still and that they could not attain. In absolute terms, their conditions had improved - but relatively, they had not. Some people had prospered but the majority of the poorer people had not' (Benedict and Benedict, 1982, p. 104). Overseas ties, and especially remittances, have threatened IMS by, as the Prime Minister of Cape Verde has remarked, 'creating illusory needs that are impossible to satisfy within the country's means' (quoted by Meintel, 1983, p. 159). For most IMS, movement towards greater self-sufficiency would be difficult and painful. Thus, the self-sufficiency and economic nationalism that continues to be widely recommended as a solution to dependence is steadily being eroded. At the same time, the alternative, a more adequate interdependence (for example, through better terms of trade for Third World produce or more employment opportunities in the rich world), is no more likely.

In the present century, IMS economies have become even more open: subsistence-oriented agricultural systems have disintegrated (e.g. Frucht, 1968; Marshall, 1982), and despite a post-independence plea for greater self-reliance, this rhetoric has failed to disguise the reality of increasing incorporation. IMS have become increasingly dependent on the external environment for capital (aid, loans, remittances and private invesment), technology and commodities, expertise, and even cultural change. In the Pacific and Indian oceans, these problems are further emphasised by the massive distances between extremely small nations, the tenuous links between them (born of their colonial histories), and fragmentation.

The decline of the productive sector (especially agriculture) and the growth of imports, offset by aid, remittances and tourist revenue, in a situation where much employment is concentrated in the public sector, has led to the conceptualisation of the smallest Pacific micro-states as MIRAB economies, dependent on migration, remittances and aid, thus sustaining the bureaucracy (Bertram and Watters, 1985). The urban bias of MIRAB economies, in aid delivery, bank loans and urbanisation (especially of the bureaucracy) and the demise of agriculture and fisheries, suggests that a better acronym might be MURAB. Whereas Brookfield has gone beyond this to suggest that since it is government employment that predominates in the bureaucracy, the micro-state economy might best be conceived of as a MIRAGE, that is scarcely a sustainable economy at all (cited in ibid., p. 497; cf. Tisdell and Fairbairn, 1984), and, independently, Dolman has commented that, compared with aid and remittances, only tourism is 'a legitimate economic activity' (Dolman, 1985, p. 46). Bertram has argued that in such small IMS, the thrust of most development planning, with its focus on production, is misplaced, since 'aid, philately and migrant remittances are not merely supplements to local incomes, they are the foundations of the modern economy' (Bertram, 1986, p. 809).

Consequently, 'the central economic problem ... is the preservation and enhancement of their status as rentier societies' (ibid., p. 309) - hence the manner in which the consolidation of the global political economy has hastened the end of the era of decolonisation. Moreover, practices once regarded in a largely negative light, such as tourism and emigration, have widely become enshrined as policies: industrialisation is by imitation and invitation, tax havens are created, and territorial waters are leased out as dependence is increasingly negotiated. Ecological degradation on land and sea reduces historical economic options. Where rentier economies have diversified, and islands occupy strategic locations, real prospects for maintaining and enhancing rentier status exist.

Two decades ago, Benedict concluded that 'the best

solution for small territories is to look for some form of economic integration with their neighbours' (Benedict, 1967, p. 8). Though the IMS have invariably officially sought greater self-reliance, in the post independence era they have been integrated into the economies of metropolitan states. Moreover, micro-states (and colonies) that have a 'special relationship' with a metropolitan power may be better off than those which do not (cf. Faber, 1984, p. 376): Dommen bluntly states that 'the particularly poor island countries are those which have failed to establish sufficiently intimate relations with a prosperous protector' (1980, p. 195). Harden's belief that 'paucity of resources may theoretically affect the criterion of independence' (1985, p. 52) is equally practical. The smallest colonies and IMS, by choice, and larger IMS, for want of a superior option, have increased their ties with metropolitan powers and moved from productive to rentier status.

The exotic image of tropical islands contrasts with a reality of struggle for survival, and the erosion of sovereignty, as self-reliance becomes no more than a mythical state. Some 20 years ago, Ward commented, in the context of small Polynesian islands, 'it certainly appears that many of the smaller islands will cease to be viable socio-economic units as present trends in cultural change continue' (Ward, 1967, p. 96). Even in much larger countries, attempts to achieve self-reliance often appear no more than reflections of the aspirations that must suffice if growth cannot easily be achieved: as Joseph puts it, in a Nigerian context, self-reliance is 'little more than a ritual for exorcising the devil of dependence' (1978, p. 223). Though even the smallest IMS have policy options that would enable greater self-reliance (Seers, 1983; Connell, 1986a, p. 35-6), there seems little real alternative to a future of economic and cultural dependence that would result from fluctuating strategies alternating between different ideologies and different internal and external sources of support - which are a function of the democratic process. Dolman has concluded one review of island states by querying: 'Is it too much to suggest that small islands, for all the problems and constraints that confront them, could become the laboratory in which alternative development strategies, shaped by the notion of self-reliance, first see the light of day?' (1985a, p. 63). Unfortunately for populist notions, it is much too late.

REFERENCES

Axline, W.A. (1986) Agricultural policy and collective self-reliance in the Caribbean. Westview Press, Boulder and London

Barry, T., Wood, B. and Preusch, D. (1984) The other side of paradise. Foreign control in the Caribbean. Grove Press, New York

Beasant, J. (1984) The Santo Rebellion. Heinemann, Melbourne

Beckford, G. (1972) Persistent poverty: underdevelopment in plantation economics of the Third World. Oxford University Press, New York

Bedford, R.D. (1980) Demographic processes in small islands: the case of internal migration. In H.C. Brookfield (ed.), Population-environment relations in tropical islands: the case of eastern Fiji, Paris, UNESCO, pp. 27-59

Benedict, B. (1967) Problems of smaller territories. Athlone Press, London

Benedict, M. and Benedict, B. (1982) Men, women and money in Seychelles. University of California Press, Berkeley

Bertram, G. (1986) 'Sustainable development' in Pacific micro-economies. World Development, 14, 809-22

Bertram, I.G. and Watters, R.F. (1985) The MIRAB Economy in South Pacific microstates. Pacific Viewpoint, 26, 497-519

Blazic-Metzner, B. and Hughes, H. (1982) Growth experience of small economies. In B. Jalan (ed.), Problems and policies in small economies, Croom Helm, London, pp. 85-101

Brierley, J.S. (1985a) Idle land in Grenada: a review of its causes and the PRG's approach to reducing the problem. Canadian Geographer, 29, 298-309

------ (1985b) A review of development strategies and programmes of the People's Revolutionary Government in Grenada, 1979-83. Geographical Journal, 151, 40-52

Brookfield, H.C. (1972) Intensification and disintensification in Pacific agriculture. Pacific Viewpoint, 13, 30-48

------ (1979) Land reform, efficiency and rural income distribution: contributions to an argument. Pacific Viewpoint, 20, 32-52

------ (1984) Population, development and environmental relations in planning in eastern Fiji and the eastern Caribbean. In F. di Castri, F.W.A. Baker and M. Hadley (eds), Ecology in practice, Tycooly, Dublin, pp. 144-79

Bryden, J.M. (1973) Tourism and development: a case study of the Commonwealth Caribbean. Cambridge University Press, Cambridge

Caldwell, J.C., Harrison, G.E. and Quiggin, P. (1980) The demography of micro-states. World Development, 8, 953-67

Caulfield, M.D. (1978) Taxes, tourists and turtlemen: island dependency and the tax-haven business. In A. Idris-Soven, E. Idris-Soven and M.K. Vaughan (eds), The world as a company town. Multinational corporations and social change, Mouton, The Hague and Paris, pp. 345-74

Clarke, C.G. (1976) Insularity and identity in the Caribbean. Geography, 61, 8-16

Cleland, J. and Singh, S. (1980) Islands and the demographic transition. World Development, 8, 969-93

Commonwealth Consultative Group (1985) Vulnerability: small states in the global society. Commonwealth Secretariat, London

Connell, J. (1980) Remittances and rural development, migration dependency and inequality in the South Pacific. ANU, Development Studies Centre Occasional Paper (Canberra), 22

------ (1984a) Diets and dependency. Food and colonialism in the South Pacific, 2nd edn. Freedom from Hunger Ideas Centre Occasional Paper (Sydney), 1

------ (1984b) Islands under pressure - population growth and urbanization in the South Pacific. Ambio, 13, 306-12

------ (1986a) Population, migration and problems of atoll development in the South Pacific. Pacific Studies, 9 (2), 41-58

------ (1986b) Small states, large aid: the benefits of benevolence in the South Pacific. In P. Eldridge, D. Forbes and D. Porter (eds), Australian overseas aid: future directions, Croom Helm, Sydney and London, 57-78

------ (1986c) Banks and bias in the Kingdom of Tonga. Pacific Economic Bulletin, 2

------ (1987a) From New Caledonia to Kanaky? The political economy of a French colony. Canberra (in press)

------ (1987b) Trouble in Paradise. Australian Geographical Studies, 24 (in press)

------ (1987c) Population growth and and emigration:

Achilles heel, cultural survival or political expediency? In M.J. Taylor (ed.), Fiji: growth without change, Allen and Unwin, Sydney (in press)

------, Dasgupta, B., Laishley, R. and Lipton, M. (1976) Migration from rural areas: the evidence from village studies. Oxford, University Press New Delhi

Cross, M. (1979) Urbanization and urban growth in the Caribbean. Cambridge University Press, Cambridge

de Vries, B.A. (1975) Development aid to small countries. In P. Selwyn (ed.), Development policy in small countries, Croom Helm, London, pp 164-84

------ (1984) Industrial policy in small developing countries. Finance and Development, 21 (2), 39-41

Diggines, C.E. (1985) The problems of small states. The Round Table, 295, 191-205

Dolman, A.J. (1984) Islands in the shade. The performance and prospects of small island development countries. Institute of Social Studies Advisory Service, The Hague

------ (1985a) Paradise lost? The past performance and future prospects of small island developing countries. In E. Dommen and P. Hein (eds), States, microstates and islands, Croom Helm, London, pp. 40-69

------ (1985b) Small island developing countries, the new law of the sea and the development potential of exclusive economic zones. Institute of Social Studies, The Hague

Dommen, E.C. (1980) External trade problems of small island states in the Pacific and Indian Oceans. In R.T. Shand (ed.), The island states of the Pacific and Indian oceans, ANU Development Studies Centre Monograph no. 23, Canberra, pp. 179-99

------ (1985) What is a microstate? In E. Dommen and P. Hein (eds), States, microstates and islands, Croom Helm, London, pp. 1-15

------ and Hein, P.L. (1985) Foreign trade in goods and services: the dominant activity of small island economies. In E. Dommen and P. Hein (eds), States, microstates and islands, Croom Helm, London, pp. 152-84

ESCAP (1985) Transnational corporations and the developing Pacific island countries. ESCAP/UNCTC, Publication Series B, no. 9, Bangkok

Eyre, L.A. (1986) The effects of political terrorism in the residential location of the poor in the Kingston urban region, Jamaica, West Indies. Urban Geography, 7, 227-

42
Faber, M. (1984) Island micro states: problems of viability. The Round Table, 292, 372-6
Feinberg, R. (1986) The 'Anuta Problem': local sovereignty and national integration in the Solomon Islands, Man, 21, 438-52
Fieldhouse, D.K. (1982) The colonial empires. A comparative survey from the eighteenth century, 2nd edn. Macmillan, London
Francis, C.Y. (1985) The offshore banking sector in the Bahamas. Social and Economic Studies, 34, 91-109
Frucht, R. (1968) Emigration, remittances and social change: aspects of the social field of Nevis, West Indies. Anthropologica, 10, 193-208
Galbis, V. (1984) Ministate economies. Finance and Development, 21 (2), 36-9
Gaspart, C. (1983) Les survivances coloniales aux Comores. In R. Cohen (ed.) African islands and enclaves, Sage, Beverly Hills and London, pp. 217-48
Gomes, A. (1986) Cape Verde, making its own rain fall. Africa Report, 31, 21-3
Gould, P. and Lyew-Awee, A. (1985) Television in the Third World: a high wind on Jamaica. In J. Burgess and J.R. Gold (eds), Geography, the media and popular culture, Croom Helm, London, pp. 33-62
Gumbs, F. (1981) Agriculture in the wider Caribbean. Ambio, 10, 335-9
Halberstein, R.A. and Davies, J.E. (1979) Changing patterns of health care on a small Bahamian island. Social Science and Medicine, 13B, 153-67
Harden, S. (1985) Small is dangerous. Micro states in a macro world. Frances Pinter, London
Hein, P. (1985) The study of microstates. In E. Dommen and P. Hein (eds), States, microstates and islands, Croom Helm, London, pp. 16-29
Henry, P. (1985) Peripheral capitalism and underdevelopment in Antigua. Transaction Books, New Brunswick and Oxford
Hope, K.R (1986) Urbanization in the Commonwealth Caribbean. Westview Press, Boulder and London
Howard, M.C. (1983) Vanuatu: the myth of Melanesian Socialism. Labour Capital and Society, 16, 176-203
------ (1986) Transnational corporations and the island nations of the South Pacific. Transnational Corporations Research Project, Occasional Paper no.

10, Sydney

Joseph, R.A. (1978) Affluence and underdevelopment: the Nigerian experience. Journal of Modern African Studies, 16, 221-39

Kaplinsky, R. (1983) Prospering at the periphery: a special case - the Seychelles. In R. Cohen (ed.), African islands and enclaves, Sage, Beverly Hills and London, pp. 195-215

Kapur, H. (1986) Great powers and the Indian Ocean. A non-aligned perspective. The Round Table, 297, 50-9

Kearney, R. (1980) Some problems of developing and managing fisheries in small island states. In R. T. Shand (ed.), The island states of the Pacific and Indian Oceans, ANU Development Studies Centre, Monograph no. 23, Canberra, 41-60

Kelly, D. (1986) St Lucia's female electronics factory workers: key components in an export-oriented industrialization strategy. World Development, 14, 823-38

Knapman, B. (1986) Aid and the dependent development of Pacific Island states. Journal of Pacific History, 21, 139-52

Lambert, B. (1975) Makin and the outside world. In V. Carrol (ed.), Pacific atoll populations, University of Hawaii Press, Honolulu, pp. 212-85

Lea, D. (1972) Indigenous horticulture in Melanesia. In R.G. Ward (ed.), Man in the Pacific Islands, Oxford University Press, Oxford, pp. 252-74

------ (1980) Tourism in Papua New Guinea: the last resort. In J.M. Jennings and G.J.R. Linge (eds), Of time and place, ANU Press, Canberra, pp. 211-31

Legarda, B. (1984) Small island economies. Finance and Development, 21 (2), 42-3

Lipton, M. (1977) Why poor people stay poor. A study of urban bias in world development. Temple Smith, London

------ (1980) Migration from rural areas of poor countries: the impact on rural productivity and income distribution. World Development, 8, 1-24

Lowenthal, D. and Clarke, C.G. (1980) Island orphans: Barbuda and the rest. Journal of Commonwealth and Comparative Politics, 18, 293-307

Macdonald, S.B. and Demetrius, F.J. (1986) The Caribbean sugar crisis. Journal of Interamerican Studies, 28, 35-58

McGee, T.G. (1975) Food dependency in the Pacific: a preliminary statement. ANU Development Studies

Centre, Occasional Paper (Canberra), 2
Maingot, A.P (1983) Caribbean migration as a structural reality. Dialogues (Latin American and Caribbean Center, Miami), 13
------ (1986) Coming to terms with the 'Improbable Revolution'. Journal of Interamerican Studies, 27, 177-90
Marjoram, T. and Fleming, S. (1983) Levers in the Solomon Islands. Raw Materials Report, 2 (1), 35-7
Marshall, D. (1982) Migration as an agent of change in Caribbean island ecosystems. International Social Science Journal, 34, 451-67
Meintel, D. (1983) Cape Verde: survival without self-sufficiency. In R. Cohen (ed.), African islands and enclaves, Sage, Beverly Hills and London, pp. 145-64
Meltzoff, S.K. and LiPuma, E. (1983) A Japanese fishing joint venture: worker experience and national development in the Solomon Islands. ICLARM Report no. 12, Manila
Momsen, J.H. (1984) Caribbean conflict: cold war in the sun. Political Geography Quarterley, 3, 145-51
------ (1986) Migration and rural development in the Caribbean. Tijdschrift v. Economische en Sociale Geografie, 77, 50-8
Moss, J. and Ravenhill, J. (1983) Trade between the ACP and EEC during Lomé 1. In C. Stevens (ed.), EEC and the Third World: a survey, vol. 3, Hodder and Stoughton, London
Nanton, P. (1983) The changing pattern of state control in St Vincent and the Grenadines. In F. Ambursley and R. Cohen (eds), Crisis in the Caribbean, Heinemann, Kingston, pp. 223-46
Newitt, M. (1984) The Comoro Islands, struggle against dependency in the Indian Ocean. Westview Press, Boulder
Peet, R. (1980) The consciousness dimension of Fiji's integration into world capitalism. Pacific Viewpoint, 21, 91-115
Potter, R.B. (1983) Tourism and development: the case of Barbados, West Indies. Geography, 68, 46-50
------ (1985) Urbanisation and planning in the Third World. Croom Helm, London and Sydney
------ (1986) Spatial inequalities in Barbados, West Indies. Transactions of the Institute of British Geographers, New Series, 11, 183-98

Premdas, R. and Steeves, J. (1984) Decentralisation and political change in Melanesia: Papua New Guinea, the Solomon Islands and Vanuatu. South Pacific Forum Working Paper, no. 3, Suva

Robertson, M. (1986) The South Pacific Regional Trade and Economic Cooperation Agreement: a critique. In R. Cole and T.G. Parry (eds), Selected issues in Pacific island development, Canberra, pp. 147-75

Rodman, M. (1984) Masters of tradition: customary land tenure and new forms of social inequality in the Vanuatu peasantry. American Ethnologist, 11, 61-80

Rubenstein, H. (1975) The utilization of arable land in an eastern Caribbean valley. Canadian Journal of Sociology, 2, 157-67

------ (1983) Remittances and rural underdevelopment in the English-speaking Caribbean. Human Organization, 42, 295-306

Sanguin, A-L. (1981) 'Small is not beautiful': la fragmentation de la Caraibe. Cahiers du Géographie du Quebec, 25, 343-59

Seers, D. (1983) The political economy of nationalism. Oxford University Press, Oxford

Selwyn, P. (1975) Industrial development in peripheral small countries. In P. Selwyn (ed.), Development policy in small countries, Croom Helm, London, pp. 77-104

------ (1978) Small, poor and remote: island at a geographical disadvantage. Institute of Development Studies, Discussion Paper (Brighton), 123

Sevele, F. (1983) Regional trade: limited potential. Pacific Perspective, 11, 23-7

Shankman, P. (1976) Migration and underdevelopment: the case of Western Samoa. Westview Press, Boulder

------ (1978) Notes on a corporate 'Potlatch': the lumber industry in Samoa. In A. Idris-Soven et al., (eds), The world as a company town. Multinational corporations and social change, Mouton, The Hague, pp. 375-404

Shaw, B. (1982) Smallness, islandness, remoteness and resources: an analytical framework. Regional Development Dialogue (Special Issue), 95-109

Shears, R. (1980) The coconut war. Cassell, Sydney

Simmonds, K.C^ (1985) The politicization of bureaucracies in developing countries: St Kitts Nevis, a case study. Phylon, 46, 58-70

Sunshine, C.A^ (1985) The Caribbean survival, struggle and sovereignty. EPICA, Boston

Sutherland, W. (1986) Micro-states and unequal trade in the South Pacific: the Sparteca Agreement of 1980. Journal of World Trade Law, 20, 313-28

Taylor, M.J. (1984) The changing pattern of Australian corporate investment in the Pacific Islands. In M.J. Taylor (ed.), The geography of Australian corporate power, Croom Helm, Sydney, pp. 25-46

------ (1986) Multinationals, business organisations and the development of the Fiji Economy. In M. Taylor and N. Thrift (eds), Multinationals and the restructuring of the world economy, Croom Helm, London and Sydney, pp. 49-85

Thaman, R.R. and Thomas, P.M. (1985) Cassava and change in Pacific island food systems. In D.J. Cattle and K.H. Schwerin (eds), Food energy in tropical ecosystems, Gordon and Breach, New York and London, pp. 189-223

Thomas, P. (1984) Through a glass darkly. Some social and political implications of the spread of television and video in the Pacific. In C. Kissling (ed.), Transport and communication for Pacific microstates, Institute of Pacific Studies, Suva, pp. 61-76

Thompson, R. (1985) Towards agricultural self-reliance in Grenada: an alternative model. In P.I. Gomes (ed.), Rural development in the Caribbean, Hurst, London, pp. 123-53

Tisdell, C. and Fairbairn, T.I. (1984) Subsistence economics and unsustainable development and trade: some simple theory. Journal of Development Studies, 20, 227-41

Townsend, D. (1980) Disengagement and incorporation: the post colonial reactions in the rural villages of Papua New Guinea. Pacific Viewpoint, 21, 1-25

Underwood, R.A. (1985) Excursions into inauthenticity: the Chamorros of Guam. In M. Chapman (ed.), Mobility and identity in the island Pacific (Pacific Viewpoint, 26), pp. 160-85

Wace, N. (1980) Exploitation of the advantages of remoteness and isolation in the economic development of Pacific Islands. In R.T. Shand (ed.), The island states of the Pacific and Indian Oceans, ANU Development Studies Centre Monograph, no. 23, Canberra, pp. 87-118

Waddell, R. (1982) The effect of national development plans on the village: a case study of the Western Solomons. In A. Utrecht (ed.), Transnational corporations in South-East Asia and the Pacific (vol. 4), TNC Research Project, Sydney, pp. 161-77

Ward, R.G. (1967) The consequences of smallness in Polynesia. In B. Benedict (ed.), Problems of smaller territories, Athlone Press, London, pp. 81-96

------ and Hau'ofa, E. (1980) The demographic and dietary contexts. In R.G. Ward and A. Proctor (eds), South Pacific agriculture: choices and constraints, ANU Press, Canberra, pp. 157-80

Watson, H. (1985) Transnational banks and financial crisis in the Caribbean. In G. Irvin and X. Gorostiaga (eds), Towards an alternative for Central America and the Caribbean, Allen and Unwin, London, pp. 126-53

Watters, R.F. (1970) The economic response of South Pacific societies. Pacific Viewpoint, 11, 120-44

World Bank (1986) World Development Report, 1986. Oxford University Press, New York

Yusuf, S. and Peters, R.K. (1985) Western Samoa. The experience of slow growth and resource imbalance. World Bank, Staff Working Paper (Washington), 754

NOTES ON CONTRIBUTORS

Dr M Bell, Department of Geography, Loughborough University, Loughborough, England

Dr J.A. Binns, School of African and Asian Studies, University of Sussex, Brighton, England

Dr C. Birkbeck, Centre for Penal and Criminological Studies, University of the Andes, Merida, Venezuela

Dr R. Bromley, Department of Geography and Regional Planning, State University of New York, Albany, USA

Dr J. Connell, Department of Geography, University of Sydney, Sydney, Australia

Dr S. Corbridge, Department of Geography, Syracuse University, New York, USA

Dr D. Drakakis-Smith, Department of Geography, University of Keele, England

Mr J. Gray, Institute of Development Studies, University of Sussex, Brighton, England

Dr R.N. Gwynne, Department of Geography, University of Birmingham, England

Dr G.P. Hollier, Department of Geography, University of Strathclyde, Glasgow, Scotland

Dr R.I. Lawless, Centre for Middle Eastern and Islamic Studies, University of Durham, England

Professor W.B. Morgan, Department of Geography, King's College, University of London, England

Dr A. O'Connor, Department of Geography, University College, University of London, England

Dr M. Pacione, Department of Geography, University of Strathclyde, Glasgow, Scotland

Dr R.B. Potter, Department of Geography, Royal Holloway and Bedford New College, University of London,

England
Dr J. Soussan, Department of Geography, University of Reading, England

INDEX